Martin Jänicke   Helmut Weidner   National Environmental Policies

A study prepared for UNU/WIDER

UNU World Institute for Development Economics Research (UNU/WIDER) was established by the United Nations University as its first research and training centre and started work in Helsinki, Finland, in 1985.
The purpose of the Institute is to undertake research and policy analysis on structural changes affecting the developing and transitional economies, to provide a forum for the advocacy of policies leading to robust, equitable and environmentally sustainable growth, and to promote capacity strengthening and training in the field of economic and social policy-making. Its work is carried out by staff researchers and visiting scholars in Helsinki and through networks of collaborating scholars and institutions around the world.

The United Nations University
World Institute for Development Economics Research
Katajanokanlaituri 6 B
FIN-00160 Helsinki
Finland

Springer

Berlin
Heidelberg
New York
Barcelona
Budapest
Hong Kong
London
Milan
Paris
Santa Clara
Singapore
Tokyo

Martin Jänicke Helmut Weidner (Eds.)

National Environmental Policies

A Comparative Study of Capacity-Building

In Collaboration with H. Jörgens

With a Data Appendix: International Profiles of Changes Since 1970

Springer

Professor Dr. MARTIN JÄNICKE
Dipl.-Pol. HELGE JÖRGENS

Free University of Berlin
Environmental Policy Research Unit
Schwendenerstraße 53
D-14195 Berlin
Germany

Dr. HELMUT WEIDNER

Science Center Berlin (WZB)
Research Unit "Norm Building and Environment"
Reichpietschufer 50
D-10785 Berlin
Germany

With 13 Figures

ISBN-13: 978-3-642-64435-1 e-ISBN-13: 978-3-642-60507-9
DOI: 10.1007/978-3-642-60507-9

Library of Congress Cataloging-in-Publication Data

National environmental policies : a comparative study of capacity
-building : with a data appendix, international profiles of change
since 1970 / M. Jänicke, H. Weidner, eds., in colloboration with H.
Jörgens.
p. cm
Includes bibliographical references.
ISBN-13: 978-3-642-64435-1 -- ISBN-13: 978-3-642-64435-1

1. Environmental policy--Political aspects--Case studies.
2. Environmental management--Political aspects--Case studies.
I. Jänicke, Martin. II. Weidner, Helmut. III. Jörgens, Helge..
GE170.N36 1996
363.7'05--dc20 96-30956

Cover design: design & production GmbH, Heidelberg
Typesetting: Camera ready by authors
SPIN 10534629 32/3137 5 4 3 2 1 0

PREFACE

This book is a collection of systematically prepared case studies describing the environmental policy of thirteen countries in terms of capacity-building. Capacity for environmental policy and management, as the concept is used in this volume, has been defined broadly as a society's "ability (...) to devise and implement solutions to environmental issues as part of a wider effort to achieve sustainable development" (OECD). Since the late 1960s capacity-building in environmental policy and management can be observed across the world. It may have made insufficient progress as yet from an environmentalist point of view, but it has produced some remarkable results, and not only in the industrialised world.

In the first chapter we present the conceptual framework that underlies the national case studies. In the course of our research project the authors of the book met together twice to discuss this framework in the light of the national experiences and to harmonise their approaches. In this way we have tried to offer more than a collection of individual and incoherent case studies, focusing only on specific environmental problems, institutions, actors, or instruments. The idea behind this book is to give a systematic, comparative overview of the fundamental conditions under which environmental policies is practised in selected countries. The wide range of countries included in this study shows up cross-national similarities in environmental policy capacity-building as well as unique developments in some countries. It highlights promising domestic approaches to environmental protection as well as potentially threatening tendencies. Last but not least, and seen from a comparative perspective, the systematic approach to the analysis of national environmental policies provides a valuable insight into the processes of cross-national policy learning and the diffusion of many of the main elements of environmental policy.

In developing the analytical approach of this book the editors could rely on a number of evaluation studies of successful environmental policy and management. Carried out at the Environmentel Policy Research Unit of the Free University of Berlin and at the Social Science Research Center Berlin. From these studies it can be seen that the success or failure of an environmental protection measure cannot be comprehensively explained as a result of the choice of particular instruments or even strategies. Environmental policy analysis has rather to take into account the background conditions for taking action: the strength, competence and configuration of organised proponents of environmental interests and their systemic opportunity structure. Both the capability of organised proponents of environmental protection and the institutional, informational, and economic background conditions (or resources) for their actions describe the capacity of environmental policy and management.

The countries represented in this volume were selected on a global basis: from the most advanced frontrunners (such as the Netherlands or Denmark) through a country plagued by economic crisis, political instability, and problems of systemic transition (Russia) to Chile, a post-dictatorial latecomer in Latin-America, and finally the extreme case of Nigeria, which exhibits nearly all forms of restriction, from underdevelopment, economic dependency and crisis to a military dictatorship with brutal suppression of environmental activists. In this, the selection follows the principle of "most different systems design".

The over-representation of highly developed countries is intentional, in order to show as many of the more advanced variants of environmental policy as possible. Without doubt, this group has most strongly influenced the process of global environmental policy development that can be observed from a comparative perspective. It was nevertheless interesting to learn – through the study of less developed countries like China and especially South Korea – how quickly environmental policy innovations, from the creation of environmental protection agencies to national environmental planning approaches have spread all over the world even into the less advanced countries.

The presentation of case studies starts with countries which, by creating environmental policy institutions and passing comprehensive legislation on a variety of pollution problems, influenced global environmental policy development early on. These countries include the United States, Sweden and Japan, but also the United Kingdom (an early pioneer). The second group includes countries which can at present be considered to be setting the pace in environmental policy. This group comprises the Netherlands, Germany, Denmark and Switzerland, along with South Korea which – much like Japan in the 1970s – has become an interesting special case of creative imitation of the frontrunners. Chile, China and Nigeria as developing countries, along with Russia, make up the third group, which is characterised by a number of severe restrictions on effective environmental protection.

The studies assembled in this volume, except for Chile[1], all tend to follow a similar design. Each chapter begins with an assessment of the main environmental problems, focusing especially on changes in important problem areas. The second section describes the development and main characteristics of environmental policy, its institutions, instruments and policy style. Following this, the main actors in environmental policy, their strength, strategies and resources are

1 In the special case of Chile it proved more useful to describe the period of military rule from 1973 to 1989 and the following democratic rule in separate sections.

analysed while in a fourth section attention is given to qualitative aspects of capacity-building, such as the dominant environmental policy paradigm, changing actor constellations, or attempts to more strongly integrate ecological aspects into other relevant policy areas. Finally, an evaluation of the national process of capacity-building in environmental policy is made.

In the last chapter of this book we summarise the findings from the national case studies and thus describe the process of capacity building in environmental policy from a cross-national point of view. The book ends with an appendix, presenting basic quantitative information about institution-building and changes in environmental quality in different countries. The data is presented in a concentrated form, which makes it possible to assess comparatively a country's performance. However, it can only give a rough picture, since the reliability of cross-national environmental statistics is still generally poor. Time series of emissions are often unreliable or incomplete. While the advanced industrial countries certainly have more environmental data than developing nations, the information is a long way from being satisfactory. Statistical information on developing countries is usually diffuse or even completely non-existent. Nevertheless, the data presented can give some indications as to which environmental problems have in general been successfully handled, and which other problems remain by and large unsolved.

This project could not have been realised without the generous financial support of the United Nations University/World Institute for Development Economics Research (UNU/WIDER). At WIDER, the project was supervised excellently by Professor Mihály Simái, Dr. Reino Hjerppe and Mrs. Barbara Fagerman. A part of this project which is particularly worthy of mention here is the second workshop held on 24 and 25 November 1995 in Berlin, which was assisted financially by the German Federal Ministry for Education, Science, Research and Technology (BMBF). We would like to offer our sincere thanks to both institutions. In addition, we wish to thank Dipl.-Pol. Harald Mönch and Dipl.-Pol. Manfred Binder of the Environmental Policy Research Unit at the Free University of Berlin (FFU) for preparing the environmental data appendix, Professor Udo Ernst Simonis for his commitment to the realisation of the project, and Alex Sawyer for English language editing and for translating parts of this book. We are also deeply indebted to Dagmar Kollande (Social Science Research Center Berlin) for her organisational support.

June 1996

MARTIN JÄNICKE
HELMUT WEIDNER
HELGE JÖRGENS

CONTENTS

LIST OF CONTRIBUTORS

ANDERSEN, M. S., Dr.
Department of Political Science
Aarhus University
Universitetsparken
8000 Aarhus.
Denmark

ANDREWS, R. N. L., Prof. Dr.
School of Public Health
Department of Environmental
Sciences and Engineering
University of North Carolina
at Chapel Hill
CB# 7400 Rosenau Hall
Chapel Hill, NC 27599-7400
USA

BRESSERS, H. Th. A., Prof. Dr.
Center for Clean Technology
and Environmental Policy
University of Twente
P.O. Box 217
7500 AE Enschede
The Netherlands

IMURA, H., Prof. Dr.
Faculty of Engineering
Institute of Environmental Systems
Kyushu University 36
6-10-1 Hakozaki, Higashi-ku
Fukuoka
Japan

JÄNICKE, M., Prof. Dr.
Environmental Policy
Research Unit
Department of Political Science
Free University of Berlin
Schwendenerstr. 53
14195 Berlin
Germany

KNOEPFEL, P., Prof. Dr.
Institut des Hautes Etudes en
Administration Publique
Université Lausanne
Route de la Maladière 21
1022 Chavannes Près Renens
Switzerland

LUNDQVIST, L. J., Prof. Dr.
Department of Political Science
University of Göteborg
Sprängkullsgatan 19
411 23 Göteborg
Schweden

MAO, Y., Prof. Dr.
Unirule Institute of Economics
2-1-1 Nan Sha Gou
San Li He
Beijing 100045
China

NAM, Y.-S., Dr.
Korea Environmental Technology
Research Institute (KETRI)
9-2 3 Samsung-Dong
Kangnam-Ku
Seoul 135-090
Korea

PLETTENBURG, L. A.
Center for Clean Technology
and Environmental Policy
University of Twente
P.O. Box 217
7500 AE Enschede
The Netherlands

POTRAVNYI, I., Dr.
Russian Academy
of Economics Plechanow
Stremyanyi Pereulok, 28
113054 Moscow
Russia

SALAU, F. K., Dr.
Auwaldgasse 5
80939 München
Germany

SILVA, E., Prof. Dr.
Department of Political Science
University of Missouri-St. Louis
8001 Natural Bridge Road
St. Louis, Missouri, 63121-4499
Chile

WEALE, A. P., Prof. Dr.
Department of Government
University of Essex
Wivenhoe Park
Colchester, CO4 3SQ
Great Britain

WEIDNER, H., Dr.
Science Center Berlin
for Social Research
Reichpietschufer 50
10785 Berlin
Germany

WEIßENBURGER, U., Dr.
Abteilung Osteuropa
Deutsches Institut
für Wirtschaftsforschung
Königin-Luise-Str. 5
14195 Berlin
Germany

THE POLITICAL SYSTEM'S CAPACITY FOR ENVIRONMENTAL POLICY

Martin Jänicke

> "The ability of a country to follow sustainable development paths is determined to a large extent by the capacity of its people and its institutions as well as by its ecological and geographical conditions. Specifically, capacity-building encompasses the country's human, scientific, technological, organisational, institutional and resource capabilities".
> (Agenda 21, Chapter 37)

1 Capacity for Environmental Policy and Management

What is capacity for environmental protection? The OECD defines it broadly as "a society's ability to identify and solve environmental problems" (OECD, 1994: 8). More specifically, the concept points to the objective limits to (and necessary preconditions of) successful solutions of a given type of problem, limitations beyond which failure sets in, even in cases of good luck, skill and highly motivated actors. Lack of ecological, technological or administrative knowledge, lack of material or legal resources, the weakness of environmental organisations or institutions in relation to vested interests are well-known examples of such limitations. The main implication for the debate on sustainable development is that long-term strategies must include concepts for improving the conditions of environmental action (capacity-building).

Generally speaking, the capacity problem is not specific to the field of environmental policy. It can be found also in other policy fields. Environmental policy, however, seems to be the most prominent field for the discussion of limitations on government intervention (Luhmann, 1990). Some authors see such limitations – or "government failure" – as regards more preventative forms of environmental policy (Jänicke, 1990). In some areas even a "capture of the system of control (...) by those which it is meant to control" (OECD, 1992: 46) has been discussed. There is at least a capacity problem for new environmental problems: "(...) in some cases existing capacity is rendered irrelevant or made inadequate by the emergence of new challenges, and awareness of these challenges has grown faster than the ability of many institutions to solve them" (OECD, 1994: 11).

A country's capacity for environmental protection is not and cannot be restricted to government policies. Increasingly it depends on societal forces of all kinds. The author therefore prefers the formula "capacity for environmental policy and management" or the OECD formulation of a societal "capacity for the environment" (OECD, 1994).

2 Capacity Limits in the Advanced Industrial Countries

The debate on capacity-building (or development) in environmental protection is at present mainly restricted to developing countries. They are the candidates for transfer of expertise, etc. from the more successful "donor countries" (OECD, 1995). One might believe that the advanced countries provide an indubitable model to be imitated. But is this really so?

Without any doubt there has been a development of capacity in the industrialised countries that can be empirically demonstrated by cases of best practice and successful environmental policy measures in fields such as air and water pollution control or the establishment of protected areas. In some advanced democracies there has been a remarkable increase in environmental policy capacity. However, the restrictions to ecologically sustainable strategies should not be ignored. Therefore, capacity-building for the environment is no less relevant for the industrialised world.

In the advanced democracies too, success in environmental policy and management depends on the type of problem to be solved. Success is more or less restricted to problems that can be handled (mainly by "additive" technical standard solutions) without restricting markets or relevant societal routines. As regards the big environmental problems – especially the problems of long-term resource management – government failure and capacity overload can be frequently observed.

A brief look at the environmental record of the industrial countries shows that between 1970 and 1990 there was a general improvement in terms of such indicators as SO_2 or suspended particulate emissions, the proportion of protected areas, or the share of the population served by sewerage systems (see the appendix to this book). However, most industrial countries showed a deterioration with respect to domestic waste, the effects of motor traffic, or the proportion of built-up areas (Jänicke, Mönch and Binder 1996, 1996a). For soil and ground water pollution or toxic waste, the same picture would be likely to emerge, but internationally comparable data are not available here. And this is a typical situation in areas where environmental capacity is low. For important environmental problems there is no "getting rich – getting clean" (see also the appendix to this book). This points to the idea that the structure of problems is important for the evaluation of environmental policy and its capacity.

In the politically difficult field of sustainable development in particular, the conditions in advanced industrial countries primarily give rise to the question of how capacities for environmental policy and management might be developed and increased.

3 The Usefulness of the Concept of Environmental Capacity

The usefulness of the concept of environmental capacity may be summarised as follows:

The first relevant function of the concept is to stress the importance of objective limitations on successful intervention. Failure in environmental policy (or other policies) cannot be explained just by the wrong choice of instruments or other kinds of "intervention failure" (OECD, 1992) as many analysts have done. It is not only the kind of action, but also its structural characteristics which matter. Not only the right choice but also the availability of other options have to be considered if a specific policy outcome has to be explained.

A second advantage of the concept of capacity might be its close relationship to the concept of development. Political development or modernisation can be described as a process of capacity-building. This was already the case in publications on developing nations in the 1960s and 1970s (Pye, 1966; Binder et al., 1971), a tradition that has been revived by the debate on sustainable development. Political development or modernisation might be defined in general as institutionalisation and internalisation of new stages of problem-solving capacities in reaction to (or anticipation of) societal challenges or crises (Jänicke, 1973). This definition could also be applied to the industrialised world. And there is no reason to restrict the concept of political development to developing countries. The concept of capacity-building also makes sense if we analyse the development of specific policies. The relatively new field of environmental policy is a good example.

This leads to the third advantage: The concept of capacity for environment and management provides a tool for comparison between the different abilities of different countries as well as the degrees of development of these abilities within a country.

Since capacity analysis deals with the structural preconditions for solutions to specific problems, the final advantage of the concept may be its practical orientation as well as its greater realism regarding the difficulty of problems that have not yet been solved.

The main function of the concept is to change the focus of environmental policy analysis. Therefore it is firstly a heuristic analytic tool. We should not overestimate the possibilities of its operationalisation. On the one hand capacity relates to concrete problems, where operationalisation may be possible. On the other hand, if we speak for example of institutions, we use the concept for a broader field or the sum of its concrete tasks. Here operationalisation creates difficulties. We should also remember, that all kinds of evaluation – the diagnosis of success or failure – depend upon explicit criteria (Bartlett, 1994).

Another reason for caution is the difference between capacity and its utilisation by concrete actors in concrete situations. But this is an important enough point to be explained in more detail later on.

4 An Analytical Framework

4.1 Categories for the Explanation of Environmental Policy Outcomes

In a number of altogether 36 case studies on the preconditions for successful environmental policy and management carried out by the Environmental Policy Research Unit at the Free University of Berlin (Jänicke and Weidner, 1995; Conrad 1995; Jänicke, 1996) a set of categories to explain differences in policy outcomes has been developed. When reformulated, it may also be useful for the analysis of environmental capacities:

Usually, perceived environmental problems lead actors to develop and implement strategies (typically against opposing target groups) under systemic conditions and within situative contexts. The outcome, then, is influenced mainly by the following factors (see *Figure 1*): (a) actors, (b) strategies (c) systemic framework conditions, (d) situative contexts and (e) problems. The structure of problems as well as the capacity to respond to them is strongly influenced by economic performance. All factors, of course, have to be seen in an international context.

The main purpose of this systematisation is to make clear that successful environmental protection is brought about by a complex interaction of influences and not by a single, isolated factor, nor a favourite instrument, nor a single type of actor, nor a particular framework condition or institution. The literature on environmental policy and management is full of such proposals of a particular measure, regardless of its context.

The shortcomings of such a "mono-factorial" view may be illustrated by the example of the time-consuming debate on environmental policy instruments (sometimes taking on the character of a debate on the "instrument of the year"). Of course, we need assessments of the general advantages and weaknesses of our policy tools. But recent research makes clear that this has been a simplistic debate using a mechanistic top-down model of policy and ignoring the complex interaction dynamics between quite different factors (Sabatier and Jenkins-Smith, 1993; Héritier, 1993; Prittwitz, 1994; Jänicke and Weidner, 1995):

Both the structural framework conditions and the situative context in which instruments are applied cannot be ignored, and these are often more important than the concrete measure applied. Just as important is the role of instruments within a strategy and the flexible use of instruments over time. If instruments are designed to be flexible they allow for appropriate, skilful reaction to changing situations, the exploitation of polluters' weaknesses, or for learning processes. One example is something that could be called "prospective intervention": In this case a measure is only announced (thus avoiding the complications of the decision-making process) but this probably leads individual proactive firms to anticipate it, thereby improving again the situative conditions of environmental policy.

Figure 1. Model of Policy Explanation.

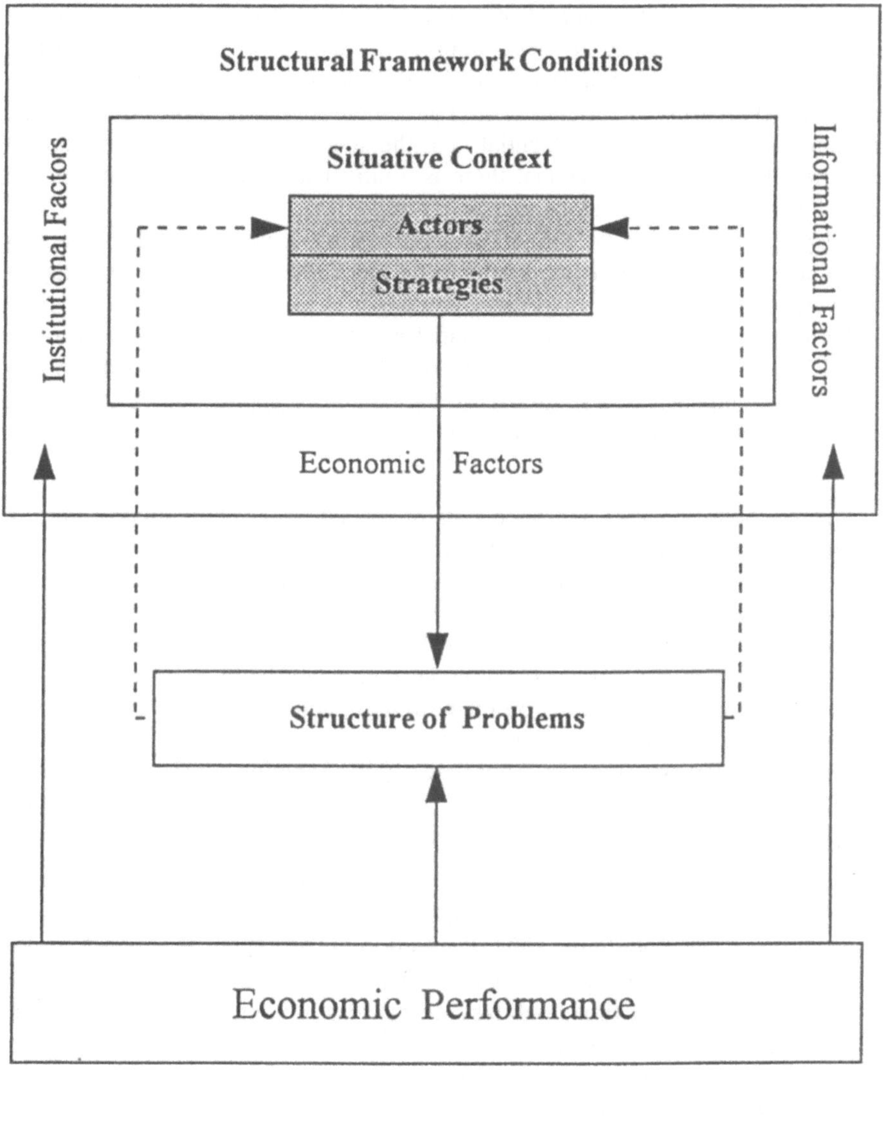

M. Jänicke / FFU (1995)

Here the mode of application is more important than the instrument itself. Command-and-control approaches may have their weaknesses regarding efficiency, innovation, or resistance from the target group. But if they remain latent and integrated in a strategy of "negotiation in the shadow of hierarchy" (Scharpf, 1991) they may be very important and leave little room for criticism. Environmental policy instruments of whatever kind often have a general informational function that may be more influential than the specific nature of the instrument concerned, telling the polluter that there is a universally perceived problem and the will to take action against it.

Today, the mechanism of cooperative goal-setting, leaving the relevant actors free to choose their instruments, seems to be more relevant than the question of policy instruments.

The list of arguments against an isolated debate on instruments – or on other possible single factors influencing policy outcomes – could be extended. To prevent this overemphasis of a single factor, the combined application of at least the following (however broad) categories seems to be indispensable to the explanation of policy outcomes:

a) *Actors*: the proponents and opponents of special issues, their support groups, but also third parties. Usually the actors are representatives of organisations and coalitions. But they are often coalitions of relevant micro-actors across organisations and institutions. They have stable general interests and "core beliefs" (Sabatier and Jenkins-Smith, 1993) as well as variable definitions of these, or situative opinions and attitudes that may change more readily. Their capacity for action depends on their strength, competence and constellation, but also on the structural framework conditions. Personal will and skill also make a difference (see also Shonfield, 1969; Schmidt, 1993; OECD, 1995: 5). Schumpeter emphasised this with respect to technical innovations. But it is also important for policy innovation.

b) *Strategy*: the general approach to the problem. Strategy is the purposeful use of instruments, capacities, and situative opportunities to achieve long-term goals (see also Bartelmus, 1994; Carew-Reid et al., 1994; Hey and Brendle, 1994; Fiorino, 1995). It includes necessary sub-goals, the flexible application of instruments over time, and, last but not least, the purposive improvement of capacities. Strategies depend on capacities such as available knowledge or the possibility of strategic and coordinated action. Environmental protection needs strategy also to compensate for the generally weak position of its proponents. For this the intelligent use of time is important. An environmental measure which is impossible to implement at present may be feasible within five years. Failure in the short term has often turned out to be a success in the long run.

c) *Structural Framework Conditions:* the systemic conditions of environmental action or the opportunity structure of the relevant actors (Kitschelt, 1986). Policy analysis generally differentiates between the influence of the political, economic and cultural framework conditions of policy action (Feick, 1992; Kern and Bratzel, 1996). The author prefers a slightly different typology:

Cognitive-informational framework conditions: the conditions under which environmental knowledge is produced, distributed, interpreted and applied. Without knowledge there is no (perceived) problem, no public awareness and consequently no policy process, and therefore, it is the first factor here. The cognitive opportunity structure (Jahn, 1993: 178), the societal interpretation of the environmental situation is the second. While culture is more of a background condition, the available knowledge about threats and options, combined with environmental consciousness, is an immediate resource for environmental actors.

Political-institutional framework conditions: the constitutional, institutional and legal structure, the institutionalised rules and internalised norms, constituting the framework for interaction. The conditions for participation, integration and long-term action should be given special emphasis in this context.

Economic-technological framework conditions: the performance, technological standard, sectoral composition, or general availability of raw materials. GNP per capita is an important determinant of existing capacities. But – as for culture – its influence is mainly indirect (contrary to its influence on the structure of problems). And – as in culture – it is no easy target for environmental capacity-building. Factors such as price structures, however, have a more immediate impact and may also be more easily influenced by (e. g. fiscal) policy.

d) *Situative contexts:* the short-term variable conditions of action (Feick, 1992; Sabatier and Jenkins-Smith, 1993). Their importance was already there to be studied with respect to the situative change that the French Revolution caused for European neighbour countries (changing the opportunities of relevant actors). In environmental policy and management, headlines sparking off public debate on a concrete pollution problem are a strong motor of environmental change. Events such as the Bhopal, Seveso or Chernobyl accidents offered sudden situative opportunities for proponents of environmental protection. In about half of the above mentioned 36 success cases analysed in 12 OECD countries, direct public pressure played an important situative role (Jänicke and Weidner, 1995; Conrad, 1995). An important situative change is also the discovery of win-win constellations, where not only environmental but also economic interests are supported by the environmental protection measure. Situative changes sometimes have an immediate influence. German industry, for example, started reducing the use of cadmium mainly in reaction to a Swedish law on the substance, provoking a public debate on the issue in Germany (Jänicke and Weidner, 1995). In times of recession, on the other hand, environmental policy and management act under restrictive conditions.

e) *The character of the problem* (Solesbury, 1976; Sabatier and Jenkins-Smith, 1993): is it easy or hard to solve? It makes a difference whether a pollution problem is urgent and experienced by the public or latent and only threatening future generations. Similarly it makes a difference if the polluter is economically relevant, has strong influence on society and no easy way out, or if the target group is weak and/or a technical or other solution is available. "The systemic or structural nature of many problems" (Carew-Reid et al., 1994) causes restrictions that may limit a given capacity.

As mentioned above, successful environmental policy measures so far have been restricted to a certain type of problem. As a first step, the advanced democracies were generally able to manage the "dilution of pollution", solid waste collection, or the creation of protected areas. Later on, end-of-pipe measures of all kinds, risk management and the substitution of dangerous materials took place. Here success in terms of reduced pollution was possible. However, there is a broad range of environmental policy failures in fields such as: waste reduction, nuclear waste disposal, soil and ground water protection, road traffic volume stabilisation, sustainable land-use and absolute reduction of resource inputs into the production process (see above). These problems have been identified, but not solved.

4. 2 Capacity for the Environment and its Utilisation

The above defined five categories of policy explanation could be used also for capacity analysis, if one differentiates between capacity, as a relatively stable condition of action, and its utilisation, which leads to the subjective and situative aspect of environmental policy and management. The above model, then, can be reformulated as follows:

The capacities for the environment are constituted by:
(1) the strength, competence and configuration of organised governmental and non-governmental proponents of environmental protection and
(2) the (a) cognitive-informational, (b) political-institutional, (c) economic-technological framework conditions.
The utilisation of the existing capacity depends on:
(3) the strategy, will and skill of proponents and
(4) their situative opportunities.
This has to be related to:
(5) the kind of the problem: its urgency as well as the power, resources and options of the target group.

Capacity, therefore, defines the necessary structural conditions for successful environmental policy as well as the upper limit beyond which policy failure sets in even in case of skilful, highly motivated and situatively well-placed proponents.

The reformulation of the analytical model described above includes an important change regarding the category of "actors": If we try to explain cases of success or failure (policy outcomes) in environmental protection, we must analyse the dynamic interaction between proponents and opponents and their conditions of action. However, in this context we are concerned with capacities for environmental protection. And here we are interested first and foremost in the possibilities available to the active part: the organised proponents of environmental policy and management (thus treating the opposing target group of polluters as a restrictive factor and part of the problem). It is true that the target groups of polluters themselves can also learn and sometimes they even become ecological modernisers. This important type of ecological innovation can be influenced not only by particular events but also by framework conditions such as a high general environmental awareness or an institutional framework of interaction that integrates ecological interests. Such framework conditions which create opportunities for proponents of environmental interests (and often originate from former environmental activities) might also explain the polluters' innovative reactions. But the structural chances of the target group of successfully opposing environmental protection measures must be analysed separately, because they have their own dynamic. This separate analysis of the target group's structural chances and the restrictions they might place on proponents is indeed necessary for evaluating whether or not the capacity of proponents is adequate to solve a certain problem.

5 Capacities in Environmental Policy and Management: Some Research Findings

Before turning to the question of capacity-building, it would be useful to summarise some earlier research findings regarding successful environmental policy and management. This may give a picture of influential factors which can be related to the presented categories of capacity for environmental policy and management. Hereby I will also refer to our own set of case studies mentioned above.

5.1 The Capacities of Proponents

Proponents of environmental policy can be specialised government environmental protection institutions as well as non-governmental green interest organisations. Comparative research in environmental policy shows that within the organisational context individual "people make a difference" also (Jacobson and Brown Weiss, 1994: 7; Jänicke and Weidner, 1995; OECD, 1995).

Following Agenda 21, studies on environmental capacity tend to stress not only the organisational and institutional strength but also the abilities of agents or proponents: the role of human capital, technical expertise, functional skills etc. (OECD, 1995).

Government institutions for environmental protection. The establishment of such institutions may plausibly be regarded as the necessary condition for successful measures. "Before pollution policy can be developed, it first has to be invented, and this in turn involves constructing the relevant institutions" (Weale, 1992: 14). Indeed, many improvements in the 1970s and 1980s can be at any rate loosely traced back to the new environmental protection institutions being created from 1970 onwards. Our 36 case studies mentioned above showed that in cases of environmental improvements at national or regional level, government regulation was by far the most important immediate factor in change (Jänicke and Weidner, 1995; Low, 1992; see also OECD, 1996). This dominant role of government is also confirmed by surveys on the factors perceived as influential by enterprises (Gothenburg Research Institute, 1995; Brinkmann and Kirchgeorg, 1995). It can reasonably be assumed that this influence could only be exerted by a competent administration (Kitschelt, 1983; Knoepfel, 1993; Jacobson and Brown Weiss, 1994; OECD, 1994). In more recent studies the importance of institutions for environmental planning is also stressed (Andrews, 1993; OECD, 1994, see also Weale, 1992; OECD, 1995a).

Unfortunately it proves difficult to test the hypothesis that early institutionalisation leads to a better environmental policy record (Jänicke, 1996): Most industrial countries created an environmental agency or ministry at nearly the same time, around the Stockholm Conference of 1972. Small countries such as Luxembourg or Norway started later, but then used the knowledge of neighbour countries. East European countries such as the GDR, Hungary and Poland also started in the early 1970s, but have been very unsuccessful in many areas. This suggests a differentiation between formal and substantive institutionalisation. In democratic systems it may be easier to create institutions in the strict sense, resulting in stable, calculable, internalised and valued patterns of behaviour (Huntington, 1965). The differing effectiveness of environmental institutions in East and West also seems to underline the importance of their opportunity structure: In the former communist countries environmental institutions neither had stringently protected jurisdiction nor support from a strong ecology movement or an independent media.

It might also be mentioned that, in empirical studies, the central institutions of representative democracy seem to play a minor role. Several cross-national studies by the Research Unit for Environmental Policy at the Free University of Berlin have come to the conclusion that the political orientation of ruling parties explains little in environmental policy (Jänicke, 1992; Fey, 1994; Binder, 1996a; Kern and Bratzel, 1996). Moreover, in our collection of success stories the influence of a political party did not show strongly as a relevant factor. Only in two cases did parliaments play a distinct role (Jänicke and Weidner, 1995; see also Häberle and Herbert, 1992). Party competition as such (Enloe, 1975; Böhmer-

Christiansen and Skea, 1991), especially the existence of green parties in the party system, does seem to be relevant at least to environmental policy formulation.

Non-governmental proponents. Today there are three important groups of non-governmental proponents of environmental protection, all of them able to influence polluters directly:

- environmental organisations (in four out of our 36 cases they immediately effected radical change),
- the media and
- ecologically innovative firms, which have become very important factors in environmental improvements within industry, albeit only in the last few years.

Environmental policy studies emphasise not only the growing direct influence of environmental organisations on target groups (e.g. Weale, 1992; Lester, 1995; Jänicke and Weidner, 1995) but also point to the increasing number of enterprises responding pro-actively to the challenge of environmentalism (e.g. Steger, 1992; Shrivastava, 1992; Fischer and Schot, 1993; Conrad, 1995; Porter and van der Linde, 1995). The situation may be illustrated by the research finding that companies in three Nordic countries (Norway, Sweden, Finland) mentioned the following principal (external) actors as exerting environmental pressure on them: Governments (90-97% of the answers), Customers (51-65%) and Environmental NGOs (26-35%) (Gothenburg Research Institute, 1995). In a similar study the media were also included, ranking second in Germany (after government and before the NGOs) and third in Norway (together with the NGOs) (Brinkmann and Kirchgeorg, 1995: 386).

5.2 The Structural Framework of Action (Systemic Opportunity Structure)

5.2.1 Cognitive-Informational Framework Conditions

Without a doubt, environmental knowledge and public awareness are important factors influencing environment policy and management. Only in 5 out of our 36 environmental success stories was public awareness not at least a relevant background factor (Jänicke and Weidner, 1995; Conrad, 1995).

Culture and the value system in a country are also highlighted in studies on environmental policy (especially OECD, 1994). A high degree of post-materialism (Inglehart, 1989) for example could plausibly constitute a good *cultural precondition* for environmental policy success. Differences in environmental awareness (e.g. if Germany and France are compared) can often plausibly be explained by cultural differences. But culture is not the only factor influencing public awareness. As mentioned above, cultural traditions are a background condition. Post-materialism, on the other hand, is statistically connected with GDP per capita, which is even more closely related to environmental improvements

(Binder, 1996a). Regarding capacity-building in environmental protection the most important task would not be to change cultural traditions but rather to improve informational and communicative capacities. There are therefore reasons not to overestimate cultural factors in this context.

Environmental knowledge and the conditions under which it is generated, distributed, interpreted and applied, however, are of the utmost importance (OECD, 1994). Jacobson and Brown Weiss (1994) emphasise inter alia the existence of an "epistemic community" (see also Haas, 1992) as a condition for successful environmental policy. The openness of the scientific community to new problems and paradigms as well as the openness of the media to new issues are important to environmental policy. The role of environmental reporting in environmental capacity has been emphasised in a comparative study by Weidner et al. (1992). Knowledge about environmental dangers is the necessary condition for public awareness. Its non-existence, on the other hand, can be seen as the functional equivalent of the suppression of respective interest articulation.

But it is not only the scope of knowledge that matters. The leading paradigm of policy actors or the "structure of available knowledge and thinking" (Vowe, 1994: 424; see also Nullmeier, 1993; Keck, 1993) is seen as being increasingly important in policy research. The leading paradigm of environmental protection has crucial influence on the strategy. If it is oriented towards visible pollution, high chimney policy may be seen as the appropriate reaction. If it focuses on emissions, then end-of-the-pipe treatment may seem to be enough. If the paradigm is sustainable environmental development, the reduction of resource input into the production process may be seen as the adequate solution. Even the availability of environmental statistics follows the leading pattern of policy orientation.

5.2.2 Political-Institutional Framework Conditions

The importance of institutional framework conditions for environmental policy outcomes has often been emphasised in the literature, though cross-national research has proven difficult (Binder, 1996, 1996a; Strom and Swindle, 1993). Three aspects seem to be most important, and may be structured as participative capacity, integrative capacity, and capacity for strategic action (Jänicke, 1992):

Participative capacity. The openness of the input structures of the policy process is an important aspect of the opportunity structure of environmental interests. Thus it is a common theme in related studies (Nelkin and Pollak, 1981; Kitschelt, 1983; Jänicke, 1990; Weidner, 1996; Weale, 1992; Zilleßen et al., 1993; Hey and Brendle, 1994; OECD, 1994; Carew-Reid et al., 1994). Decentralisation and strong local communities are seen as a favourable condition for participation. The role of cities in climate protection policy may be cited as an example (OECD, 1994a). Sometimes the pressure from environmental movements is mentioned as a relevant factor for the general opening-up and modernisation of political systems (Paehlke, 1990; Jänicke, 1993; Weidner, 1996). The openness of the legal system to "protective" interests is also mentioned as a kind of catalyst to participation (Weidner, 1996). In terms of resources, environmental laws as well as constitutional mechanisms such as plebiscites (which changed nuclear energy policy in at least three countries) can be regarded as important tools for proponents.

Integrative capacity. Participation and decentralisation require a high level of integration. In the literature volunteer cooperation – "consensual capacity" (Jänicke, 1992) – is generally stressed as an institutional condition for success in environmental policy (Lundqvist, 1980; Kelman, 1981; Kitschelt, 1983; Brickman et al., 1985; Jänicke, 1992; Zilleßen et al., 1993; Fey, 1994; OECD, 1994; Carew-Reid et al., 1994). According to Badaracco, "co-operation is a superior approach" and corporatism is good for environmental policy (Badaracco, 1985; see also Ricken, 1995). Our own research is also open to the corporatist argument: In OECD countries, low rates of unemployment correlate with improvements in the environment (Jänicke, 1992; Binder, 1996a). This may indicate that in countries with better pollution control outcomes there is no issue competition between environmental and employment policy. But it may also mean that effective steering mechanisms in employment policy – e. g. neo-corporatism, concerted action or other cooperative mechanisms – are not only helpful in industrial and employment policy (Schmidt, 1986; Traxler and Unger, 1991), but also in pollution control. This institutional hypothesis is plausible if the correlation between the reduction of SO_2 emissions and low inflation rates is considered (Binder, 1996a).

Integrative capacities are important at different levels. Firstly, intrapolicy cooperation (Knoepfel, 1993), i. e. the internal integration of the policy field, is important. This includes the coordination of environmental policy at different levels (and in respect to different jurisdictions) of the political system. The second field of integration is interpolicy coordination, the cross-sectoral integration of conflicting policies (Carew-Reid et al., 1994). Thirdly, the external integration of environmental policy institutions and non-governmental actors, including target groups is needed. "Network management" (Carley and Christie, 1994; Dutch Committee for Long-Term Environmental Policy, 1994; Glasbergen, 1995; Ricken, 1995) has today become a highly sophisticated, professional strategy.

Integrative capacities together with participative ones offer possibilities for describing types of political systems (Kitschelt, 1983). Examples might be the open, but fragmented US-American system, or the closed and highly integrated French system, or the open and integrated systems of smaller democracies such as the Netherlands, or Norway.

Capacity for strategic action. This capacity (Scharpf, 1987; Jänicke, 1992) may at present be seen as the highest stage of institutional capacity-building in environmental policy and management. In the institutional debate on sustainable development it is nearly always considered important (Andrews, 1993; Carew-Reid et al., 1994; OECD, 1994, 1995a; see also Weale, 1992). If we compare countries such as Italy or Greece on the one hand and Japan or the Netherlands on the other, we find great national differences in this capacity.

Broadly speaking, a cooperative policy style is good for policy innovation, because innovators are integrated earlier into the decision-making process than is the case in countries with a more confrontational tradition (Jänicke, 1992). Obviously, if the environmental policy goal is anchored in a broad consensus, implementation will be easier and more successful.

5.2.3 Economic-Technological Framework Conditions

According to our own comparative research, per capita GDP is most closely connected with environmental policy outcomes. But the statistical result is somewhat paradoxical or ambivalent, because GDP per capita correlates with environmental improvements in some fields, but with deterioration in others. If a standard technological solution – clean-up technology, substitutes, higher efficiency, or recycling – is available, the probability of improvement is highest in rich countries (Jänicke and Mönch, 1988; Jänicke, 1992; Binder, 1993; Jänicke, Mönch and Binder 1996). The correlation of high R&D expenditure with successful air pollution control (Binder, 1993) supports this argument. That wealthier countries both demand and can afford stronger environmental measures was also found to be the case for air protection and recycling activities in the states of the US (Ringquist, 1993: 107; Khator, 1993: 218). On the other hand, if a standard technological solution has not been developed (examples are land-use, soil contamination, or road traffic) GDP per capita is usually a good predictor of ecological deterioration (Jänicke, Mönch and Binder, 1996). The indicator "protected areas" also shows some – weaker – correlation with GDP per capita. Here it is not technology but – possibly – the availability of public resources for buying the land that may give an explanation.

As mentioned above, economic performance is a difficult aspect of environmental capacity. This is because it has not only a strong, but a contradictory impact on the environment, influencing both the structure of problems and the capacity to solve them. Its influence on capacity for environmental protection is complex and more indirect, implying a connection to research, educational, communicative or administrative capacities. And the general state of the economy is not a reasonable target for environmental policy.

However, the technological standard of a country's economy may be influenced more easily and directly, e.g. by transfer of technology and expertise (OECD, 1995). The prices of environmentally sensitive goods are another economic factor that can be influenced by policy (OECD, 1993, 1996; Potier, 1994). This important aspect of economic framework conditions is well-known and need not be discussed in detail here.

6 Capacity-Building

If capacity in environment relates to the ability of societies to identify and solve ecological problems, "capacity development in environment relates to the 'process' by which those abilities are developed" (OECD, 1994: 9). This includes "the capacity for capacity development" (OECD, 1995: 46). And, of course, capacity development can only be conceived of as a comprehensive strategy.

Political science has its own tradition of the concept 'capacity'. The idea of development through overcoming crises (formulated by Erikson in 1968) played a certain role in the debate. Along these lines political modernisation can be defined as the institutionalisation of a higher capacity to identify and solve problems. This is usually promoted by crises (i. e. a threat of severe loss or even collapse) and made possible by already existing capacities and former policy experiences. Institutionalisation includes the internalisation of new routines of cognition and policy action. Crises that do not lead to development may in the long run lead to decay (Huntington, 1965). Pye (1966), Binder, Almond and Coleman (Binder et al., 1971), or Rokkan used the concept of development crisis as potentially leading to higher capacities of the political system: crises of penetration, identity, legitimacy, distribution, participation and integration. In our context the capacities of participation and integration are especially important. The capacity of a system for strategic action or the capacity to provide relevant knowledge could be added to the list.

If we examine the process of capacity-building in environmental policy since 1970, we see typical developments. As a rule, however, systemic capacity development is no steady, linear process. In the UK (under Thatcher), the US (under Reagan, and after the 1994 Congressional elections) and also in Germany (after unification) or Japan (since 1985) we observe phases of downturn in environmental policy connected partly with the reduction of institutional capacities (e. g. curtailment of financial resources or participative opportunities). However, this has sometimes strengthened the position of environmental organisations.

6.1 Proponents

The long-term process of capacity development usually begins with the institutionalisation of the new policy area. In the early 1970s most industrial countries founded an environmental agency and/or a ministry for the environment (Jörgens, 1996). This institution was again differentiated, as a rule, according to the growth of staff, budget and scientific competence. The process of capacity-building also implied policy integration throughout dispersed jurisdictions (intrapolicy coordination) and a closer cooperation with "polluting" policy areas such as transport or energy (interpolicy coordination). The existence of a pluralistic and cooperative national institution for environmental planning represents a high level of institutional development. The next step, then, might be the spread of decentralised institutions for cooperative environmental planning in cities and enterprises.

Stages in Capacity Development: *Proponents of Environmental Protection*
1. Environmental policy institutions: • no central ministry/agency for environment • weak and isolated national ministry/agency • environmental institutions at all levels (including local communities) • environmental divisions in most other ministries also • institutions for integrated environmental planning *2. Green interest organisations:* • local organisations, no clear autonomous national interest organisation • weak or non-professional national interest organisations • strong and highly competent green organisations, playing a consultative role in political decisions • green organisations playing a consultative role in industry also *3. The media:* • few, largely official reports on the environmental situation • environmental reports in some critical sections of the media, but dominant media resistance • environmental problems widely reported in all the media • frequent direct media attacks against polluters *4. The "green" business sector:* • environmental interests are scarcely articulated within the economic system • environmental interests are articulated mainly by the eco-industry (producers of clean-up technology) • environmental interests are articulated by particular green business organisations • ecological pioneer enterprises (e. g. trade or insurance companies) have a strong impact on the whole economy

A process of capacity development can also be observed with respect to non-governmental actors:

The role of environmental organisations developed from more traditional nature conservation to an active representation of environmental interests, usually increasing their membership, financial resources and competence. They develop from scattered, lay grass-roots organisations to a kind of national institution (Roth, 1994) which often plays a consultative role in government decisions and finally increasingly in enterprise decisions also. Today just the five largest environmental organisations in the EU (of the 12) have about 10 million members, with nearly 20% of the adult population in the Netherlands belonging to some environmental organisation (Hey and Brendle, 1994: 642).

In the media a few newspapers usually start by publicising some critical information, then the issue becomes established in a larger segment of the media, with an opposing segment of media remaining friendly to polluters. In most advanced democracies today the environmental problem is by and large accepted as an issue by most publishers. The media, however, also represent an area from where it can be seen that capacity development is not always a linear process. Highly concentrated commercialised mass media with close ties to economic power may in recent years have worsened conditions for environmental awareness in some countries.

Capacity development can also be observed in the universities, which were for a long time important generators of knowledge for the polluting sections of the economy, and in some countries made a relatively late changeover to new ecological paradigms, often following government initiatives (and the associated research budgets).

The development of a green business sector – at first formed only by the eco-industry, but then also by consultancy firms, insurance companies and ecological pioneer enterprises with their own green organisations – can also be described as part of the process of capacity-building (see box "stages in capacity-building").

6.2 The Changing Constellation of Actors

In recent times capacity-building in environmental policy and management also has been described in terms of actor constellation (Knoepfel, 1993). This can be shown for the development in the United States (Hoffman, 1994):

In a first stage (1960-70) industry acted relatively autonomously within its institutional sphere. In a second stage (1970-82) the state intervened in this sphere, under pressure from environmental organisations. In a third stage (1982-88) – under the presidency of Reagan – state and industry cooperated under the common pressure of green organisations. In the fourth stage (since 1988) the cooperation between state and industry partly includes environmental organisations. In addition, an intra-market proponent of environmental protection has become influential: insurance companies, "green" capitalists or pioneer enterprises.

This kind of capacity-building by the improvement of actor constellation could also be shown for the case of Germany, Switzerland or other advanced countries.

6.3 The Systemic Framework of Action

The opportunity structure, or the sum of systemic resources available to the proponents of environmental protection also shows typical lines of development.

The participative capacity of the political system as well as its integrative capacity have usually increased during the process of environmental policy development. The typical development of participative capacities can be described as follows:

Development of Participative Capacities
• Participation is restricted to general elections • Participation rights are instituted regarding licensing of local sites and urban planning • Participation exists in the formulation of national environmental policy • Participation is possible in policy formulation in fields such as energy or transport • Participation in ecologically relevant long-term planning (Agenda 21)

The crucial stage is (or would be) the opening of the closed policy networks which make de facto decisions in ecologically problematic policy fields, before proposals are decided upon by the representative institutions. The pluralistic structures of Public Utilities Commissions in the US are an example of such high-level participation.

The process of developing integrative capacities within government may be described by the following typical steps:

Development of the Integrative Capacities of Government
• Only additive environmental institutions, fragmentation of jurisdictions • Intrapolicy cooperation • Selected ad-hoc interpolicy cooperation • Environmental subdivisions in all relevant ministries and administrations • Strategic interpolicy cooperation, • Long-term intersectoral planning on a broad societal basis

The improvement of the scope as well as the structure (paradigm) of knowledge is another level of capacity development. Some indicators of this process are shown in the following table:

Selected Indicators of Capacity-Building: Environmental Knowledge and Awareness
1. Statistics (as indicators of the leading paradigm) • Scattered or irrelevant data • Mainly air and water quality data • National emissions and solid waste data also • Data on relevant material flows (such as heavy metals) also • National eco-balances, green GDP *2. Science and Research* • No systematic research on pollution problems • Few independent environmental research institutes typically with disputed professional standards • Erosion of "productionist" research paradigms and priorities • Environment becomes a relevant theme in universities and research institutes

7 Capacity-Building for What?

This introduction has tried to make clear that the concept of capacity-building could be a useful heuristic tool in the analysis of environmental policy and management. But how can such capacities be developed beyond their present levels? And where do we need additional capacities?

We do need much more research on that, in particular research on the conditions restricting development of a sustainable environmental policy.

Policy failure can be regularly observed in respect of industrial target sectors with a broad basis within both society and government, such as:

- the construction complex (the construction industry, local government, or institutions interested in increasing the value of land); this sector uses the largest share of materials and land, and generates the most solid waste and goods transportation,
- the road traffic complex (car producers and their suppliers, the service network, the mineral oil industry, the road construction industry, etc.),
- the energy complex (the multinational primary energy industries, the utilities, closely associated with the powerful energy-intensive basic industries), or
- the agro-industrial complex.

Ecologically sustainable development is more than end-of-pipe treatment or even ecological modernisation of those sectors. It means a structural change of their societal role and importance. And this is a form of intervention for which capacity is still very low. Scepticism as to whether it ever will be developed seems plausible. But who could have imagined the environmental policy capacity of today in 1970?

The following case studies on the national environmental policies of different countries will show how far this policy field has already developed, which factors have contributed to progress in environmental protection and which restrictions are still to be overcome.

References

Andrews, R. N. L,. 1993: Long-Range Planning in Environmental and Health Regulatory Agencies, Ecology Law Quarterly 20, 515-582.

Badaracco, J. L., 1985: Loading the Dice. A Five-Country Study of Vinyl Chloride Regulation. Boston and London: MIT Press.

Bartelmus, P., 1994: Environment, Growth and Development – The Concepts and Strategies of Sustainability. London, New York: Routledge.

Bartlett, R. V., 1994: Evaluating Environmental Policy – Success and Failure, in: N. J. Vig and M. E. Kraft (eds.), Environmental Policy in the 1990s. Toward a New Agenda (second edition). Washington, D.C.: CQ Press.

Binder, L., J. S. Coleman, J. LaPalombara, L. W. Pye, S. Verba and M. Weiner, 1971: Crises and Sequences in Political Development. Princeton, N. J.: Princeton University Press.

Binder, M., 1996: Die Operationalisierung umweltpolitischen Erfolges: Probleme und Lösungsansätze, in: M. Jänicke (ed.), Umweltpolitik der Industrieländer. Entwicklung – Bilanz – Erfolgsbedingungen. Berlin: edition sigma, 133-152.

Binder, M., 1996a: Mögliche Erfolgsbedingungen der Schwefeldioxidminderung im internationalen Vergleich, in: M. Jänicke (ed.), Umweltpolitik der Industrieländer. Entwicklung – Bilanz – Erfolgsbedingungen. Berlin: edition sigma, 153-201.

Böhmer-Christiansen, S. and J. Skea, 1991: Acid Politics: Environmental and Energy Policies in Britain and Germany. London, New York: Belhaven.

Brickman, R., S. Jasanoff and T. Ilgen, 1985: Controlling Chemicals: The Politics of Regulation in Europe and the United States. Ithaca, New York: Cornell University Press.

Brinkmann, J. and M. Kirchgeorg, 1995: Umweltorientiertes Unternehmensverhalten, *Zeitschrift für Umweltpolitik und Umweltrecht* 18, 377-390

Calvert, J. W., 1989: Party Politics and Environmental Policy, in: J. P. Lester (ed.), Environmental Politics and Policy – Theories and Evidence. Durham, London: Duke University Press.

Carew-Reid J., R. Prescott-Allan, S. Bass, and B. Dalal-Clayton, 1994: Strategies for National Sustainable Development. A Handbook for Their Planning and Implementation. IUCN/IIED, London: Earthscan.

Carley, M., I. Christie, 1994: Managing Sustainable Development. London: Earthscan.

Conrad, J. (ed.), 1995: Successful Environmental Management in European Companies, Forschungsstelle für Umweltpolitik/Freie Universität Berlin (final report of a project for the European Commission, to be published in 1996).

Dutch Committee for Long-Term Environmental Policy, 1994: The Environment: Towards a Sustainable Future. Dordrecht, Boston, London: Kluwer Academic Publishers.

Enloe, C. H., 1975: The Politics of Pollution in a Comparative Perspective. Ecology and Power in Four Nations. New York: David McKay.

Feick, J., 1992: Comparing Comparative Policy Studies – A Path Towards Integration, *Journal of Public Policy* 12, 257-285.

Fey, A., 1994: Umweltsituation und Umweltpolitik in der ausgehenden Sowjetunion. Frankfurt/M. u. a.: Peter Lang.

Fiorino, D. J., 1995: Making Environmental Policy. Berkeley, Los Angeles, London: University of California Press.

Fischer, K. and J. Schot (eds.), 1993: Environmental Strategies for Industry – International Perspectives on Research Needs and Policy Implications. Washington, D. C.: Island Press.

Glasbergen, P. (ed.), 1995: Managing Environmental Disputes – Network Management as an Alternative. Dordrecht, Boston, London.

Gothenburg Research Institute, 1995: The Nordic Business Environmental Barometer. Oslo: Bedriftsoekonomens Forlag.

Haas, P. M., 1992: Introduction: Epistemic Communities and International Policy Coordination, *International Organization* 46, 1, 1-36.

Häberle, T. and W. Herbert, 1992: Umweltbewußtsein bei Experten und Bevölkerung (Zwischenbericht zum Projekt). Mannheim: Forschungsstelle für gesellschaftliche Entwicklungen der Universität Mannheim.

Héritier, A. (ed.), 1993: Policy-Analyse – Kritik and Neuorientierung, *Politische Vierteljahresschrift*, Special Issue 24. Opladen: Westdeutscher Verlag.

Hey, C. and U. Brendle, 1994: Umweltverbände und EG. Strategien, Politische Kulturen und Organisationsformen. Opladen: Westdeutscher Verlag.

Hoffman, A. J., 1994: The Environmental Transformation of American Industry: An Institutional Account of Environmental Strategies in the Chemical and Petroleum Industries. Unpublished paper, MIT, Dep. of Civil and Environmental Engineering.

Huntington, S. P., 1965: Political Development and Political Decay, *World Politics* 17, 3, 386-430.

Inglehart, R., 1989: Kultureller Umbruch – Wertwandel in der westlichen Welt. Frankfurt/M., New York: Campus.

Jacobson, H. K. and E. Brown Weiss, 1994: Strengthening Compliance with International Environmental Accords: Some Preliminary Observations from a Collaborative Project. Paper presented at the XVI World Congress of the International Political Science Association Convention, 21-25 August, Berlin.

Jahn, D., 1993: The Rise and Decline of New Politics and the Greens in Sweden and Germany, *European Journal of Political Research* 24, 177-194.

Jänicke, M., 1973: Die Analyse des politischen Systems aus der Krisenperspektive, in: M. Jänicke (ed.), Politische Systemkrisen. Köln: Kiepenheuer & Witsch.

Jänicke, M., 1990: State Failure. Cambridge: Polity Press.

Jänicke, M., 1992: Conditions for Environmental Policy Success: An International Comparison, *The Environmentalist* 12, 1, 47-58.

Jänicke, M., 1993: Über ökologische und politische Modernisierungen, Zeitschrift für Umweltpolitik und Umweltrecht 16, 159-175.

Jänicke, M. (ed.), 1996: Umweltpolitik der Industrieländer. Entwicklung – Bilanz – Erfolgsbedingungen. Berlin: edition sigma.

Jänicke, M. and H. Mönch, 1988: Ökologischer und wirtschaftlicher Wandel im Industrieländervergleich. Eine explorative Studie über Modernisierungskapazitäten, in: M. G. Schmidt (ed.), Staatstätigkeit. International und historisch vergleichende Analysen, *Politische Vierteljahresschrift*, Special Issue 19. Opladen: Westdeutscher Verlag, 391-405.

Jänicke, M., H. Mönch and M. Binder, 1996: Umweltindikatorenprofile im Industrieländervergleich: Wohlstandsniveau und Problemstruktur, in: M. Jänicke (ed.), Umweltpolitik der Industrieländer. Entwicklung - Bilanz - Erfolgsbedingungen. Berlin: edition sigma, 113-131.

Jänicke, M., H. Mönch and M. Binder, 1996a: Getting Rich - Getting Clean? Umweltindikatorenprofile im Industrieländervergleich, Zeitschrift für angewandte Umweltforschung (ZAU) 9, 1, 41-55.

Jänicke, M. and H. Weidner (eds.), 1995: Successful Environmental Policy. A Critical Evaluation of 24 Cases. Berlin: edition sigma.

Jörgens, H., 1996: Die Institutionalisierung von Umweltpolitik im internationalen Vergleich, in: M. Jänicke (ed.), Umweltpolitik der Industrieländer. Entwicklung - Bilanz - Erfolgsbedingungen. Berlin: edition sigma, 59-111.

Kelman, S., 1981: Regulating America, Regulating Sweden: A Comparative Study of Occupational Safety and Health Policy. Cambridge, Mass.: MIT Press.

Khator, R., 1993: Recycling: A Policy Dilemma for American States? *Policy Studies Journal*, 21, 2, 210-226.

Keck, O., 1993: Information, Macht und gesellschaftliche Rationalität. Das Dilemma rationalen kommunikativen Handelns, dargestellt am Beispiel eines internationalen Vergleichs der Kernenergiepolitik. Baden-Baden: Nomos Verlagsgesellschaft.

Kern, K. and S. Bratzel, 1996: Umweltpolitischer Erfolg im internationalen Vergleich: Zum Stand der Forschung, in: M. Jänicke (ed.), Umweltpolitik der Industrieländer. Entwicklung – Bilanz – Erfolgsbedingungen. Berlin: edition sigma, 29-58.

Kitschelt, H., 1983: Politik und Energie. Frankfurt/M., New York: Campus.

Kitschelt, H., 1986: Political Opportunity Structures and Political Protest: Anti-Nuclear Movements in four Democracies, *British Journal of Political Science* 16, 1, 57-85.

Knoepfel, P., 1993: New Institutional Arrangements for the Next Generation of Environmental Policy Instruments: Intra- and Interpolicy Cooperation. Cahiers de l'IDHEAP no 112. Lausanne: Institut des Hautes Etudes en Administration Publique.

Lester, J. P., (ed.) 1995: Environmental Politics and Policy – Theories and Evidence. 2nd edition. Durham, London: Duke University Press.

Low, P., 1992: International Trade and the Environment: An Overview, in: P. Low (ed.): International Trade and the Environment, World Bank Discussion Papers 159, Washington, D. C.

Luhmann, N., 1990: Ökologische Kommunikation (third edition). Opladen: Westdeutscher Verlag.

Lundquist, L. J., 1980: The Hare and the Tortoise: Clean Air Policies in the United States and Sweden. Ann Arbor: University of Michigan Press.

Nelkin, D. and M. Pollak, 1981: The Atom Besieged. Extraparliamentary Dissent in France and Germany. Cambridge and London: MIT Press.

Nullmeier, F., 1993: Wissen und Policy-Forschung. Wissenspolitologie und rhetorisch-dialektisches Handlungsmodell, in: A. Héritier (ed.), Policy-Analyse. Kritik und Neuorientierung, *Politische Vierteljahresschrift*, Special Issue 24. Opladen: Westdeutscher Verlag, 175-196.

OECD, 1992: Market and Government Failures in Environmental Protection: The Case of Transport. Paris: OECD.

OECD, 1993: Taxation and Environment: Complementary Policies. Paris: OECD.

OECD, 1994: Capacity Development in Environment. Paris: OECD.

OECD, 1994a: Urban Energy Management. A Handbook of Good Local Practices, Heidelberg Conference, Sept. 7-9, 1994. Paris: OECD.

OECD, 1995: Developing Environmental Capacity. A Framework for Donor Envolvement. Paris: OECD.
OECD, 1995a: Planning for Sustainable Development. Paris: OECD.
OECD, 1996: Environmental Performance in OECD Countries. Paris: OECD.
Paehlke, R., 1990: Democracy and Environmentalism: Opening a Door to the Administrative State, in: R. Paehlke and D. Torgerson (eds.), Managing Leviathan. Environmental Politics and the Administrative State. Ontario, Lewiston, NY: broadview press, 35-51.
Porter, M. and C. van der Linde, 1995: Green and Competitive: Ending the Stalemate, *Harvard Business Review*, September-October 1995, 120-134.
Potier, M., 1994: Agreement on the Environment, *OECD-Observer* 189, 8-11.
Prittwitz, V. v., 1994: Politikanalyse. Opladen: Westdeutscher Verlag.
Pye, L. W., 1966: Aspects of Political Development. Boston: Little, Brown and Co.
Ricken, C., 1995: Nationaler Politikstil, Netzwerkstrukturen sowie ökonomischer Entwicklungsstand als Determinanten einer effektiven Umweltpolitik – Ein empirischer Industrieländervergleich, Zeitschrift für Umweltpolitik und Umweltrecht, 18, 4, 481-502.
Ringquist, E. J., 1993: Environmental Protection at the State Level – Policies and Progress in Controlling Pollution. London: Armonk.
Roth, R., 1994: Demokratie von unten. Neue soziale Bewegungen auf dem Wege zur politischen Institution. Köln: Bund-Verlag.
Sabatier, P. A. and H. Jenkins-Smith (eds.), 1993: Policy Change and Learning: An Advocacy Coalition Approach. Boulder/CO: Westview.
Scharpf, F. W., 1987: Sozialdemokratische Krisenpolitik in Europa. Frankfurt/M., New York: Campus.
Scharpf, F. W., 1991: Die Handlungsfähigkeit des Staates am Ende des zwanzigsten Jahrhunderts, *Politische Vierteljahresschrift* 32, 4, 621-634.
Schmidt, M. G., 1986: Politische Bedingungen erfolgreicher Wirtschaftspolitik. Eine vergleichende Analyse westlicher Industrieländer (1960 – 1985), *Journal für Sozialforschung* 26, 3, 251-273.
Schmidt, M. G., 1993: Theorien in der international vergleichenden Staatstätigkeitsforschung, in: A. Héritier (ed.), Policy-Analyse – Kritik and Neuorientierung, *Politische Vierteljahresschrift*, Special Issue 24. Opladen: Westdeutscher Verlag, 371-393.
Shonfield, A., 1969: Modern Capitalism. The Changing Balance of Public and Private Power. London et al.: Oxford University Press.
Shrivastava, P., 1992: Corporate Self-Greenewal: Strategic Responses to Environmentalism, *Business Strategy and the Environment* 1, 3, 9-21.
Solesbury, W., 1976: Issues and Innovation in Environmental Policy in Britain, West Germany, and California, *Policy Analysis* 2, 1-38.
Steger, U. (ed.), 1992: Handbuch des Umweltmanagements – Anforderungs- und Leistungsprofile von Unternehmen und Gesellschaft. München: Beck.
Strom, K. S. Swindle, 1993: Political Parties, Institutions and Environmental Reform. University of California, Center for German and European Studies: Working Paper 2.17.
Traxler, F. and B. Unger, 1991: Institutionelle Erfolgsbedingungen wirtschaftlichen Strukturwandels, *Wirtschaft und Gesellschaft* 16, 189-223.
Vowe, G., 1994: Politische Kognition. Umrisse einer kognitionsorientierten Ansatzes für die Analyse politischen Handelns, *Politische Vierteljahresschrift* 35, 423-447.

Weale, A., 1992: The New Politics of Pollution. Manchester, New York: Manchester University Press.

Weidner, H., 1996: Basiselemente einer erfolgreichen Umweltpolitik – Beschreibung, Analyse und Evaluation der Instrumente der japanischen Umweltpolitik. Berlin: edition sigma.

Weidner, H., R. Zieschank and P. Knoepfel (eds.), 1992: Umwelt-Information. Berichterstattung und Informationssysteme in zwölf Ländern. Berlin: edition sigma.

Zilleßen, H., P. C. Dienel and W. Strubelt (eds.), 1993: Die Modernisierung der Demokratie. International Ansätze. Opladen: Westdeutscher Verlag.

UNITED STATES

Richard N. L. Andrews

Introduction

Environmental problems in the United States reflect deeply rooted historical incentives promoting exploitation of natural resources for economic purposes. Distinctive features of U.S. environmental policy include a system of government that is constitutionally limited and divided, and imposes major restrictions on government power to restrict environmental exploitation by private property owners. However, it also provides unusually open citizen access to government decisions, including rights even to bring lawsuits against government itself as well as against business polluters to compel environmental protection. These features produce policies that are perhaps more accountable but also more unstable than in many other countries. Rather than a progressive maturation of policy capacity to intervene in environmental and other social problems, one sees periodic "swings of the pendulum" between policy control by advocates and opponents of such a capacity, between initiatives to create and initiatives to destroy such a capacity.

In 1970, the United States took a world leadership position in instituting far more authoritative national regulatory policies for environmental protection. These policies produced some clear benefits, but also imposed heavy burdens of compliance, and by the mid-1990s they were facing intensive political counter-attack, as part of a far broader "pendulum swing" from support to opposition towards government regulations and spending.

It is possible that the present debate will produce moderate reforms that are widely acknowledged to be needed. It is equally possible, however, that it will produce merely a lingering erosion in policy effectiveness, or even a severe reduction of national environmental policy capacity if some of the most radical proposals should be enacted. Whether such changes would be replaced by equal or better environmental results from policies at the sub-national or international level, or would in fact produce a fundamental historic reversal of commitment to environmental protection, remains to be seen.

1 Main Environmental Problems

The United States entered the "era of environmental protection" with widespread and serious environmental problems. These problems included:

- High rates of fossil fuel combustion per capita and per unit of economic output, resulting from a history of cheap and abundant domestic supplies of fossil fuels, and from policies promoting continued availability of cheap energy.
- Serious air and water pollution in industrial regions – despite significant displacement of coal combustion by cleaner-burning natural gas and electric energy – due especially to massive increases in industrial production during World War II. These increases continued throughout the 1950s and 1960s due to Cold War military subsidies, major increases in civilian consumption per capita, and growing export markets.
- Uncontrolled disposal practices for industrial wastes, which were usually discarded into unlined pits and lagoons, dumped on rural lands, or discarded with urban household and commercial wastes in unlined general-purpose dumps and landfills.
- Generally uncontrolled wastewater and waste management practices by cities and towns, including discharge of untreated or partially treated sewage into surface waters, ocean dumping of wastes by coastal cities, and open dumping and burning of wastes on nearby land by most other local governments.
- High and rising rates of air pollution from motor vehicles, due to high per-capita ownership and use of automobiles, large-scale public subsidies for highway construction and low-density suburban development, and resulting displacement of both passenger and freight transport from railroads to automobiles and trucks.
- Widespread water pollution from agricultural soil erosion, fertilisers, pesticides, and animal wastes, including both "non-point" agricultural sources in rural areas and concentrated animal confinement operations.
- Steady losses of natural landscapes, wetlands and other natural ecosystems, and endangered species, due especially to widespread low-density urbanisation (especially in attractive coastal regions), large-scale public infrastructure projects (especially dams and highways), increasingly intensive agricultural and forest products operations, and policies that promoted and reinforced these trends.

These problems became increasingly visible to the public during the 1960s, particularly as peace, prosperity, suburbanisation, and increased education shifted public values toward greater appreciation of nature; as the problems themselves worsened; as television made nation-wide patterns of environmental damage visible to more people; and as citizen advocacy groups mobilised to combat them.

Changes Since 1970. The United States since 1970 has achieved significant improvements in many of the problems on which it has concentrated environmental policy attention (U.S. Council on Environmental Quality, 1993). These include major reductions in most air pollutants, substantial reductions in urban and industrial water pollution, and significant increases in officially protected natural lands. The U.S. has ended the disposal of hazardous industrial wastes into urban solid waste landfills; it has reduced their quantities substantially, and required far more secure measures for treating the remainder; and it has substantially increased the proportion of wastes that is recycled. It also has significantly reduced rates of energy consumption per unit of economic production.

Emissions of most air pollutants except nitrogen oxides, for instance, have dropped by one-third to one-half since 1970, and of lead almost totally. Emissions per capita of volatile organic compounds (VOCs), highest among the OECD countries in 1970, were reduced by almost 50% from 1970 to 1990. The "pollution standards index" (PSI) for major U.S. urban areas, perhaps the best integrated indicator of air pollution exposure, also improved by 50% from 1980 to 1991. Releases of reportable toxic chemical pollutants decreased by over 30% in just four years after reporting requirements were instituted. For water quality, violation rates for bacteria decreased from 36% in 1975 to 20% in 1989.

U.S. industrial energy consumption per constant US$ GDP has declined by almost 45% since 1970, and per capita energy consumption declined despite a growing economy. The use of agricultural chemicals has levelled off since about 1980, after continuous prior annual increases since the 1940s. Park and wilderness areas increased by 160% from 1969 to 1987, wildlife areas more than tripled, and nearly 10,000 miles of rivers were protected as wild and scenic rivers; but urban areas also nearly doubled during the same time period.

On the negative side, however, some environmental problems remain stubbornly unsolved, some fundamental trends remained problematic, and policies that had been reasonably effective for dealing with large industrial sources were proving distinctly less successful for dealing with the large numbers of smaller and more dispersed sources that remained. It is possible therefore that some of the successes to date have merely slowed or postponed problems rather than producing permanently sustainable solutions.

NO_x air emissions, for instance, remained stubbornly constant, offsetting air pollution reductions by industry and per vehicle, and the number of vehicle miles travelled continued to increase more rapidly than gains in fuel efficiency (68% versus 60% respectively, from 1970 to 1991). Data for most water pollutants other than bacteria remained ambiguous: most monitoring stations showed no trend, but of those that did show any trend, somewhat more showed increases in pollution than decreases. An estimated 40% of water pollution in U.S. rivers comes from agriculture, yet it remained unregulated; shellfish bed closures increased dramatically in some regions; and many fisheries declined to their lowest levels ever, due to both overfishing and pollution (Safina, 1994).

Recycling increased significantly, from 7% in 1970 to over 17% in 1990, but these rates were still far below those of many other countries; generation of solid waste also continued to increase far more rapidly than recycling, both absolutely (by 60%) and per capita (from 3.27 to 4.3 pounds per day).

More generally, the actual rate of energy use in the U.S. remains high, as does its production of polluting chemicals; production of pollutant chemicals continued undiminished: organic chemical pesticide production declined significantly from its peak production year (1981), but is still almost 11% higher than in 1971, and petroleum production was over 40% higher in 1990 than in 1971. Urbanisation of the landscape, and related destruction of wetlands and other natural habitats, also continues with little policy guidance or control.

2 Development and Main Characteristics of Environmental Policy

The United States has many decades' experience in the protection of natural lands by reserving them as national parks, public forests, wildlife refuges, wilderness areas, and other special designations. Its experience with national policies to control pollution, however, dates only to the past several decades.

Recent U.S. environmental policy has developed in approximately five "stages". Stage 1 (before 1970) was a *pre-federal* period, in which responsibility for pollution control was left almost exclusively to sub-national governments. Some of the most polluted cities made significant progress in reducing SO_2 and particulates from coal combustion, but few states developed the capability or the political will to control pollution more generally.

In Stage 2 (1970-1980) this pattern was superseded by *federal primacy*, in which a national institutional capacity was created (the Environmental Protection Agency in particular) and authorised to establish minimum ambient standards, explicit control measures, and mandatory elements of sub-national environmental policies and programs. A broad policy statute, the National Environmental Policy Act (1970), established a nation-wide policy goal of harmonising human activities with environmental conditions, and required "environmental impact statements" for all major federal actions; it also created a Council on Environmental Quality to advise the President and oversee implementation.

The Clean Air and Clean Water Acts (1970, 1972) established national goals for air and water quality and required permits – based on "best practicable technology" for each sector – for major point sources of pollutant emissions. They also set percentage reduction requirements for motor vehicle emissions, and directed EPA to set standards for toxic air and water pollutants, but they did not control "non-point" sources such as agriculture. Several additional laws authorised EPA to regulate individual toxic chemicals in pesticides, drinking water, and industrial production; and two more – the Resource Conservation and Recovery Act (1976) and "Superfund" Act (1980) – regulated waste disposal facilities and practices, particularly for industrial hazardous wastes, and financed cleanup of contaminated sites (Portney, 1990).

In Stage 3 (1980-1983), the Reagan administration attempted to unilaterally *dismantle* the national institutional capacity that had been developed over the previous decade, in the name of freeing businesses from excessive government regulation. Major actions included appointment of ideological loyalists rather than competent professionals to senior positions in the agencies; severe budget reductions (33% of EPA's operating budget); rewriting or rescinding more than 50 regulations, and delaying most new ones; reducing enforcement by 50%; and reducing environmental grants to state and local governments by more than one-third. This initiative ultimately produced a major political backlash, and reaffirmation of national environmental policies as well as dismissal of nearly all the senior environmental officials appointed by the Reagan administration. However, it also caused lasting damage to U.S. environmental policy capacity, including serious disruption of EPA's programs, undermining of more moderate reform initiatives, and enactment of even more rigid and prescriptive laws to prevent any repeat of such administrative reversals (Vig and Kraft, 1984).

Stage 4 (1984-1994), then, was a period of *legislative micro-management*, in which the Congress reacted against this presidential counter-revolution by enacting far more prescriptive statutory requirements. It was also a period of increased enforcement actions by the EPA, including for the first time criminal indictments, and increased lawsuits against polluters by environmental groups (U.S. EPA, 1992). These pressures produced significant increases in regulatory activity, compliance, and environmental results, but also imposed far more burdensome requirements on small businesses and local governments as well as other pollution sources.

Finally, Stage 5 began with the 1994 Congressional elections, which transferred control over the national legislature not only from the Democratic to the Republican party, but to an exceptionally ideological group of new legislators whose central priority was reducing the power and capacity of the national government. In this current stage of environmental policy, traditional opponents of regulation were joined by many small businesses – and ironically, by many local governments – that had been burdened by the prescriptive environmental laws passed in reaction to the Reagan attacks (Andrews, 1996). In effect, U.S. environmental policy is now repeating the political battles of the Reagan era only with roles reversed, with its opponents now in control of the legislature and its supporters in the White House. Opinion polls clearly showed continued public support for environmental protection, and opposition to weakening government capacity to assure it; but the polls also showed widespread public dissatisfaction with government regulation and spending in general, and environmental regulation was one of the most visible examples of such regulation. The future of U.S. environmental policy capacity therefore will depend heavily on which forces are most successful in framing the political choices between these preferences.

One possible outcome of this fifth stage could be a set of moderate reforms, which would increase flexibility in priorities and means of compliance but still maintain effective federal authority to assure that national environmental policy goals are achieved. A series of such reforms was recommended in early 1995 in a report commissioned by the congressional appropriations committees from the National Academy of Public Administration. This report recommended (among others) that EPA should offer "accountable devolution" of major responsibilities to state governments, and greater operating flexibility to target companies that were willing to go "beyond compliance" in their own environmental improvements – somewhat like the Dutch covenants approach (National Academy of Public Administration, 1995).

These recommendations were well received, and EPA itself is now implementing some of them and seriously considering others. This "moderate reform" scenario, if supported by the Congress, would probably lead to continued progress by those states and target firms that had both the commitment and the resources to do so, but it might also lead to stagnation or regress in those that lacked either the resources or the will to continue without strong national pressure (Lester, 1995). Even leading states and firms might also be undermined by opportunistic competition from others.

A second possible outcome of the fifth stage, however, could be a radical weakening of national environmental policy capacity, if some of the more radical congressional proposals should be enacted. In early 1995, for instance, the Congress enacted a budget amendment which reopened many national forests to far more intensive logging, and which included sweeping "supermandate" language stating that such logging is considered automatically in compliance with all environmental laws. Similarly damaging provisions were proposed to weaken other environmental policies, particularly the Endangered Species Act and wetland protection regulations. Others would require government compensation for any reduction of property values resulting from environmental regulations ("takings"); require compensation for any additional federal requirements on state and local governments ("unfunded mandates"); restrict EPA from regulating and enforcing against particular industries; and in some cases, rewrite and weaken the basic environmental laws themselves.

Finally, a third possible outcome could be simply a continued political stalemate, which would leave most of the current institutional capacity in place but reduce its effectiveness by cutting budgets and increasing paperwork burdens. Proposed budget appropriation bills, for instance, would reduce EPA's budget more than any other major agency; and concurrent proposals would require burdensome analyses of risk and cost for new regulatory proposals, and open them to increased legal challenges by target groups.

The continuing political volatility of the U.S. experience suggests caution in attributing any consistent direction or linear process of "capacity-building" to these stages. Policy-making in the United States is more commonly described as a "pendulum," swinging periodically back and forth between pro- and anti-government dominance, rather than as a steady process of capacity development. The stages are sequential in time, but by no means inevitable, and represent oscillation in political power and institutional character.

3 Main Actors

The chief actors in U.S. environmental policy-making include not only legislatures and administrative agencies, but also the courts, voluntary organisations, philanthropic foundations, and the news media.

A distinctive feature of the U.S. governance system is its constitutional fragmentation: between national and sub-national governments, among separate branches of the national government (legislative, administrative, and judicial) and even within these branches (Senate vs. House, president vs. administrative agencies). This fragmentation was purposeful, intended to create "checks and balances" against strong concentrations of government power in particular factions or elites. In recent years it has often produced divided governments, in which the majority of the Congress are of a different political party than the President. It also produces unusually open opportunities for political mobilisation and influence, making its policies more responsive to political pressures but also far less stable than in other systems. It is also vulnerable to political stalemate ("gridlock"), or to divergence between symbolic and actual results, when powerful interests conflict.

The United States has a distinctively two-party rather than multi-party political system. Both of its dominant parties have generally functioned as broad alliances, differentiated only by somewhat general leanings favouring business or labour and lesser or greater government action: the Republican party has included both social and fiscal conservatives as well as more moderate progressives, and the Democratic Party has included southern conservatives as well as classic liberals and supporters of more socialist policies. Each party includes both supporters and opponents of environmental policy, though in recent years these issues have become more partisan, with Democratic leadership predominantly supportive and Republicans more consistently hostile; and the ideological lines between the parties appear to be sharpening more generally, at least at the national level and in some states. The U.S. has no effective "Green Party," nor any other influential third party in recent decades.

The U.S. Congress plays a far stronger and more direct role in environmental policy than do most parliaments. In many parliamentary democracies, legislation takes the form of broad principles – the "precautionary" principle, the "polluter pays" principle, the "subsidiarity" principle – whose details are worked out in administration by the ministries and in more specific statutes by sub-national governments. Many U.S. environmental laws, in contrast, consist of detailed requirements to be carried out by administrative agencies and state governments. In recent years these mandates have been further tightened by adding deadlines and "hammer clauses," which specify automatic (and often severe) statutory consequences if the deadlines are missed.

This approach severely limits the discretion of administrative agencies, subnational governments, and even the president, to set overall environmental policy goals and priorities and to plan and implement measures to achieve them. Since the 1970s the Congress has also proliferated large numbers of subcommittees, thus dispersing power to advance legislation among more members (and thus in a sense increasing policy capacity), but at the same time fragmenting this capacity and burdening the agencies with accountability to more and more subcommittees. EPA alone, for instance, is answerable to as many as 70 Congressional subcommittees.

In the Executive branch, there is no single Ministry of Environment. The Environmental Protection Agency exercises regulatory authority over pollution problems under at least eight separate laws, and administers subsidies for public wastewater treatment plants and cleanup of contaminated sites, but it has no general authority to coordinate overall environmental policy. Policies of other mission-oriented, sectoral agencies – such as the Departments of Agriculture, Defence, Energy, Housing, Interior, and Transportation, and their subsidiary agencies – often have far greater impacts on environmental problems, both positive and negative, than does EPA; and each of these agencies has its own history of brokering the political demands of competing environmental user constituencies (Graham, 1976).

The courts are a particularly important element of U.S. environmental policy capacity, since they are politically independent of the legislature and executive and can rule on both the constitutionality of the laws and the administrative agencies' regulations interpreting them. Since 1970 many environmental statutes have established specific rights for environmental advocates to challenge not only polluters but government agencies as well, to compel them to protect the environment, and this right of access has greatly increased both access and influence for green organisations in regulatory proceedings as well. The courts in turn have shown themselves willing to review in detail the basis of the agencies' decisions, and to overrule them in response to challenges by both industries and green organisations.

State and local governments are also important actors in U.S. environmental policies, though U.S. federalism differs from that of other government systems such as Germany and the European Union. On the one hand, the constitutional supremacy of the U.S. national government has allowed it to enact not just broad principles, but detailed mandates that supersede the laws of state and local governments, and to require state and local governments to implement them if they wish either to receive funds from the federal government or to avoid direct federal regulation of their industries. On the other hand, the national government does not have the capacity to implement most of these mandates without active state and local cooperation, and so the actual outcomes are in effect negotiated. The results vary widely: some states have both strong resource capacity and political will to assure environmental protection, and have repeatedly produced innovations that have been imitated by others; other states have the resources or the political will but not both, and some have neither (Lester, 1995). Primary authority over land use, environmental services such as water supply and waste management, and many other policies also remains with state and local governments.

The United States is distinctive in both the large number and diversity of its "green" organisations (Gottlieb, 1994; Dunlap and Mertig, 1992). Many of these organisations are also far more active and effective in policy advocacy than their counterparts in other countries. One major class of such organisations is national membership groups, such as the Sierra Club, Wilderness Society, National Wildlife Federation, Greenpeace, and others. Some of these have active local chapters or affiliates, while others operate mainly through "chequebook membership" to support professional staffs at the national level. A second class includes groups that have smaller memberships but greater support from foundations and other donors to analyse and advocate environmental policy proposals: examples include several environmental law groups in their early years, such as the Environmental Defence Fund and Natural Resources Defence Council (both of which have since developed larger memberships), as well as "think tanks" such as Resources for the Future, the Rocky Mountain Institute, and the Environmental Law Institute.

A third class of green organisations includes many local and *ad hoc* groups, often created to protect their own communities or to oppose specific proposals such as waste disposal facilities or major construction projects. Some of these have formed national networks as well, such as the Citizens' Clearinghouse for Hazardous Wastes, the environmental justice movement, and the Student Environmental Action Coalition. Finally, U.S. "green" groups include many organisations of environmental scientists and professionals – the Air and Waste Management Association, Water Pollution Control Federation, National Association of Environmental Professionals, and others – which advocate environmental protection based on their professional principles and interests.

Other organisations have also been important "green" influences. Civic organisations such as the League of Women Voters, and associations of municipal governments themselves (National League of Cities, Conference of Mayors, and others) have all played key roles in promoting environmental policy in the U.S. Often overlooked, but enormously influential, have been the many philanthropic foundations – Ford, Hewlett, Joyce, MacArthur, Mellon, Pew, Rockefeller, and many others – that have supported U.S. green organisations with millions of dollars of resources.

By the early 1990s, however, conservative businesses and foundations had also become increasingly effective in copying some of the tactics of pro-environmental groups, such as by funding their own ideologically-defined advocacy groups and "think tanks" to counter the green organisations. These organisations often had civic-, populist-, or green-sounding names masking their business support and anti-green agendas. Examples included the Heritage Foundation, Competitive Enterprise Institute, and Mountain States Legal Foundation; and the "Alliance for America," "Endangered Species Co-ordinating Council," and "Wise Use" movement. To these tactics, anti-regulatory groups also added sophisticated use of mass-mailing, advertising, electronic communication techniques, and computerised "fax attacks" to generate political pressure.

The U.S. has a growing but poorly organised community of businesses that benefit from environmental protection. These include environmental technology and services vendors, tourism, recreation, and outdoor equipment businesses, businesses that have made costly fixed investments in compliance, and a growing number of others that have created market niches as "green" companies. A few labour unions have also played influential roles (the Oil, Chemical and Atomic Workers, for instance), though others have allied themselves with management in protecting existing jobs at the expense of environmental improvements. Banks and insurance firms have also been drawn inadvertently into roles as environmental protection auditors, since they may otherwise find themselves liable for the costs of their clients' practices.

While some businesses have taken strongly pro-environmental positions, however, many have not. U.S. businesses have deeply rooted adversarial attitudes toward both labour and government, in contrast to the corporatist cooperative cultures that are more typical of western Europe and Japan, and these attitudes are often reciprocated in anti-corporate populist attitudes of environmental advocates and the general public.

The strongest business opposition typically comes from economically powerful, high-pollution sectors such as mining, logging, agribusiness, and some chemical firms; from some business and trade associations, which typically represent the lowest common denominator rather than the leading firms in their sectors (for instance the U.S. Chamber of Commerce, National Association of Manufacturers, and others); from ideological media opponents of government regulation, such as *The Wall Street Journal* and some columnists; and from the National Federation of Independent Businesses, representing many small firms which are more heavily burdened by regulations than larger ones and often lack the capital and expertise to modernise.

The mass media have played a vital role in building public awareness and support for U.S. environmental policy. Media influence includes not only newspapers, news magazines and television news programs but also televised documentaries, popular science and outdoor magazines (*Environment, National Geographic, Scientific American*, and others), and magazines published by green organisations themselves. By the early 1990s, however, anti-green organisations had become increasingly effective in disseminating revisionist critiques and "oped" arguments, both to try to discredit environmental issues directly and to reframe them as issues of government over-regulation (Brick, 1995). At some influential newspapers, including *The New York Times*, strongly pro-environmental reporters were also replaced by more sceptical successors. By late 1995, however, the extreme nature of some of the anti-environmental legislative agenda had begun once again to generate widespread mass-media support for environmental policy.

4 Capacity-Building

Environmental policy capacity in the United States has developed in patterns that are far more uneven, adversarial, and vulnerable to political instability than is often recognised.

Within government, the federal Environmental Protection Agency was created in 1970 in a reorganisation which brought into one agency the functions of regulating pollution and waste management, subsidising municipal wastewater treatment, and related research and development programs. It has since been given sweeping statutory powers to expand these functions, and by 1995 it received an annual budget of approximately US$7 billion. It has never been given an overall statutory mission, however, and most of its programs therefore are not well integrated with one another. Its priorities are determined mainly by annual budget decisions, which have always been far less than necessary to implement all its official responsibilities; and more than half its budget in any case is earmarked for cleanup of contaminated sites and construction of wastewater treatment plants (Andrews, 1993).

The U.S. has developed probably the most extensive scientific and technical capacity of any country to support environmental policy making. This is mainly due to its requirements that government agencies explicitly justify their regulatory proposals in public and in the courts, including documentary records providing "substantial evidence" in support of proposed regulations. As a result, extensive capacity for scientific, engineering, and economic analysis of proposed regulations has developed both in government agencies themselves – such as EPA's Office of Research and Development and Science Advisory Board – and in consulting firms, regulated businesses and trade associations, applied research institutes, green organisations, and universities.

Many U.S. states have also greatly increased their capacity for environmental policy, in response both to public pressure and to federal requirements. Some leading states have acted far beyond federal requirements, serving as "laboratories of innovation" to be copied by others. Examples include California's tough air quality standards and carcinogen labelling requirements, New Jersey's "community right-to-know" and mandatory recycling laws, Massachusetts' Toxic Use Reduction Act, Wisconsin's negotiated siting law, and the Oregon Benchmarks program for environmental policy planning. Others, however, have done only what was required of them, and some state legislatures have even explicitly prohibited any regulatory initiatives beyond those required by federal legislation.

Since 1970 there has also been an unprecedented increase in the number, strength, professionalisation and effectiveness of proponents of environmental protection. Many environmental advocacy groups have been created, and the leading ones have developed substantial staff capabilities – lawyers, economists, scientists, and publicists as well as political lobbyists – to promote environmental protection. This capacity has fluctuated substantially over time, however, and is now in a period of great uncertainty (Downs, 1971; Dunlap and Mertig, 1992; Gottlieb, 1994).

Environmental groups grew rapidly in the early 1970s, then eroded somewhat as the Arab oil embargo and economic fears partially displaced environmental concerns from the political agenda; mass mobilisation tactics were partially replaced by staff-led initiatives. Green groups' support grew again as fears of toxic chemicals and hazardous wastes re-ignited public concern, then again softened as a more pro-environment President was elected (Carter) and government policies appeared to address the problems and economic worries again dominated public attention. Membership grew dramatically again in the 1980s, in opposition to the Reagan administration's attempts to undermine national environmental policy mandates, but softened once more when these threats were averted. A serious danger emerged with the very success of the major environmental groups in institutional capacity-building, as their organisational balance subtly shifted from personal awareness and substantive commitment to larger-scale public relations, fund-raising and bureaucratic management (Gottlieb, 1993).

In the 1990s environmental proponents appeared poised for another major increase in effectiveness, with the rising visibility of global environmental issues (the ozone hole, global warming, the Rio Environmental Summit, and others) and appointment of many strong environmental leaders to influential positions in the Clinton Administration. Instead, however, their memberships and media influence declined once again, and they found themselves increasingly on the defensive against the most effectively organised anti-environmental forces they had yet faced. By the end of 1995 it was not yet clear whether they would succeed once again in mobilising mass public pressure to block major reversals of environmental policy.

4.1 Resources for Proponents and Opponents

Key resources for proponents of environmental protection in the U.S. include both money and information. Primary sources of money include membership contributions, which have fluctuated over time with the degree of public fears about environmental issues; grants from foundations, some of which have been among the most durable sources of support for environmental proponents; government grants for particular projects and studies; and penalties awarded by the courts in lawsuits against polluters. Environmental opponents under the Reagan Administration and in the present Congress have sought to cut off government and other sources of funding for environmental proponents – "defunding the Left," in their slogan – including even those which protect natural areas through market transactions, such as The Nature Conservancy.

One of the most powerful resources available to U.S. green organisations, however, is their extensive formal rights of access to government information and decision processes. The Administrative Procedures Act (1946) requires government agencies to justify their decision proposals publicly and in writing, and the Freedom of Information Act guarantees public access to most government documents on which decisions are based. The National Environmental Policy Act further requires the agencies to produce information on the environmental effects of their proposals ("environmental impact statements"), which opponents as well as other agencies and state and local governments can then review and challenge. Many environmental statutes also explicitly guarantee both citizen access to administrative decision processes, and citizen lawsuits to compel government agencies as well as businesses to protect the environment. These rights of access are not always sufficient to defeat opposing interests, which enjoy the same access and often have far greater economic resources, as well as traditional American concepts of property rights, in their favour. However, they have been a powerful tool for equalising access among economic and other interests such as green organisations.

In contrast to many other countries, U.S. capacity-building has not been strongly influenced by international events, except for a few specific treaties such as the Montreal Protocol (CFC phase-out), CITES (protection of endangered species), and trans-boundary agreements along its Canadian and Mexican borders. One finds little institutionalisation in the U.S., for instance, of many influential international ideas such as the "precautionary" or "polluter pays" principles, and even the principle of "sustainable development" remains primarily symbolic rhetoric – with competing interests each redefining it to suit their own conflicting agendas – rather than an influential basis for national policy development.

Some U.S. presidents have taken strong positions of international environmental leadership: President Nixon on controlling ocean dumping and trade in endangered species, the Stockholm Conference, creation of a World Heritage Trust, and limits on whaling, President Carter at least in his sponsorship of the *Global 2000* report. Others, however, have sharply retreated from such leadership: President Reagan withdrew U.S. support for the Law of the Sea Treaty and for international family planning programs, for instance, and President Bush's hard-line positions at the 1992 UN Conference on Environment and Development – the United States refused to consider any reduction in its consumption of fossil fuels and other resources, and was the one nation that refused to sign the Biodiversity Convention – left the U.S. increasingly isolated from international environmental leadership (although the Bush Administration did lead in supporting accelerated phaseout of CFCs).

An important emerging question is how U.S. environmental policies will be influenced by the current globalisation of finance and trade, and by related institutional forces including trade agreements (such as NAFTA and GATT) and other initiatives to harmonise international environmental standards (such as European Union eco-audit and product standards, and ISO-14000 certification of environmental management procedures). These are already having some influence on major U.S. businesses, and may perhaps force correction of some environmentally harmful practices even without further improvements in formal government policies. However, they may also clash with current U.S. proposals to devolve more environmental policy capacity to sub-national levels of government. International harmonisation will almost certainly lead to requests for exemptions from stricter U.S. requirements by firms complying with the international standards, and may also create new barriers to stricter or more innovative policies by state or local governments.

4.2 Degree of Policy Integration

U.S. environmental policy is not well integrated (Andrews, 1993). A National Environmental Policy Act was passed in 1970 to try to promote environmental policy integration: it included a visionary statement of national policy to promote harmonisation of human actions with the integrity of natural systems, as well as "action-forcing mechanisms" to promote implementation (the "environmental impact statement" requirement, for example), and a Council on Environmental Quality to oversee implementation and advise the President. However, it set no operational goals or "green plans" for the agencies, and provided no clear authority for policy integration superseding the agencies' existing mandates. Environmental impact statements helped block some of the worst specific proposals, but did not redirect the agencies toward environmental goals more generally.

The U.S. has no single department equivalent to a Ministry of the Environment, and even the Environmental Protection Agency has no single "organic act" establishing its mission and priorities and integrating its programs. The Council on Environmental Quality was somewhat influential in its early years, but its staff was eliminated by the Reagan Administration, and it has never regained significant influence over the many departments and agencies whose actions have significant environmental consequences. Environmental policy integration therefore remains largely personalised, in the initiatives of individual political leaders such as William Ruckelshaus and William Reilly at EPA and Vice President Al Gore and others in the Clinton Administration, rather than institutionalised.

In the 1980s, important new environmental protection incentives were successfully inserted into the legislation of several sectors, such as the agriculture and energy laws and the inter-modal transportation act, but it is not clear whether these initiatives will survive the new anti-environmental coalition in the Congress. Some current legislative proposals, in fact, might either explicitly remove these provisions or fundamentally change the more basic sectoral policies to which they were tied. Examples included proposals to radically reduce agricultural subsidies, which might also eliminate the incentives for environmental practices that were conditions of those subsidies; and proposals to break up energy utilities into separate economic units for power production and distribution, which might eliminate the distribution companies' incentive to promote conservation and might also increase price competition either from "greener" sources or from dirtier, less regulated coal-burning plants. Proposals to privatise many functions, or to devolve them from federal to state and local governments, might also exacerbate competitive market pressures at the expense of environmental protection.

Beginning in the early 1990s, EPA staff undertook a National Environmental Goals Project to propose specific environmental results that should be sought, and define operational objectives and benchmarks for tracking them; several states also developed similar projects. These initiatives could in principle provide a basis for a more integrated form of multi-sectoral "green plan," similar in principle to those of the Netherlands and elsewhere, but there was no evidence yet of the broader political support that would be necessary to implement such a process at the national level.

4.3 Development of Policy Paradigm

The primary U.S. paradigm for pollution control policy has been national minimum standards and program requirements, largely based on technological controls of industrial and municipal pollution sources and on substance-by-substance standards for particularly hazardous contaminants. Its primary policy tools have been federal minimum standards for purity in ambient air and drinking water; permit requirements forcing industries to implement the "best practice" of existing control technologies for air emissions, water pollution discharges, and waste disposal facilities; restrictions on the production and use of pesticides and a few other toxic chemicals; and federal subsidies for municipal wastewater treatment and cleanup of contaminated sites.

The paradigm of technology-based, "best-practice" regulations has been generally successful in reducing emissions from automobiles, as well as from many major industrial and municipal pollution sources. It has also been costly and often economically inefficient, however, and is encountering increasing political resistance as it is applied to smaller, more dispersed, and less affluent businesses. In recent years, therefore, EPA administrators and others have been among the leaders in advocating fundamental changes in this paradigm.

One proposed reform is to use "risk-based decision-making," which would base priorities and regulatory and enforcement decisions on detailed quantitative estimates of risk to public health and the environment. A second proposal is to shift from "end-of-pipe" control technologies – which are often costly and inefficient, and merely remove pollutants from one medium to be deposited in another – to pollution prevention, which would reduce the underlying causes of pollution at their source: by using less harmful materials in production, recycling more of them, making production processes themselves more efficient, and reducing consumer use of environmentally damaging products. A third proposal is to use more "market-oriented incentives" – such as pollution taxes or charges, marketable emissions allowances, liability principles, and information disclosure requirements – in place of the more rigid traditional "command-and-control" regulations (U.S. EPA, 1990).

Some of these initiatives have been successfully introduced, but most only on a limited scale due to the constraints of existing laws. EPA has implemented risk-based decision-making in setting enforcement priorities and in recommending budget priorities, for instance, but it must still operate within the statutory limits and budget priorities set by the Congress. EPA advocates pollution prevention in its official statements, but its legal requirements for control technology are still in effect. EPA has pioneered several market incentives, but most of these remain experiments at the margins of the traditional paradigm rather than replacing it.

The U.S. has pioneered, for instance, in the use of marketable emissions allowances to increase economic efficiency in pollution reduction. These began with EPA's "bubble policy" in the 1970s, which allowed flexibility for firms to over-control some sources and undercontrol others; it was then extended to allow trading of emissions among different firms and sources as well, even for instance buying and destroying old high-polluting cars when that was cheaper for the same environmental benefit than controlling industrial emissions. Trading of production allowances was also allowed for phase-outs of leaded gasoline and CFCs, thus minimising excessive economic costs of phaseout; more recently, it is being used to create a nation-wide market for sulphur emission allowances for electric utilities, and potentially for other air and water pollutants as well.

Another of the most effective U.S. innovations was a 1986 law requiring businesses to report publicly each year the actual amounts of several hundred toxic chemicals they release into the environment (the "Toxics Release Inventory"). This information requirement dramatically increased both external media pressure and internal corporate awareness of the linkage between pollution and economic waste, encouraging pollution prevention (Gottlieb, 1995).

Other promising policy innovations, however, have been promoted far more widely than they have actually been implemented. Despite a large body of literature advocating unit-based pollution taxes and charges, for instance, the U.S. actually has almost none of them, except at the level of local water and waste authorities, and even these are more often set at levels simply to generate revenues or to pay for environmental services rather than to promote waste reduction.

Some leading businesses, however, have also made significant improvements in pollution prevention by their own decisions, either because they discovered unexpected market benefits or because it was simply cheaper to reduce the costs of regulatory procedures (Gottlieb, 1995). In reality, the most powerful market incentives in U.S. environmental policy, for better or worse, are not deliberate "policy tools" such as pollution taxes or emissions allowances, but the far greater costs of economic liability for pollution cleanup (especially for hazardous waste contamination), of waste disposal (due to EPA regulations requiring safer design and operation of waste facilities), and of the regulatory process itself, especially reporting requirements and the opportunity costs of regulatory delays in competitive industries.

Some of the most far-reaching innovations in the policy paradigm are now being attempted by the Clinton Administration, as part of its broader initiative to "reinvent government." These new approaches promote integrated sectoral and facility-wide approaches to replace fragmented permitting of air, water, wastes, and individual pollutants (the "Common Sense Initiative" and "Project XL"). They also offer the option of "performance partnership agreements" with state governments, which would allow more flexibility in priorities and approaches in exchange for agreement to achieve specific environmental results. Finally, they include experiments in integrated "place-based" policy-making for local and regional ecosystems, instead of uncoordinated permitting of each separate facility. The Clinton initiatives also include significant streamlining of EPA staff and regulatory requirements, and reorganisation of several of its ten regional offices from media-specific (air, water, wastes) to integrated units for pollution prevention and enforcement. These initiatives have not yet been widely implemented, let alone evaluated for their environmental and economic results.

5 Policy Evaluation

If we were to adopt as a principle that good environmental policy would gradually evolve from piecemeal toward more integrated approaches, and from cleanup and end-of-pipe control toward pollution prevention, ecological modernisation and sustainable development, how well has the United States done so far in developing its environmental policy capacity?

Overall, U.S. environmental policies have corrected some of the most acute problems, and slowed the worsening of some others. Their successes have been uneven and often limited, however, and many fundamental trends – in energy and resource use, in production and consumption of pollutants, and in transformation of ecosystems and landscapes, for instance – may only have been slowed somewhat rather than redirected into sustainable patterns.

The United States has pioneered in developing many innovative concepts and instruments of environmental policy, which have been influential in other countries and international agreements as well. Examples include environmental impact statements, quantitative and comparative risk assessments, tradable emissions allowances, information disclosure requirements, and the use of citizen lawsuits to compel environmental protection measures by both businesses and government – among others.

The basic structure of U.S. environmental policy, however, has remained entrenched in the paradigm of detailed federal standards and technology-based permits, rather than evolving further toward pollution prevention and ecological modernisation principles. Most of the reforms remain experiments at the margin of the regulatory mandates; the main force driving further progress often seems to be the increasing burden of conventional regulation itself rather than evolution of the policies into more advanced paradigms.

U.S. environmental policy initiatives have produced impressive results for some environmental problems, but have left others relatively untouched. They have been dramatically successful, for instance, in improving waste management practices, but far less so in controlling individual toxic chemicals. They have been relatively successful in controlling point sources of air and water pollution, but far less so in reducing non-point sources, so that their success in achieving overall environmental results has often fallen far short of their effectiveness in controlling some sources. They have reduced pollution dramatically from each individual motor vehicle, but have not stopped the constant increase in vehicle miles travelled and related transport and land use patterns, which will continue to erode the benefits of the technical measures in the future. Finally, most policies remain fundamentally focused on pollution control requirements despite rhetoric promoting pollution prevention, and some even pose serious disincentives to prevention alternatives.

Many of the measures that have been reasonably successful in reducing pollution from large point sources, moreover, are not easily transferable to pollution by other sources – such as small businesses, households, and individual behaviour – that are increasingly recognised as important contributors to the remaining problems. EPA requirements for effective inspection and maintenance programs for motor vehicles, for instance, aroused immediate and effective political opposition. There is simply far less political will to impose serious restrictions on farms, land development, small businesses, local governments, households, and individual behaviour such as motor vehicle use, than on large manufacturing industries. There is even less will to adopt such restrictions – or to increase foreign aid expenditures – in order to reduce these problems elsewhere in the world.

More generally, many of the strengths of the U.S. approach are also the sources of its most serious limitations. One is its fragmentation, which allowed rapid policy development and implementation but also burdened it with institutional rigidities and inefficiencies that are stubbornly resistant to further improvement. A second is its highly adversarial culture, which has given environmental proponents a more "level playing field" in regulatory politics but has not produced a more effective consensual process or stable results.

Perhaps most fundamentally, U.S. political culture has always harboured deep strains of self-interested individualism, on behalf of both individuals and business corporations, and of grass-roots populism. These forces have always been distrustful of "big government," but in a few periods have subordinated this distrust to an even greater distrust of "big business:" for instance, during the Progressive Reform era of the early 20th century, the New Deal era of the 1930s, and most recently the Environmental era of 1970 through the early 1990s.

With the 1994 elections, U.S. environmental policy has reached an important cross-roads. Anti-environmental business interests appear to have succeeded in re-establishing their traditional alliance with anti-government populists in opposing government regulation to protect the environment. It remains to be seen how successful they will be in actually reducing national environmental policy capacity or blocking its use, or whether more moderate reform advocates will succeed in forging compromise strategies that correct the most serious regulatory burdens while maintaining or even increasing its effectiveness.

References

Andrews, R. N. L., 1993: Long-Range Planning in Environmental and Health Regulatory Agencies, Ecology Law Quarterly 20, 3, 515-582.

Andrews, R. N. L., 1996: Risk-Based Decision Making, in: N. Vig and M. Kraft (eds.), Environmental Policy in the 1990s, 3rd edition. Washington, D.C.: CQ Press, Chapter 10 (in press).

Brick, P., 1995: Determined Opposition: The Wise Use Movement Challenges Environmentalism, Environment 37, 8: 17-20, 36-42.

Downs, A., 1971: Up and Down With Ecology: The Issue-Attention Cycle, The Public Interest (May).

Dunlap, R. and A. Mertig, 1992: American Environmentalism: The U.S. Environmental Movement, 1970-1990. Philadelphia, PA: Taylor & Francis.

Gottlieb, R., 1993: Forcing the Spring: The Transformation of the American Environmental Movement. Washington, D.C.: Island Press.

Gottlieb, R. (ed.), 1995: Reducing Toxics: A New Approach to Policy and Industrial Decisionmaking. Washington, DC: Island Press.

Graham, O. L. Jr., 1976: Toward A Planned Society: From Roosevelt to Nixon. New York: Oxford University Press.

Lester, J., 1995: Federalism and State Environmental Policy, in: J. P. Lester (ed.), Environmental Politics and Policy, 2nd edition. Durham, NC: Duke University Press, 39-60.

National Academy of Public Administration, 1995: Setting Priorities, Getting Results: A New Direction for EPA. Washington, D.C.

Portney, P., 1990: Public Policies for Environmental Protection. Washington, D.C.: Resources for the Future.

Safina, C., 1994: Where Have All the Fishes Gone?, Issues in Science and Technology (Spring), 37-43.

U. S. Council on Environmental Quality, 1993: Environmental Quality: 23rd Annual Report of the Council on Environmental Quality. Washington, D.C.

U. S. Environmental Protection Agency, 1992: Enforcement Accomplishments Report, FY 1991, Report No. 300-R92-008. Washington, DC: U. S. Government Printing Office.

U. S. Environmental Protection Agency, Science Advisory Board, 1990: Reducing Risk: Strategies for Environmental Protection. Washington, D.C.

Vig, N. and M. Kraft, 1984: Environmental Policies in the 1980s: Reagan's New Agenda. Washington, D.C.: CQ Press.

SWEDEN

Lennart J. Lundqvist

1 Main Environmental Problems and Developments Since 1970

With an area of almost 450,000 km^2, Sweden is the third largest nation in what was before 1989 called Western Europe. The distance from the southernmost shore to the northernmost border is greater than that between Copenhagen and Rome. Lakes and streams cover 9% of the country. Agricultural land is 8%, while 3% is populated. The rest – 80% – is forest, marshland, and mountains. If Sweden's total area were equally distributed among all inhabitants each Swede would have about 50,000 m^2 at his or her disposal (Cabinet Bill 1990/91:90, Part A, p. 70).

Abundant and pristine as this may look, there have been, and still are, some serious problems for the Swedish environment. Rapid urbanisation and industrialisation have concentrated more than four fifths of the population in cities and suburbs. The rapid economic growth in the post-war period, and the consumption patterns associated with the expanding welfare state, brought pressure to bear on the natural environment. Water pollution forced the closing of beaches in the 1950s and 1960s. The agricultural use of pesticides led to prohibition of fishing in lakes around the country in the 1960s. Domestic SO_2 emissions increased rapidly with urbanisation and increased combustion of oil for heating purposes. The prevailing south and southwesterly winds carry long-range sulphur emissions to Sweden; imported emissions cause acidification of water and soil which is above the critical level for permanent damage in large parts of southern Sweden. Today, nitrogen emissions from Sweden's more than 3.5 million cars, and from intensive use of fertilisers in agriculture, cause trouble in populated areas and along the country's southern shores.

Since the advent of a "modern" environmental policy in the late 1960s, much progress has been made to improve Sweden's environmental quality. Emissions to water of phosphorous and BOD7 from municipal sewers have come down to pre-World War II levels (Lundqvist, 1995a: 44). BOD7 emissions from the

Swedish pulp and paper industry are down by two thirds since the 1960s (SCB, 1993: 66). CO_2 emissions from fuel combustion have been cut by half, but those from transportation have increased (SCB, 1994c). Emissions of heavy metals to air and water are in many cases down to a trickle compared to the situation in the early 1970s (cf. SCB, 1994b: 14 f.). The total consumption of CFC's decreased by 75% between 1986 and 1994. VOC emissions have continued to fall since 1988, and will have decreased by 45% by the end of the 1990s. In its 1994 evaluation of the implementation of nine environmental policy objectives, the Swedish Environmental Protection Agency (SEPA) stated that while most of the ambitious goals for the year 2000 will be achieved, some thorny problems will remain. Emissions of CO_2 will not become stable at the 1990 levels without further strong interventions. The decrease in NO_x emissions will not meet its target without draconian measures against traffic in larger urban areas. Because of insufficient funding, the programme for preserving natural and cultural values in agricultural areas will only reach halfway towards the decided policy goals. Nature conservation in forested areas "still has a long way to go" before the present objectives are reached (SEPA, 1994a: 7 f.)

2 The Organisational and Instrumental Development of Environmental Policy

2.1 Policy Institutions: A Specialised "Sector"

Environmental policy and administration were given low priority in Sweden until the 1960s. The situation was characterised by institutional *fragmentation*. The 1963 Nature Conservation Committee (*Naturskyddsnämnden*) was set up to secure public access to natural resources for leisure activities and to defend nature conservation in cases concerning the large-scale use of land or water. A tiny Water Inspectorate *(Vatteninspektionen)* set up in the late 1950s was expected to examine all plans for extending the water and sewage systems in Sweden, and to supervise industrial pollution. At the same time, however, the National Highways and Water Construction Agency *(Väg- och Vattenbyggnadsstyrelsen)* subsidised sewage pipe lines, but *not* the construction of sewage treatment plants! An Air Quality Management Committee *(Luftvårdskommittén)* was set up in 1964.

The debate leading up to policy integration centered around the character of the "new" environmental issue. One emerging, albeit numerically small line of reasoning was *ecological*; as an intrinsic value, environment quality should supersede other societal concerns. The other line was *traditional*; environmental quality is one "interest" among others, and should be given "accurate" weight in policies and decisions on the use of resources in society.

The traditional view clearly won out. The new Swedish Environmental Protection Agency (SEPA; *Naturvårdsverket*) set up in 1967 consolidated the central administrative units just mentioned. The SEPA gained a broad area of responsibility for policy implementation, ranging from protection of rare species and management of national parks to promulgation of pollution-control guidelines and payment of state subsidies to sewage treatment plants. SEPA's competence expanded over the first years of its history. An Environmental Research Board was established in 1968, headed by the SEPA Director General and served on by SEPA personnel. The 1969 Environmental Protection Act gave SEPA responsibility for developing the general principles of the Act into more precise emission guidelines and other criteria for implementative measures against polluters. This work was carried out with vigour during the first few years, providing SEPA with important links to both environmental researchers and industrial branches (cf. Lundqvist 1971).

A special environmental "court", the National Environmental Licensing Board (*Koncessionsnämnden för miljöskydd)* was set up in 1969 to handle applications from polluters for permits required under the Environmental Protection Act. Appointed by the Cabinet, the Licensing Board has an experienced judge as Chairman. The three ordinary members have environmental, technological, and industrial experience. When municipal permits are at issue, there is a fourth member with municipal experience. The total number of employees is about 30. The permit cases handled by the licensing board are those on the so-called A-list of the Environmental Ordinance, i.e., cases which could possibly cause serious environmental disruption (Loftsson et al., 1993: 76).

National environmental policy-making remained a matter for the Ministry of Agriculture. An Environmental Advisory Council *(Miljövårdsberedningen)* was set up in 1968 to provide advice and information to the Prime Minister and the Cabinet on broad, long-term environmental issues involving several sectors in society. The Council held many meetings on such issues as climate change and acidification (cf. Lundqvist, 1971: 148f.; Loftsson et al., 1993: 77), but its importance for policy seemed to diminish somewhat in the 1980s. Neither the 1987/88 Cabinet Bill on the Environment, nor the one of 1990/91 mentions the Advisory Council or its place and function within Swedish environmental policy. More recently, the Council has been given important functions as an information centre and source of inspiration for the Local Agenda 21 process (cf. e.g. SOU, 1994:128).

By the late 1960s, the municipal Health Protection Committees (*hälsovårdsnämnderna* – politically appointed by the Municipal Councils*)* already had responsibility for supervision and control of sanitary issues, including local water and air quality as well as waste management. However, the County Administrations (*länsstyrelserna* – the regional administrative arm of central government) became responsible also for the supervision and control of local polluters requiring permits under the 1969 Environmental Protection Act.

2.2 Policy Instruments: A Regulatory Emphasis

Sweden's environmental policy started out with a strong *regulatory* emphasis. The principal idea of the 1969 Environmental Protection Act was to promote *end-of-pipe* solutions through legal means. All potentially harmful activities conducted on a property or in some permanent facility must be licensed by means of a permit spelling out the conditions under which they may be conducted. The polluter must take such precautions and tolerate such restrictions "as may *be reasonably* demanded" to prevent or remedy damages to the environment. Assessments of what is "reasonable" should be based on what is "technically feasible" and "economically reasonable"; particular attention should be paid to the balance between disturbing effects on the one hand, and the usefulness of the activity and costs of protective measures on the other. The policy organisation just described was to determine the permissibility of the activities, set the conditions for polluters, and supervise the compliance with these conditions.

The 1985 Act on Chemical Products and subsequent ordinances regulate the manufacturing, import, handling, packaging and sale of chemical products. The Act stresses the principle of *substitutability*; whenever there is a less dangerous product which performs as well as the more dangerous one, it must be chosen. The Act also emphasises *environmental impact assessment*; all manufacturers or distributors of chemical products must present a satisfactory assessment of the environmental and health hazards of the product (cf. Michanek, 1991: 242 ff.).

Regulation of land use has been employed since the turn of the century to protect and conserve different natural areas. By the early 1990s, there were 22 national parks, 1,363 nature reserves, 88 conservation areas, and 1,425 conservation objects in Sweden. The parks, reserves and areas comprised over 31,000 km^2, or 6%, of Sweden's land area (Sveriges Nationalatlas, 1991: 152 ff.). The Natural Resources Act of 1987 stipulates that geographical areas of "national (environmental) interest" may be subjected to industrial or other polluting activities only under very special circumstances. Almost 1,400 such areas have been statutorily defined through the system of national physical planning, totalling 101,000 km^2, or 22% of Sweden's total land area. About one-quarter of these areas are physically protected as nature reserves (cf. above), which means that industrial activities cannot take place there. The other 75% may be secured through (a) prohibitions in municipal overview plans; (b) the permit procedures in the Environmental Protection Act, or (c) the County Administrations' decisions (Sveriges Nationalatlas 1991: 164 ff.).

2.3 Policy Style: Consensus and Cooperation

The policy principles promulgated in the late 1960s described the public sector's role as one of promoting a "fair adjustment between competing demands on natural resources", while "the preservation of natural resources for coming generations" came second. The intense cooperation in the early years between controller and polluter in the development of emission guidelines established a very special style of Swedish environmental policy. The relations between controllers and polluters are primarily based on *cooperation*, aiming at *consensus* on the objectives and methods of pollution abatement. The environmental field thus closely resembles what the Swedish Power Study concluded about parts of the "Swedish Model": it is an "extremely deliberative, rationalistic, open and consensual" political culture (SOU, 1990:44: 187).

This consensual and rationalistic approach was actively pursued as the implementation strategy of SEPA by its legendary Director General during the first 25 years, Valfrid Paulsson. He wanted a relationship between controllers and polluters which could be characterised by prudence and "reason". A major vehicle was *information*; if only each side knew exactly what the other wanted and why, this would lead to rational and balanced decisions. The keyword was *trust;* polluters should be relied upon to implement agreed pollution control programmes and prescribed control measurements *without* day-to-day interference from regional or local environmental officers (cf. Lundqvist, 1971: 127 ff.).

3 Main Environmental Policy Actors: Strength and Resources

3.1 Policy Institutions: Centralised Intelligence and Decentralised Implementation

It took 20 years for "modern" Swedish environmental policy to get a ministry of its own. The 1987 Ministry for Energy and Environment was to have "an offensive and coordinating role within the Cabinet"; it should install environmental aspects into other policy sectors, and thus also into other Cabinet Ministries (Cabinet Bill 1987/88:85, p. 28). Energy issues were brought back under the Ministry for Industry in 1990, when the Ministry of Environment took over issues of national physical planning from the Ministry of Housing and Physical Planning. There were even aspirations towards making this "specialised" Ministry for Environmental Affairs a "super-ministry", as coordinator of environment-related issues at the Cabinet level (Loftsson et al., 1993: 71).

The Ministry of Environment and Natural Resources today handles such policy issues as regulation for environmental protection, nature conservation, control of chemical products and waste management. It is responsible furthermore for developing economic instruments for environmental policy. The management of Sweden's natural resources is within the Ministry's remit, as is protection against radiation and the control and supervision of nuclear technology with regard to its potential effects on man and the environment. The Ministry also has some implementative functions. Under the Natural Resources Act, permit cases which concern particularly large projects with extensive environmental implications can be decided only by the Ministry (Cabinet Bill 1991/92:100, part 15, p. 30 f.). Still, the Ministry is a small body with only about 150 employees, three-fifths of whom have an academic background. Moreover, governmental expenditure for environmental purposes is far less than one percent of the total state budget.

Figure 1. State Grants to Environmental Policy as a Share of the Total State Budget, Fiscal Years 1973/74-1994/95.

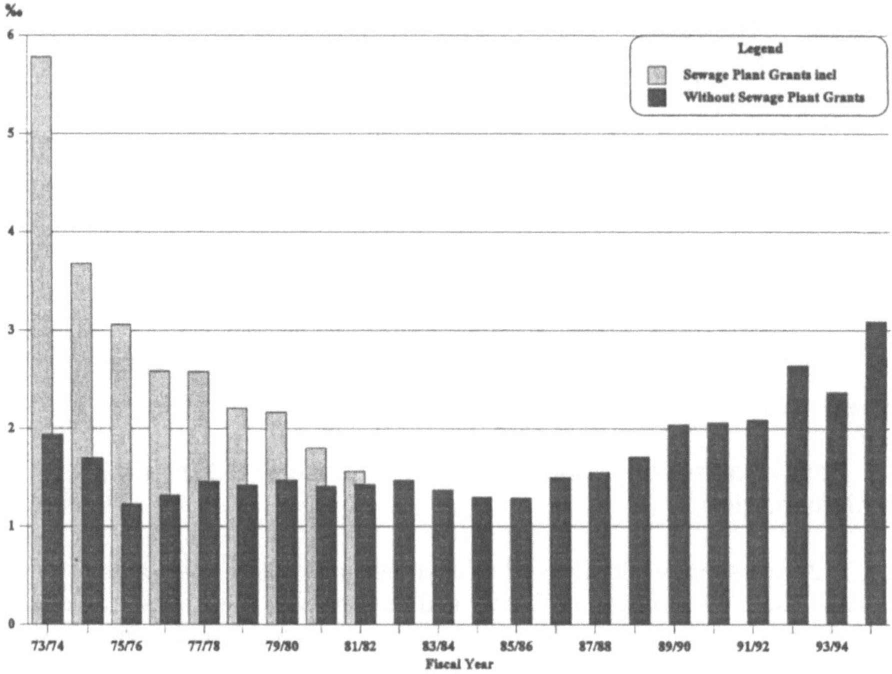

Source: Cabinet Budget Bills for respective years.

The Swedish Environmental Protection Agency's area of responsibility has become increasingly dominated by research planning and funding, information gathering, advising implementing units and other national agencies, and giving advice to government on long-term policy issues. Since 1978 SEPA has been responsible for the Programme on Measurement and Control of Environmental Quality (*PMK*). Through systematic monitoring, data-gathering and inventories, the Agency tries to create a firm basis for its policy advice and priority recommendations. Later decisions confirm and further this SEPA development from "regulation" to "intelligence". A 1987 Parliament decision gave the agency responsibility for providing environmental information pertinent to decisions under the new Acts on Planning and Building and on Natural Resources Management. The 1988 revision of the Environmental Protection Act instructed the SEPA to develop guidelines, action programmes, general advice and instructions, as well as evaluation criteria for activities within its area of competence.

The 1991 Cabinet Bill on Environment further strengthened the environmental responsibilities of the national agencies for road, air and railway traffic, agriculture, fisheries and forestry. Their charters now require them to formulate sectoral plans and programmes with precise environmental objectives to "implement nationally determined goals in the most cost efficient way." The central environmental agencies – SEPA and the Chemical Inspectorate (*Kemikalieinspektionen*) – are to provide information and expertise to the sectoral agencies, and to actively cooperate with them in formulating, following up and evaluating sectoral agency plans (Cabinet Bill 1990/91:90, p. 66).

The SEPA's mission is presently described as securing "a healthy environment and biological diversity". The SEPA mainly functions as an environmental policy "think-tank"; its emphasis is on documentation of threats to and trends in environmental quality, and on development of programmes and action plans which function as a crucial input in the Cabinet's formulation of environmental policy (see, e.g., SEPA, 1987, 1990, 1993). The Agency's staff is presently about 470 people. Of those with academic training, about 110 have a background in technology, while 140 are natural scientists. There are 30 social scientists, 20 economists and 10 jurists (SEPA, 1994b).

The 24 County Administrations (*länsstyrelserna)* function as the regional arm of central government, implementing national policies and controlling regional development. In addition to the County Administrations' duties under nature conservation, water, and planning legislations, the Environmental Protection Act of 1969 empowered the County Administrations to issue permits for a variety of polluting activities which are potentially harmful to the environment and to supervise and control such activities within the county borders.

The 1988/89 decentralisation of environmental administration set out to create a coherent administrative system for both permit decisions and supervision. The polluting activities listed in the Environmental Protection Act and its Ordinance were regrouped according to their duration, scope, and effects on humans and the environment. This cut in half the number of permit cases handled by the National Licensing Board. County Administrations would take on permit decisions for these cases, in addition to a large number of medium-sized plants and other facilities; the total was estimated at 4,200 plants and industrial activities. The County Administrations would continue to be responsible for supervising their performance.

A 1992 analysis of the County Administrations and their Environmental Units showed that the County Administrations' environmental staff is the second largest among the specialised sub-units, accounting for 635 of the total 5,500 man years within the County Administrations in 1991/92. They took the largest share of money among the sub-units, or SEK 250 million out of the totally SEK 2.3 billion used by the County Administrations in 1991/92 (Cabinet Bill 1992/93:100, Part 14:34 f.). The decentralisation of the late 1980s has been gradually changing the regional role towards one of functional specialisation and integration. The CA Environmental Units are actively developing new environmental policy instruments for the regional/local level. They also work with actors across sectoral boundaries to develop regional environmental analyses for Regional Environmental Action Programmes (Cabinet Bill 1992/93:100, Part 14:55 ff.).

The Swedish constitution acknowledges a general right of the currently 288 municipalities (*kommuner*) to act independently of central government on issues of "common concern" to the citizens living within their borders. They also have the power to raise taxes from their inhabitants. The new 1991 Municipal Act enables Swedish local governments to organise their activities as they see fit. Previous mandatory boards and committees can be replaced by a political board and committee organisation tailored to the municipalities' needs.

Local environmental policy has been affected by both environmental and constitutional reforms. A 1981 reform enabled municipalities to take over – by agreement – some or all supervision of polluting activities within the municipality. Eight municipalities had taken over supervision across the board of polluting activities by 1987, while about 40 had limited their participation to minor polluting activities (SOU, 1987:32: 200). The 1989 environmental reform transferred regional permit power over 6,000 medium-sized and small plants and facilities to the municipal Environmental and Health Protection Committees (EHPCs), in addition to the more than 4,000 activities requiring prior notification of the EHPCs (SOU, 1987:32: 309 f.). Local environmental inspectors are also responsible for supervising the environmental performance of some 93,000 farming units, 115,000 units handling chemicals, 44,000 places where food is marketed or handled, as well as some 3,000 water works.

Since there was no economic compensation from central government, municipalities felt obliged to fire support staff to be able to afford more inspectors to handle this supervisory workload. Total manpower in Sweden's municipal environmental inspectorates is now about 2,100, of which 1,500 are directly engaged in the field. Since they have become burdened with administrative work, the inspectors spend only about 25% of their time on direct field supervision of environmentally harmful activities (SOU, 1993:19: 73 ff.). This makes for some discrepancy between what is expected under environmental and health laws and what is actually achieved. Many of the environmentally harmful industries are only visited once every three years. Two-thirds of 305 randomly surveyed firms had not implemented required measures since previous supervisions, or were actively obstructing action through administrative or court procedures (SOU, 1993:19: 85 f.). The environmental policy's share of local government expenditure oscillated around 0.5% in the1990s. This means that local governments spend on average SEK 90 per capita on the implementation of environmental and health policies (SOU, 1993:19: 73; Lönn, 1992).

The 1991 Municipal Act increased municipal freedom to organise local affairs. Within one year of the Act, 57 local governments abandoned their Environmental and Health Protection Committees in favour of amalgamating them with the Building and/or Fire and Rescue Committees (SOU, 1993:19: 41f.). As many as 73 municipalities integrated their environmental protection officers into other municipal administrations, such as the Real Estate and/or Building Administrations, the Technical and Infrastructure Offices, or the Municipal Central Chancellery. This process of horizontal fragmentation of environmental administrative capacity has been particularly visible in smaller municipalities (SOU, 1993:19: 43).

3.2 Green Organisations

The emergence and formation of the Green Party (*Miljöpartiet – de gröna*) in 1981 was very much a result of the referendum on nuclear power in 1980. Many of the party founders were active in the "People's Campaign: Say No To Nuclear Power". The 1988 election campaign was totally dominated by such environmental issues (Djerf, 1989; Asp, 1990; Bennulf, 1989, 1994). This no doubt helped the Green Party to become the first really new party in 70 years to enter Parliament (Bennulf, 1990; Vedung, 1988, 1989; Gilljam and Holmberg, 1990).

The Green Party is far more active on environmental issues than other parties in Parliament; 23% of all Green Party MP bills to Parliament in 1991, and 35% of all Green Party MP questions and interpellations to Ministers in that year concerned the environment. The other "greenish" parties – the Left and the Centre Parties – were way behind (Parliamentary Record, 1991).This activity did not help, however. The electoral relevance of the environmental issue was much lower in the 1991 autumn elections, and the Green Party did not achieve beyond the 4% needed to remain in Parliament (Gilljam and Holmberg, 1993). However, they gained over 5% of the votes in the 1994 elections, thus becoming the first party ever in Sweden to return to Parliament after having been ousted in an earlier election. In the autumn 1995 elections to the European Parliament, the Greens ran a very energetic "No to EU" campaign, winning 17% of the votes and becoming the third largest of the 11 contending parties (*Från Riksdag & Departement,* 1995: 13). Party membership is presently 5,500 (*Vem gör vad i Miljösverige,* 1994: 114).

Environmental interest groups had little influence on the formation of the new environmental policy in force from 1967 onwards. No new environmental organisations with radical programmes and strategies emerged alongside the venerable Swedish Society for Nature Conservation (*Svenska Naturskyddsföreningen,* founded in 1909). The environmental awakening of the 1960s did, however, lead to a short outburst of ecologism in the 1970s, as new "alternative" movements arose. These groups lacked both a strong common organisation and a large number of followers. Their quite "unorthodox" tactics most probably precluded easy access to the then tightly knit policy networks of Swedish central government. After the 1973 oil crisis, environmentalists were very much active in the formation and campaigning of the anti-nuclear movement up to the nuclear referendum of 1980. From there, the environmentalists split up, some going to the Green Party, others remaining at the margins, and still others becoming incorporated into the professional environmental organisations. Although a "significant political force" in Sweden, the Swedish environmental movement has been "subjected to a range of incorporation pressures that have made it difficult for an autonomous social movement to develop." The very strong and early response to environmental issues from the political parties led to an "ideological incorporation of environmentalism as part of the established parties. The author contends that because of these incorporation pressures and the "instrumental orientation of the national culture, environmentalism has had difficulty carving out a public space of its own." (Jamison et al., 1990: 13ff.).

Table 1. The Context of Capacity-Building: National Public Authorities, Research Organisations, Private Interest Organisations, and Media in the Environmental Policy Field in Sweden (1994).

Organisations According to Activity/Purpose	*Number of Organisations Registered*[1]	
National Public Authorities[2]		*46*
Specifically environmental	3	
Sectoral, with statutorily defined environmental duties	10	
Public health	4	
Technological development	8	
Transportation, energy, resource development	9	
Research, information, monitoring	8	
Fiscal, judicial	4	
Research Organisations		*48*
Funding agencies, environmental	4	
Funding agencies, general	5	
Funding agencies, sectoral	5	
Public/semi-public research institutes with environmental emphasis	11	
Branch research institutes with environmental aspects involved	18	
Science academies and programmes	5	
Interest Organisations		*113*
Ecotechnology, ecoinformation, ecoproduction	24	
Environmental professionals	14	
Environmental "movement" organisations	17	
Industrial production	35	
"Traditional" interest organisations (labour market, people's movements, etc.)	10	
Professionals in other sectors of society	6	
International contacts with some environmental content	7	
Media[3]		*67*

Sources: Vem gör vad i Miljösverige1994, passim. Svensk Periodicaförteckning 1993-94, passim.

1 It should be noted that inclusion in the *Vem gör vad* is based on self-reporting, which means that the table should be read merely as a rough indicator of the contextual framework of environmental capacity-building.

2 Ministries are not included.

3 Includes perodicals with "alternative", "earth", "eco-", "enviro-", "nature", or "recycle/reuse" in titles or names.

The "interest" arena of environmental policy-making today is quite "crowded". In 1994, more than 50 Swedish interest organisations could be characterised as pro-environmental. About half of them are active in eco-technology, eco-information, and eco-products. One quarter consists of associations of environmental professionals. There are 35 business and industrial environmental organisations, as well as environmental bureaus or units within trade union and other labour market organisations. As many as 67 special periodicals with some sort of "green" label are available. All leading newspapers, as well as the news programmes on radio and television, have special reporters working exclusively on environmental issues.

Of the more than 100 organisations in Table 1, 17 may be classified as environmental "movement" organisations. The membership of these "movement" organisations is rather small except for the three largest organisations – the Society for Nature Conservation, Greenpeace, and the WFN. They have together 460,000 members, or 5% of the Swedish population. A majority of their followers are only "checkbook members" (cf. Andrews, in this volume). The smaller "movement" organisations which opt for a more active or activist profile do not gather a large membership.

Table 2. The Swedish Environmental Movement: The Largest Organisations and their Membership, 1994/95.

Organisation, group	*Membership*
Swedish Society for Nature Conservation (*Naturskyddsföreningen)*	200,000[1]
World Wide Fund for Nature (*Världsnaturfonden)*	150,000[1]
Greenpeace Sweden	110,000[1]
JAK Association for Economic Liberation *(JAK Riksföreningen för ekonomisk frigörelse)*[2]	40,000
Swedish Ornitological Society (*Sveriges Ornitologiska Förening)*	20,000
The Field Biologists (Fältbiologerna, *Naturskyddsföreningen*'s youth organisation)	5,500
Swedish Anti Nuclear Movement (*Folkkampanjen mot Kärnkraft)*	3,000
The Natural StepFoundation (*Stiftelsen Det Naturliga Steget*)[3]	n.a.
Environmental Federation (*Miljöförbundet)*	2,000
Friends of the Earth Sweden (*Jordens Vänner)*	1,600

Sources: *Vem gör vad i Miljösverige1994*, passim. Personal information to the author from different organisations, April 1995.

1 Paying members; number of active members or "activists" not given.

2 Lends money without interest to "provide an economy in balance with nature".

3 Supports networks of people from differnent vocations who are prepared to "use their competence to serve the nature"; no exact figure on people engaged is available.

Still, we know from many opinion surveys that most citizen's groups share a concern for the environment. Generally stated proposals to save the environment gain strong support, while proposals to restrict private car use are unpopular across broad segments of the Swedish population. Environmentalist support is generally strongest in the so-called new middle class (well educated public sector employees with medium-level incomes). Young and middle-aged people in urban areas also tend to score high on support for the Green Party and other environmental organisations (Holmberg and Gilljam, 1987; Westholm, 1991; Bennulf and Holmberg, 1992; Bennulf, 1994; Gilljam and Holmberg, 1993). Most Swedes contribute to environmental protection by "ecologising" part of their lifestyles, but some groups are more environmentalist than others; Green Party sympathisers tend to be more active in recycling, composting, and using collective transportation systems. Women and elderly people also show a greener lifestyle than do other citizens (Bennulf and Gilljam, 1991; Bennulf and Holmberg, 1991; Bennulf, 1992).

The influence of environmentalist groups on policy and implementation is varied. A case study on the siting of hazardous nuclear waste facilities concludes that local action was not just a NIMBY (not-in-my-backyard) form of local protest. Local activist groups contested both the *location* and *production* of such waste, thus contributing to the debate on the nationwide (nuclear-based) energy policy as such (Lidskog and Elander, 1993). However, other studies imply that environmentalist groups are not an integral part of environmental policy implementation. A survey of the Chief Environmental Inspectors in Swedish municipalities found that environmentalists are very seldom sought after by the Inspectors for comments and advice. If local environmentalists want to get into the decision-making process, they themselves have to initiate contacts with the local environmental administration. Other groups such as local enterprises, business organisations, and other "established" interests are very often brought into the process through the Inspectors' initiatives (Lundqvist, 1994: 1752).

3.3 The Target Groups of Environmental Policy

As noted earlier, the traditional political style brought polluting industries very early on into the development of emission guidelines for implementing the 1969 Environmental Protection Act. This was in turn probably also instrumental in the emergence of environmental units or divisions within 35 industrial branch organisations in Sweden (cf. Table 1). The comprehensive business and industrial federations, most individual branch organisations, and individual enterprises of some size now have environmental divisions. These divisions serve as research and intelligence units and as information and lobbying vehicles on environmental issues *(Vem gör vad i Miljösverige,* 1994). They participate in state commissions to investigate environmental problems and propose solutions (cf. below). Individual firms frequently use environmental auditing as a means of monitoring and evaluating their environmental performance. There is now an Association of

Swedish Environmental Auditors (*Miljörevisorer i Sverige*) with the aim of improving theories, methods and skills in environmental auditing (*Vem gör vad i Miljösverige,* 1994: 115).

What about the resources used by Swedish business and industry for environmental purposes? Reductions in pollution from industrial and mobile sources may have been triggered by government interventions, but they have been made possible by technological innovation and economic investment in more effective pollution control equipment. The level of investment in pollution control technologies proper can be estimated from official statistics. However, other investments with possibly beneficial environmental "side effects" are very difficult to assess. With these caveats in mind, we can say that total environmental investment in Swedish industry rose from SEK 400 million in 1981 to 1,200 million in 1985, and to 2,600 million in 1988, after which it fell back to SEK 2,300 million in 1991 (annual prices). Their share of total investment seems to have risen from 2% in 1981 to 8% in 1988 (a year with a peak in the business cycle). By 1991, the average branch share was just under 6% of all industrial investment (SCB, 1994c). It is even more difficult to estimate environmental investment in the transportation sector. One indication is that after catalytic converters were made mandatory for 1989 and later car models, the share of cars with such exhaust control equipment rose to one-third of the total stock of cars by 1994 (Cabinet Bill 1994/95:4, p. 6).

4 The Processes and Features of Capacity-Building

4.1 Policy Content: Towards a New Paradigm

The 1967/69 policy established an environmental "sector" with a mandate to protect the environmental "interest" against other sectors, where "protection" was a euphemism for "cleaning up". The major 1987/88 Environmental Bill spoke of a new strategy of integrating environmental policy into all other policy sectors: "A successful environmental management presupposes that care for the environment is integrated into the development plans for different sectors of society". These sectors all have "a responsibility to prevent new environmental damage", but the costs should be borne by those causing the problems (Cabinet Bill 1987/88:85, p. 35 f.).

The 1990/91 Environmental Bill reveals a further shift in scope: "Responsibility and care for the environment should permeate all walks of life and society. The mission of the 1990s is to readjust all societal activities in an ecological direction (...) [in order to] (...) protect human health, to preserve biological diversity, to manage natural resources for sustainable use, and to protect the natural and cultural landscape" (Cabinet Bill 1990/91:90, p. 11). The "Polluter-Pays" and "Precautionary" principles should be cornerstones of future policy, whose main features will be:

- Intensified international cooperation to limit transnational and global envi-

ronmental problems, including support for an environmentally adjusted development in the Third World and Eastern Europe;
- Improved production of knowledge and monitoring of the state of the environment in order to obtain a better basis for preventive environmental action;
- Increased sectoral responsibilities and decentralisation to secure broad support for environmental management, including individual, business, and municipal action as well as participation by public agencies and units in all sectors of society. A further emphasis on evaluation of policy measures and instruments (Cabinet Bill 1990/91:90, p. 13).

The Right-Centre Coalition government of 1991-1994 saw the objective of environmental policy as "a long-term sustainable development (...) where future generations can take over national goods, including environmental and natural resources, of at least the same size as they are today". The major policy statement deserves to be quoted:

> "Concern for the environment must influence decision-making at all levels and in all sectors of society. This means giving prior consideration to the environmental impact of every decision which might have a major impact on the environment. It also means that the 'precautionary principle' must be more widely applied, i.e. measures to prevent environmental destruction must not be postponed owing to lack of scientific certainty (...) The objective is to preserve 'biodiversity', i.e. the complete range of animals and plant species found in our country" (Cabinet Bill 1991/92:150, Annex I:12, p. 1).

The principles of Sweden's environmental policy have thus shifted since the "Adjustment" era of the 1960s and 1970s. The scope of the policy also changed during the 1980s and 1990s. The policy of the late 1960s was mainly concerned with pollution control (air, land and water) and nature conservation. Biocides, herbicides and waste emerged on the agenda, and there was some connection of environmental policy to national physical planning and environmental research. Today's environmental policy comprises not only these features, but is also directed towards agriculture and forestry, traffic and communications, energy provision, industrial activities, hazardous as well as other products, waste, environmental recovery, education to promote environmental awareness, and environmental impact assessment and evaluation (Cabinet Bill 1990/91:90, Table of Contents, pp. 578 ff.)

4.2 Policy Instruments: A Widening Choice

A broader choice of policy instruments seems to have come as a response to changing "problem structures" in environmental policy. With point-source pollution gradually under control, and with the problems of diffuse and long-range pollution gradually coming to the fore, traditional regulatory instruments have been supplemented by economic and information measures aimed at discouraging or stimulating a broad spectrum of behaviours related to the environment. Official documents emphasise that economic instruments in conjunction with persuasion and exhortation are becoming more and more central to the government's strategy for reaching a more sustainable society (Cabinet Bill 1991/92:100, Part 1:5, p. 5; cf. ibid., Part 15, p. 7).

A Commission Report in May 1993 proposed amalgamating more than a dozen environment-related acts and ordinances into one single Environmental Code. One indicator of how regulatory measures are changing is the proposed shift from indivudual source control to "environmental quality standards". The state should have powers to issue general, and binding, quality standards, criteria and guidelines to secure environmental and public health (SOU, 1993:27, Part I: 58). Such ambient quality norms should be supported by measures such as physical planning, environmental adaptation of the infrastructure and other interventions to ensure that the quality standard is maintained. Urban areas suffering from air pollution from transportation are prime targets for the application of quality norms (SOU, 1993:27, Part I: 363 ff.).

The other principle is that of "cyclicality and producer liability". A rule "will apply to all handling of products, to the effect that the full life expectancy of a product is to be taken into account when assessing the hazards it involves to human beings and the environment." The state authorities should be "empowered to make it the duty of anyone commercially producing, importing or selling a product or a packaging (a producer) to ensure that the waste resulting from such activities is removed, re-used, recovered or disposed of in an environmentally acceptable manner" (SOU, 1993:27, Part I: 60). By the end of 1995, a new commission was still working to refine and fill out the original 1993 proposal.

The "cyclicality and producer liability" principle has sparked some interesting mixes of regulatory, self-regulatory and economic instruments. A producer liability clause included into the Public Sanitation Act states that if negotiations with different packaging industries do not result in voluntary compliance with the principle, the government may force these industries to comply (Cabinet Bill 1992/93:180). The Eco-cycle Bill of February 1993 and subsequent decisions amount to what could be labelled a system of *ecological corporativism*. The cooperation and consensus strategy of earlier environmental policy is now elaborated further. Under this Bill, manufacturers are held responsible for all packages and packaging materials used in the production and distribution of manufactured goods during the whole life cycle of the package or material. Manufacturers are also required by law to achieve certain levels of re-use or recycling of packages and materials by 1997 (SFS, 1994:1235). In return, the producers are left free to organise and run the processes which would close the eco-cycle for the products and materials in question. In fact, a self-regulatory oganisation is already being formed. The largest manufacturers in the branches affected by the producer liability regulations have formed special "materials companies" to be run as "non-profit businesses", with the actual collection and recycling done by outside contractors. The materials companies will control the re-use and recycling activities of their member producers, while non-member producers will have to account for their re-use and recycling to the SEPA (REPA-registret AB, 1994).

As for *economic* instruments proper, Sweden has adhered to the *Polluter-Pays* principle, except for a limited period in the 1970s. Subsidies were then given to both local government and to private firms to stimulate investment in environmental clean-up and protection (cf. Lundqvist, 1995a). Subsidies are now down to a trickle, and used exclusively for the development of new technology, liming of acidified lakes and streams, and conservation of the "cultural" landscape. Economic measures such as fees, charges and taxes are increasingly used. The Commission on Environmental Charges saw several of its proposals (cf. SOU, 1990:59) accepted by the Cabinet and the Parliament as part and parcel of the major tax reform of 1990/91. As much as SEK 18 billion of the cut in income taxes were to be financed through environmental and energy taxation. The "*green taxation*" on fuels and environmentally hazardous products seems to have been accepted without any political clashes in Parliament. By 1993/94, state income from effluent charges, from direct environmental taxes and from environment-related energy taxation, totalled around SEK 40 billion, or about 12% of total state income from taxation (Cabinet Bill 1994/95:120, pp. 52 ff.). Several environmental and energy taxes were raised in 1995 to help finance Sweden's accession to the European Union (Cabinet Bill 1994/95:203, pp. 30 ff, 48).

Information plays an ever-increasing role as an environmental policy instrument for both detection and protection. There are nationwide monitoring programmes and a host of regional and local measurement programmes to aid in environmental decision-making. Local governments run environmental information campaigns on recycling, composting, sorting of household waste, etc. Information is also of central concern in the local activities relating to AGENDA 21. All local governments are presently engaged in a process of developing local plans for sustainable development. Special working groups have been set up at the political and/or administrative level in local government, and these groups function as nexuses in a rapidly growing "discursive" network of municipal politicians/administrators, affected local interests, environmentalist groups, and interested individuals. The national government and central environmental agencies envisage that the information obtained in the local strategic plans coming out of this locally based endeavour will form the basis for a national plan for sustainable development (SOU 1994:128). However, the realism of this vision remains to be proven (cf. Lundqvist, 1995b). The crucial role of information in policy implementation is shown by the requirement for environmental impact assessment procedures in decision-making processes on a wide variety of undertakings, involving a widening circle of actors.

Voluntary *environmental labelling* of industrial products, using the inter-Nordic *Swan* label, was introduced in 1989. After having had their products tested by Standardisation in Sweden (SIS), producers can apply to the Environmental Labelling Board for a license to use the *Swan* symbol. The board consists of representatives of environmental, industrial, wholesale and retail as well as labour market organisations. The Chairman is appointed by the Cabinet. As many as 84% of Swedish consumers prefer *Swan*-labelled products to those without environmental labelling (*SIS Miljömärkning*, 1994). There are also several other labelling programmes. The Swedish Society for Nature Conservation (cf. below) runs *Good Environmental Choice* together with the main wholesale grocery chains in Sweden (SNF, 1994; Ds 1992:58: 118 f.). An NGO called the Control Association for Ecological Cultivation runs the *KRAV* (literally "demand") programme. Farmers, horticulturists and others marketing ecologically grown and organic agricultural products are allowed to use the *KRAV* label if they pass the Association's inspection and test (Råd & Rön 1994:10: 30 f.). The *Seal* and *TCF* (Totally Chlorine Free) labels are used by pulp and paper producers and distributors (Swedma, 1994: 4).

4.3 A "Swedish Model" for Capacity-Building: Special Investigative Commissions

Swedish ministries are quite small by international standards, but they have a very central role in policy formulation. This is due to the separation of policy-making from implementation through the division between the ministries on the one hand, and independent agencies and boards on the other. The prime vehicle for policy formulation has been the system of *ad hoc commissions*, set up by ministries to investigate a particular problem and come up with suggested solutions. These investigative commissions often include representatives from the parties in the Parliament, from affected interests, and from relevant administrative agencies and boards.

Three aspects are crucial to these commissions' role in Swedish politics: (a) their production of policy-relevant *knowledge*, (b) their role as an arena for political *compromise*, and (c) the possibilities for government to *control* the formulation of new policies. The last point is related to the Cabinet's setting the terms for the commissions through directives, and the Cabinet's appointment of members on the commissions. Some notable changes have occurred in the importance of these three aspects. Upon returning to power in 1982, the Social Democrats radically cut down on the number of commissions, and restricted the lead time from appointment to final report. While there is less generation of knowledge, and there are fewer compromises made than before within the commissions, the Cabinet's control over the commissions, and thus over policy formulation, has increased (Johansson, 1992: 115). However, since 1989, the number of Commission Reports (SOU) has increased and is now on a par with the situation prevailing before 1982.

What is the role of investigative commissions in the formulation of Sweden's environmental policy? Of the more than 1,850 SOU's published from 1973 to 1993, a total of 67 can be classified as falling within the domain of environmental policy. If we include about a score of reports on other policy areas which contain discussions of related environmental aspects, we find that just under 5% of the SOU reports in the last 20 years have dealt with environmental policy problems. Another vehicle for policy formulation is represented by the so-called ministerial investigations (*departementsutredningar).* There were 2,659 reports from such investigations between 1973 and 1993. Less than 4% – or 97 – touched on environmental issues.

An illustration of the role of investigative commissions in relation to other capacity-enhancing measures is provided by the Omnibus Environment Bill sent to Parliament by the Social Democratic Government in 1991. No less than 39 different reports, action plans, memoranda etc. are mentioned as the basis for the bill. The Environmental Ministry itself produced the overall description of the environmental situation. There were seven (7) Commission Reports, and two (2) from departmental investigations. The SEPA provided 10 of its own action plans and/or memoranda, and was involved in another four (4) basic documents together with other state agencies. The two other "environmental" agencies, i.e., the Licensing Board for Environmental Protection and the Chemical Inspectorate, provided three (3) items. Agencies with "sectoral" environmental responsibilities produced 12 input reports to the Bill (Cabinet Bill 1990/91:90, p.). Several different means and vehicles are thus used for interpolicy and intrapolicy cooperation and for overall policy integration.

4.4 Environmental Research and Monitoring for Policy Capacity

Swedish environmental research has been closely integrated with policy. The Research Board is closely integrated with the SEPA, whose Director General is chairman of the Board. Applications to the Board for grants are examined by Research Committees, consisting of active researchers as well as representatives of sectoral agencies, municipalities, and sometimes industry. Personnel from the SEPA Research Department are secretaries on the Committees. When a particular environmental problem is seen as being in need of special attention, a so-called "Programme Area" is formed. All (interdisciplinary) research on this area is then coordinated by a Programme Area Officer from SEPA's Research Department, ultimately responsible to the Research Board. The environmental research budget is decided upon together with all other state-supported research in the Cabinet's Omnibus Research Bill sent to the Parliament every third year (cf. Cabinet Bill 1989/90:90, p. 459 f.).

The research budget has grown steadily over the years, from about SEK 20 million in the early 1970s to about SEK 150 million in the mid-1990s. The total environment research grant – the major part of which is given to SEPA's Research Board – has kept pace with the overall environmental policy budget ever since the mid-1970s, usually constituting about one-tenth of total state expenditures on environmental policy.

Apart from building up scientific competence on environmental problems, SEPA and the Research Board stress the practical application of research results in policy-making and implementation. Organisational and procedural changes have

Figure 2. Governmental Expenditure on Environmental Policy; Total Grants and Grants to Environmental Research, Fiscal Years 1973/74 – 1994/95; Running Prices.

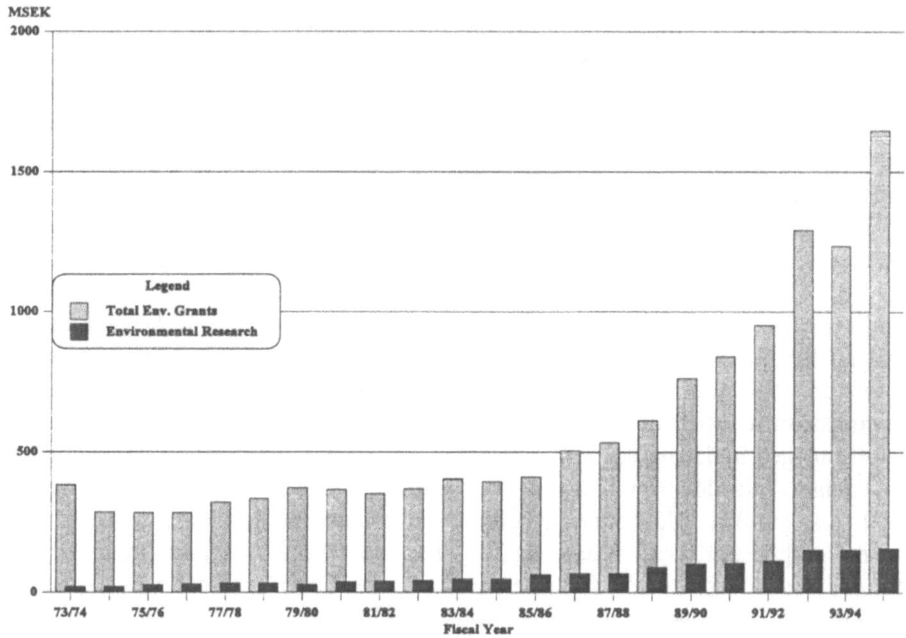

Source: Swedish Parliamentary Record.

been undertaken to ease contacts between researchers and practitioners. Information and reports about Board-sponsored research results are easily available through several series of SEPA research reports. The objectives set up in the SEPA Action Plans during the 1980s and 1990s have to a large extent been based on the results of environmental research.

The 1983 Parliament decision to create so-called Wage Earners' Funds highlighted the traditional Left-Right cleavage in Swedish politics. Upon winning the 1991 elections one of new Centre-Right Government's first priorities was to dissolve these funds. In the 1993 Cabinet Bill on "Frontline Research", the Centre-Right Coalition argued that part of the funds should be "used for long-term knowledge expansion though basic research and education of researchers in areas chosen on the basis of judgements concerning their importance for the development of Sweden's future competitiveness". This included strategic environmental research as well as research in medicine (Cabinet Bill 1992/93:171, passim).

Of the SEK 10 billion thus released from the dismantled Wage Earners' Funds, the new Foundation for Strategic Environmental Research – (MISTRA) – is charged with using the interest on SEK 2,5 billion to support research networks which "attack complex, most often international problems in a way suited to the character of the specific problem". Cooperation with industrial R&D is seen as vital; the research sponsored by MISTRA should "be of importance for finding solutions to important environmental problems", and "opportunities for achieving industrial applications shall be taken advantage of". It is made clear in MISTRA's Request for Applications that the Foundation "will give priority to research on sustainable technologies, as seen in a broad perspective." (MISTRA 1994). By the end of 1995, more than two dozen Programme Committees were busy formulating comprehensive research programmes for specifically defined environmental problem areas.

As early as 1978 SEPA became responsible for the programme on measurement and control of environmental quality. Regional and local programmes were set up for water recipient control, covering most of Sweden's freshwater bodies and streams. Regional air quality measurement programmes were established around the largest urban conglomerations (cf. Lundqvist, 1994: 1739 ff.). Problems of coordination, dissemination and retrievability soon developed. A commission revealed in its 1994 report that more than a dozen public agencies were engaged in data collection, with little exchange of information. Field studies made data collection very costly, and measurement stations were not coordinating with one another. No common central environmental data base existed. The report emphasised that remote sensing methods should be used more in the future, closely coordinated and integrated with the net of data-collecting ground-stations. An inter-agency advisory group should organise a new environmental data system, coordinated thoroughly from collection to storage and dissemination. Priority should be given to measurements which provide the basis for environmental indices (SOU, 1994:125: 91 ff., 161 ff.).

Both SEPA and Statistics Sweden regularly publish environmental statistics. The SEPA earlier ran a series called *Monitor* reporting trends in different environmental pollutants or different parts of the environment. Every third year or so, Statistics Sweden and SEPA publish *The Natural Environment in Figures,* which gives an overview of global and regional environmental problems, reports on factors influencing the national environment, and contains a lengthy statistical description of "the state of the environment" in Sweden. Statistics Sweden also runs an information bulletin, presenting annual trends of different pollutants, hazardous products etc.

4.5 Local and Regional Capacity-Building: Professional Networks

The Swedish Power Study argued that the environment is the only "sector" without a central "political" network (Petersson, 1989: 186 f.). At local level, a 1991 survey of municipal Chief Environmental Inspectors (CEI's) revealed a lively contact network. The CEI's have frequent contact with environmental administrators in other municipalities. These intermunicipal professional contacts are more intense than CEI contacts with their own local environmental politicians, local action groups, local media, and local associations. Also notable is that contacts with environmental officers and county level are almost as lively as those with the local EHPC politicians. The CEI *initiates contacts* with officers in other municipal agencies, as well as with environmental officers in other municipalities and at the county and national levels. However, he (85% of the CEI's are men) *is contacted* by local citizens, by local media, by local environmentalist groups, and by local interest associations. The active contacts with their peers in the Inspectorates in other municipalities and at the county level has as its most important purpose the collection or exchange of information (Lundqvist, 1994: 1754).

To sum up: Municipal CEI's and Inspectorates have lively contacts with their colleagues in other municipalities. There is also a close contact network between CEI's and county level Environmental Officers. We may speak of *a local and regional network of environmental professionals.* The lively contacts among professional "peers" are a strong vehicle in ecological capacity-building; their main function is to gather, assess, and exchange information.

4.6 Sweden and International Environmental Cooperation

Ever since the 1972 Stockholm Conference on the Environment, Sweden has taken pride in being an environmentally progressive country. Swedish policy-makers have expressed the opinion that their country must stand out as "the good environmental example" if Sweden's efforts to achieve more radical international solutions are to be successful. But the regard Sweden has won in this field does not just stem from some "greenish" morality. Indeed, the country's dependency on international trade, its vulnerability to long-range transport of air pollutants, and its linkage to the ecological fate of the Baltic Sea all contribute to Sweden's environmental activism.

The main issue for the 1990s has been the fate of Sweden's environmental profile after joining the European Union. The 1994 report of the Commission on the Consequences of Sweden's Participation in West European Integration was by no means unequivocal. The environmental policy advantages for Sweden of an EU membership are regarded as

> "(...) highly dependent upon our ability to present good proposals and develop close cooperation with other progressive (environmentally, my remark) countries in order to convince the more hesitant ones. An active use of this possibility may be important both with regard to chemicals and when it comes to air and water pollution. However, an EU membership (....) may limit our possibilities to develop good examples. The prohibition against tax rebates for motor vehicles in Environmental Class 1 is but one example of such a restriction in the membership alternative" (SOU, 1994:7: 281).

The report thus rests its argument on how much political will and skill Sweden can muster to convince other EU nations of the need for more stringent environmental regulations. Much of the debate on future "national environmental sovereignty" has dwelt upon the implications of the so-called "Environmental Guarantee" (Article 100a, pt. 4) of the Rome Agreement. The Swedish reaction following the European Court's spring 1994 decision on Germany's PCP regulations is indicative of this concern. The Court's decision was interpreted as a demand on the Commission to develop a more thorough argumentation for its decision to allow Germany's stricter regulations to continue. The Minister of the Environment of the Centre-Right Government therefore moved to offer Brussels access to the important research and documentation on PCP's in Sweden; which would give the Commission "scientific arguments for its renewed consideration of the German request" (Ministry of the Environment, 1994). The incoming Social-Democratic Government stated in a special report in early November 1994 (three weeks after getting into power) that its environmental policy towards the European Union would be one of *activism*. To promote "green" decisions, Sweden should make alliances with "environmentally progressive" countries, and Swedish environmental standards should not be adjusted to less stringent European ones (Ds, 1994:126).

5 Evaluation: Swedish Environmental Capacity-Building: Framing or Framed?

Viewing the Swedish policy development in the light of the theoretical and conceptual framework drawn up for this comparative endeavour (Jänicke, in this volume), what conclusions could be drawn about environmental capacity and capacity-building in Sweden? First of all, let us look at the *strength, competence and configuration of environmental proponents*. We then find that the central environmental agency – the SEPA – has played a very decisive role in the development of Swedish policy. Gradually, the basis of this role has shifted from regulatory and implementative activities to what I would like to call the Agency's *epistemic* capacity. The collection, assessment and dissemination of environmental knowledge is very much a matter for the Agency through its command over environmental research and monitoring. It should be noted that this capacity is also well spread among regional and local environmental administrations. This implies that in policy-making terms, problem discovery and definition is very much a scientific and administrative capacity; such other proponents as environmentalist organisations and political parties get much of their arguments from environmental professionals.

Thus, it seems fair to say that the *cognitive-informational* conditions for environmental capacities are very much determined by environmental science and bureaucracy. This is very much a reflection of the *political-institutional* framework. Constitutionally separated from the special agencies, and furnished with a very limited staff, the Swedish Environmental Ministry can not get – or produce

– an alternative cognitive and informational base for its policies and decisions. But what about the emergence and growth of The Green Party and the consolidation of environmental opinion in Sweden? Has not this development provided a favourable *political opportunity* for environmental policy and management, i.e., setting the agenda and transforming knowledge and principles into action?

Although the *situative context* was extremely favourable when the Greens got into Parliament in 1988, the political opportunities turned out not to be so favourable. In fact, the Greens were in a parliamentary limbo; their support was not needed as long as the Social-Democratic Government could rely on the Left Party – also quite "greenish" – for support. And when the Leftists withdrew their support in response to the draconian Social-Democratic economic and labour market policy package in early 1990, the Greens made a strategic mistake; they made their support conditional on a complete concession to all their demands by the Social Democrats, at the same time as they locked the door on a Bourgeois Coalition Government. Therefore, they had no hand in the 1990 tax reform, which established environmental taxation as one of the cornerstones of Sweden's environmental policy. In the 1994-1997 Parliament, he Greens were able to get some environmental policy concessions from the Social-Democratic minority Government early on, only to find that the Social-Democrats had then turned to the old-timer of "greenish" parties in Sweden – the Centre Party – for long-term, coalitionist cooperation.

Despite the strength of green opinion, and the increasing number of votes given to the Green Party and to the "greenish" Leftist Party, environmental issues still take up a rather small part of Parliament's legislative work (5% of bills, questions, and interpellations) or of the state budget (4 promille of state expenditures). An important factor in explaining why growing political strength has led to larger shares is the prevailing *economic framework conditions*; the Swedish economy endured its most dramatic crisis ever in the early 1990s, and the huge state budget deficit at present overshadows all other priorities in Swedish politics.

Despite these more recent political and economic complications, Sweden has managed to set up an organisation and make use of instruments which have contributed to a remarkable overall *capacity for the environment,* i.e., a capacity to understand different environmental problems and take successful action to solve them. Founded on the almost 200-year-old constitutional right of open access to government and bureaucracy files, the *participative capacity* has now been extended to ecologically relevant long-term planning through the very citizen-focused Local Agenda 21 process. *Integrative capacity* includes strategic inter-policy cooperation at ministerial levels, as well as long-term intersectoral planning involving national and regional agencies. *Epistemic capacity* is continuously being improved with respect to information gathering over the whole range of environmental issues. Policy formation is based on problem-oriented environmental research. *Implementative capacity* has been broadened through a widening variety of policy instruments, including economic, informative and organisational measures, all the way down to local level.

That this has come about may be attributed in large measure to two factors, namely the political system's *capacity for strategic action* and the *strength and behaviour of target groups.* In trying to solve the first-generation problems of large-scale industrial pollution, the government and responsible authorities could build on a tradition of close cooperation with well-organised, responsible target-group interests. The corporatist character of the welfare state was at its height, and this was brought into the environmental field. A consensual and cooperative policy style developed in the late 1960s, and has continued uninterrupted ever since. The target groups developed their own environmental organisations, with a well entrenched system of "opposite numbers" able to deal on an equal professional basis with environmental administrators at different levels of government. They have built up an impressive organisation for environmental auditing and labelling as "natural" parts of industrial strategies. Frequent discussions within the system of ad-hoc investigatory Governmental Commissions has probably been an important vehicle for furthering consensus and mutual understanding. This may be seen by some as an unnecessary accommodation of target group views on *technological capacity* limits, and by others as a precondition for increased *implementative capacity.*

Cooperation with polluters and other affected interests in order to achieve consensus is still the dominant mode in the policy-making process. In a way, this helped early on in framing the policy field in a positive way, as it worked more or less in tandem with the then dominant *character of problems* of policy. Today, the problems are different; they increasingly concern diffuse sources and thus an increasing number of likewise diffuse cooperation partners. This raises an important question: Will the present environmental policy system find itself "framed" in a way that diminishes the capacities for participation, integration and implementation? The outcome of the present Local Agenda 21 process, of the efforts of interpolicy coordination, and of using new informative and economic instruments directed at individual behaviour, will prove crucial.

References

Asp, K., 1990: *Medierna och valrörelsen* (The Media and the Election Campaign), in: M. Gilljam and S. Holmberg (eds.), *Rött blått grönt. En bok om 1988 års Riksdagsval* (Red, Blue, Green. On the 1988 Parliament Elections). Stockholm: Bonniers.

Bennulf, M., 1989: *Miljöopinion under åttiotalet*, in: S. Holmberg and L. Weibull (eds.), *Åttiotal. Svensk opinion i empirisk belysning* (The Eighties. Swedish Opinion in an Empirical Light), SOM-rapport 4. Göteborg University: Department of Political Science.

Bennulf, M., 1990: *Grönt ljus. Väljarna och miljöpartiets valframgång* 1988 (A Green Light. The Voters and the Environmental Party's Success in the 1988 Elections). Göteborg University: Department of Political Science (paper).

Bennulf, M., 1992: Members of Swedish Environmental Organisations: Social and Attitudinal Bases, paper for the Conference on Soziale Bewegungen in Westeuropa, Saarbrücken, 13-15 November.

Bennulf, M., 1994: *Miljöopinionen i Sverige* (Environmental Opinion in Sweden). Lund: Dialogos.

Bennulf, M. and M. Gilljam, 1991: *Snacka går ju – men vem handlar miljövänligt?* (Talking is OK, but Who Acts Friendly towards the Environment?), in: L. Weibull and S. Holmberg (eds.), *Politiska opinioner* (Political Opinions), SOM-rapport 7. Göteborg University: Department of Political Science.

Bennulf, M. and S. Holmberg, 1991: The Green Breakthrough in Sweden, Scandinavian Political Studies 13: 165-184.

Bennulf, M. and S. Holmberg, 1992: *Förbättras eller försämras miljön i Sverige? (Is Sweden's Environment getting Better or Worse?),* in: S. Holmberg and L. Weibull (eds.), *Trendbrott?* (Breaking the Trend?), SOM-rapport 8. Göteborg University: Department of Political Science.

Cabinet Bills, Private Bills, Standing Committee Reports and Comments, Interpellations, Questions, Debate; *Riksdagen i siffror* (The Parliament in Figures); all to be found in Swedish Parliamentary Record (*Riksdagstrycket*). Stockholm: Swedish Parliament.

Djerf, M., 1989: *Miljövalet. Mediernas dagordning i valet 1988* (The Environmental Election. The Media Agenda in the 1988 Election), Paper for the IX Nordic Conference on Mass Communication, Borgholm, 20-23 August.

Ds, 1992:58: *Varor som faror* (Dangerous Products). Stockholm: Ministry of Environment and Natural Resources.

Ds, 1994:33: *Så fungerar miljöskatter!* (This is how Environmental Taxes Work!). Stockholm: Ministry of Environment and Natural Resources.

Ds, 1994:126: *Det svenska miljöarbetet i EU* (Sweden's Environmental Strategy in the European Union). Stockholm: Ministry of Environment and Natural Resources.

Från Riksdag & Departement, 1995: *Så fördelades rösterna i EU-valet* (The Results of the EU Election) # 29:13.

Gilljam, M. and S. Holmberg (eds.), 1990: *Rött blått grönt. En bok om 1988 års riksdagsval* (Red, Blue, Green. On the 1988 Parliament Elections). Stockholm: Bonniers.

Gilljam, M. and S. Holmberg, 1993: *Väljarna inför 90-talet* (The Voters in the Nineties). Stockholm: Norstedts.

Holmberg, S./Gilljam, M. 1987: *Väljare och val i Sverige* (Voters and Elections in Sweden). Stockholm: Liber.

Jamison, A. et al., 1990: The Making of the New Environmental Consciousness A Comparative Study of the Environmental Movements in Sweden, Denmark and the Netherlands. Edinburgh: Edinburgh University Press.

Johansson, J., 1992: *Det statliga kommittéväsendet. Kunskap, kontroll, konsensus* (State Investigative Commissions in Sweden; Knowledge, Control, Consensus). Stockkholm: Department of Political Science.

Lidskog, R. and I. Elander, 1993: Reinterpreting Local Conflict. NIMBY and Nuclear Waste Management in Sweden, Policy and Politics 20: 249-264.

Loftsson, E. et al., 1993: *Svensk miljöpolitik* (Swedish Environmental Politics). Lund: Studentlitteratur.

Lundqvist, L. J., 1971: *Miljövårdsförvaltning och politisk struktur* (Environmental Administration and Political Structure). Lund: Prisma/Verdandidebatt.

Lundqvist, L. J., 1994: Environmental Cooperation among Swedish Local Governments. Professional Networks and the Evolution of Institutions for Collective Action, International Journal of Public Administration 17: 1733-1766.

Lundqvist, L. J., 1995a: Municipal Sewage Treatment in Sweden 1960-1990. From Bans on Bathing to Schools of Salmon, in: M. Jänicke and H. Weidner (eds.), Successful Environmental Policy – A Critical Evaluation of 24 Cases. Berlin: edition sigma, 43-59.

Lundqvist, L. J., 1995b: *När miljöpolitiken drabbas av starr* (When Environmental Politics is Struck from Cataract), Svenska Dagbladet, April 21.

Lönn, K., 1992: *Miljö- och hälsoskyddets resurser i kommunerna* (Local Government Resources for Environmental Protection). Stockholm: The Environmental Health Officers' Association of Sweden.

Michanek, G. (ed.), 1991: *Lagbok i miljörätt 1991*. Stockholm: Allmänna förlaget.

Miljöaktuellt, 1994: *Svårt klara återvinningsmålet. Nya system kräver mer tid* (Difficult to Reach the Recycling Demands: New Systems Need More Time), Vol. 3, March 23, p. 8.

Ministry of the Environment, 1994: *Sverige bidrar till uppföljning av PCP-målet* (Sweden Contributes to the PCP Objective). Press release June 6.

MISTRA, 1994: Invitations to Propose Research Programme (March 28).

Parliamentary Record of Sweden (Cabinet Bills, Private Bills, Questions, Interpellations, Debates).

Petersson, O., 1989: *Maktens nätverk* (The Network of Power). Stockholm: Carlssons.

REPA-registret AB, 1994: *Information om reglerna kring producentansvar för samtliga förpackningar* (Information on the Regulations Concerning Producer Responsibility for All Packages). Stockholm: REPA.

Råd & Rön, 1994: *Tre säkra miljömärken* (Three Reliable Environment Labels), nr. 10, 30-33.

SCB, 1993: *Naturmiljön i siffror* (The Natural Environment in Figures). Stockholm: SCB.

SCB, 1994a: *Utsläpp till vatten 1992* (Emissions to Water in 1992). Stockholm: SCB Na 22 SM 9401.

SCB, 1994b: *Miljösverige* (Environmental Sweden). Stockholm: SCB.

SCB, 1994c: Communication to the author, September through November, 1994.

SEPA, 1987: *Vilken miljökvalitet?* (Which Environmental Quality?). Solna: SEPA.

SEPA, 1990: *Då, nu, sedan. En resultat- och framtidsanalys av miljövårdsarbetet* (Yesterday – Today – Tomorrow. A Result- and Future-oriented Analysis of Environmental Management. Solna: SEPA.

SEPA, 1993: *Ett miljöanpassat samhälle. Naturvårdsverkets aktionsplan Miljö '93.* (An Environmentally Adapted Society. SEPA's Action Programme Environment '93). Stockholm: SEPA.

SEPA, 1994a: *Hur har det gått? Redovisning av myndigheters arbete mot 9 miljömål.* (What's the Result? Evaluation of the Activities of Responsible Authorities Towards 9 Environmental Objectives). Stockholm: SEPA.

SEPA, 1994b: Personal information to the author about staff (June 1).

SFS (Swedish Code of Statutes), 1994:1235: Förordning om producentansvar för förpackningar (Ordinance on Producer Responsibility for Packages).

SIS Miljömärkning, 1994: *8 sidor information om miljömärkningen Svanen* (8 pages with Information about the *Swan* Environmental Label), Advertising Supplement, Svenska Dagbladet, November 1.

SNF, 1994: *Märk-värdigt. Naturskyddsföreningen och miljömärkningen* (Worth Labelling. The Nature Conservation Association and Environmental Labelling). Stockholm: Swedish Association for Nature Conservation.

SOU, 1987:32: *För en bättre miljö* (For a Better Environment). Stockholm: Ministry of Environment and Natural Resources.

SOU, 1990:44: *Demokrati och makt i Sverige. Maktutredningens slutrapport* (Democracy and Power in Sweden. The Final Report of the Power Study). Stockholm: Allmänna förlaget.

SOU, 1990:59: *Sätt värde på miljön! Miljöavgifter och andra ekonomiska styrmedel* (Put a Value on the Environment! Environmental Charges and other Economic Policy Instruments). Stockholm: Ministry of the Environment.

SOU, 1993:19: *Kommunerna och miljöarbetet* (The Municipalities and their Work for a Good Living Environment). Stockholm: Ministry of Environment and Natural Resources.

SOU, 1993:27: *Miljöbalk* (Environmental Code). Final Report of the Environmental Protection Commission, Part I. Stockholm: Ministry of Environment and Natural Resources.

SOU, 1994:7: *EU, EES och miljön* (EU, EES and the Environment). Stockholm: Ministry of Environment and Natural Resources.

SOU, 1994:125: *Samordnad insamling av miljödata* (Coordination of Sweden's Environmental Data Collection). Stockholm: Ministry of Transport and Communications.

SOU, 1994:128: *Lokal AGENDA 21 – en vägledning* (Local Agenda 21 – some guidelines). Stockholm: Environmental Advisory Council.

Svensk Periodicaförteckning (Swedish List of Periodicals), 1994: Stockholm: Tidnings AB Svensk Bokhandel.

Svenska Dagbladet (Stockholm newspaper).

Sveriges Nationalatlas, 1991: *Miljön* (The Environment). Stockholm: SNA Förlag.

Swedma, 1994: *Miljömärkt direktreklam* (Environment-Labelled Direct Marketing), Swedma informerar, nr. 2.

Vedung, E., 1988: The Swedish Five-Party Syndrome and the Environmentalists, in: K. Lawson and P. Merkl (eds.), When Parties Fail. Emerging Alternative Organisations. Princeton: Princeton University Press.

Vedung, E., 1989: Sweden: The 'Miljöpartiet de Gröna', in: F. Müller-Rommel (ed.), New Politics in Western Europe. The Rise and Success of Green Parties and Alternative Lists. Boulder: Westview Press.

Vem gör vad i Miljösverige, 1994 (Who Does What in Environmental Sweden 1994): Stockholm: Swedish Environmental Protection Agency and Natlikan.

Westholm, A., 1991: The Political Heritage. University of Uppsala: Department of Political Science.

JAPAN

Hidefumi Imura

1 Introduction: Main Environmental Problems

Japan is an island nation stretching relatively far from North to South. Its land is full of topographical and climatic variety, providing a rich natural environment and much scenic beauty. But this natural wealth has been sacrificed or subjected to irreversible changes during the course of post-war economic development. Natural coastlines, in particular, were transformed to artificial ones as a result of land reclamation from the sea and rapid urbanisation.

In the mid-1950s, Japan entered a period of rapid growth. Throughout the 1960s growth averaged 10 percent. The pressure of industrial activity on the environment reached and went beyond breaking point in the mid-1960s. Environmental crises such as asthmatic diseases (Yokkaichi asthma) caused by air pollution, and organic mercury poisoning (Minamata disease) emerged in a number of parts of the country. Citizens' anti-pollution movements and media campaigns appeared, urging stronger government action to regulate polluting industrial activities. The result was a number of environmental laws and the creation of the Environment Agency in 1971.

In the 1970s, various control technologies, such as flue-gas desulphurisation and denitrification and advanced waste water treatment were made available for factories and plants, and clean air and water returned to most industrial cities. The scars of the damaging environmental pollution of the past still remain today, but one cannot use them to judge the current state of the environment in Japan. Visitors to Japan's industrial cities, once notorious for environmental disasters, find that the problems have largely been solved. However, there are persistent as well as emerging problems: air pollution in large metropolitan areas caused by increasing automobile traffic, water pollution in lakes and semi-closed coastal seas due to rapid urbanisation and the increase in household effluents, and the effects of acid rain.

Since the 1980s, the major areas of activity causing environmental degradation have shifted from the manufacturing industries to public works projects. Public investment in infrastructure, for example roads, new towns, railways, harbours and airports, has been promoted as an economic measure to increase domestic demand and stimulate economic growth. These projects are under the control of public works authorities such as the Construction and Transportation Ministries.

In the 1990s, Japanese people entered a new phase of environmental awareness. With the expanding scope of the problems, increasing attention is being paid to global issues such as climatic change, ozone layer depletion, acid rain, and the destruction of the rain forests. Drastic changes took place in people's fundamental world view as regards the relationship between the environment and the economy. Environmental issues have gained prominence from the public, and environmentalist attitudes have gradually penetrated the business community, pointing to the new direction of environmental policy, which will come to rely more on the voluntary efforts of citizens and businesses.

2 Development and Main Characteristics of Japanese Environmental Policy

For more than two decades, Japanese environmental policy and administration were governed by two laws, the Basic Law on Pollution Control and the Nature Conservation Law, enacted in 1967 and 1972 respectively. The scope of the former, in particular, was restricted to pollution control, and it proved insufficient for providing the basis of a new generation of environmental policy. In November 1993, therefore, the law was abolished and the new "Basic Environmental Law" was enacted. The new legislation embraces all Japanese environmental regulatory administration, providing a new basis and framework for national environmental policy in response to the globalisation of environmental concerns.

Japan's environmental measures were implemented with a combination of regulation and control technology. While the government assessed the feasibility and prospects of various technical solutions, it allowed industries time to develop their own methods. Then the environmental targets were raised progressively, forcing industries to introduce appropriate technical measures. At that stage, the initiatives of central and local governments and their administrative guidance for industries played an important role. The "pollution control agreements" approach has been widely used in settling disputes between companies and private residents. Thanks to these measures, Japan has achieved a drastic reduction in air and water pollution from industrial activity.

With the 1990s, Japan's environmental policy entered a new phase. Major targets of environmental policy today are global issues, and problems arising from people's high-consumption lifestyles. In October 1991, the Japanese government published an Action Plan for the Prevention of Global Warming. It set out the goal of stabilising carbon dioxide and methane emissions at 1990 levels by the year 2000.

The Basic Environmental Law of 1993 proclaims the need to establish a society in which economic development places small burdens on the environment, while at the same time changing people's behavioural patterns and lifestyles.

Here, regulatory measures would no longer be sufficient for a new environmental policy; voluntary action from businesses and citizens and market-based economic measures such as environmental charges and taxes should be considered instead. Pursuant to the Basic Law, the Basic Environmental Plan, a comprehensive national environmental plan to better integrate the contributions of the various agencies involved, was approved by the Cabinet in January 1995.

3 The Main Actors

3.1 Environmental Policy Institutions

The Environment Agency is mandated to promote policies for pollution control and nature conservation. It is assigned to the Prime Minister's Office, but its Director-General is a Minister of State and member of the Cabinet. One of its major activities is basic environmental policy planning and overall coordination of the environmental measures implemented by other ministries and agencies. It is empowered to set nation-wide environmental quality standards for air and water, as well as atmospheric emission and water effluent standards for factories and plants, based on specific laws such as the Air Pollution Control Law and Water Pollution Control Law. The Agency also implements environmental laws such as the Natural Parks Law and the Law on Compensation for Pollution-Related Damage to Health. It promotes environmental research, for which it has the National Institute for Environmental Studies (NIES). The Centre for Global Environment Research and the National Environment Training Institute are subordinate to the NIES.

The Environment Agency has two advisory councils: the Central Council for the Environment and the Central Council for Nature Conservation. Members of the councils are respected scholars, academics, representatives of different interest groups and other influential opinion-formers. The Director General of the Environment Agency asks the councils to study and discuss selected policy issues and prepare reports.

A number of other ministries and agencies carry out sectoral environmental protection activities. Among them are Ministry of International Trade and Industry (MITI), Ministry of Construction, Ministry of Health and Welfare, Ministry of Agriculture, etc.

Implementation and enforcement of laws falling under the responsibility of the Environment Agency are delegated in principle to prefectures and designated municipalities, which are invested with administrative status and competence almost equivalent to that of the prefectures. Japan is a centralised country and the central government has strong control over local governments. In the 1960s, however, local governments took initiatives in environmental policy-making and played a pioneering role at a time when there was no relevant legal framework for environmental control provided by central government. Since many environmental problems express themselves at local level, prefectures and municipalities can enact their own ordinances for environmental management, and they are authorised to set more stringent standards than those at national level. Many of them have their own research institutes and monitoring networks. They monitor air and water quality, and conduct inspections of factories and plants. The Environment Agency provides advice for local environmental authorities on general policy directions and technical expertise. It grants subsidies to local governments to improve their environmental monitoring systems, promote measures associated in the Basic Environmental Plan, draft local action plans for stabilising global climatic change, etc.

An example of the role of local governments can be demonstrated in the case of the country's environmental impact assessment (EIA) system. The draft EIA Law was presented by Environment Agency in the early 1980s, but it was dropped after strong resistance from industry. Instead of a law, central government adopted a Cabinet Decision on EIA in 1984. Meanwhile, a number of local governments established their own EIA schemes, which are generally more comprehensive and take more account of citizens' participation in decision-making than the national scheme.

Table 1. Local Ordinances and Guidelines on Environmental Impact Assessment (EIA).

Category	*Number of prefectures and designated cities with EIA schemes*
Ordinances on EIA	5 prefectures and 1 designated city
Guidelines on EIA	36 prefectures and 8 designated cities

Note: Japan has 47 prefectures, and 12 designated cities with administrative competence and a status almost equivalent to prefectures.

3.2 Green Organisations

Some environmental NGOs in Japan, particularly those devoted to nature and wildlife conservation, have a long history and a large membership. Another type of environmental NGOs originated in anti-pollution protest movements at the height of industrial expansion in the 1960s and 1970s. They include groups of pollution victims and patients suffering from pollution-related diseases.

Currently there are literally thousands of grassroots citizens' action groups and other NGOs involved in environmental activities. Most of them, however, are small groups of people working at local and community levels. They do not have enough financial resources or manpower.

3.3 The Media

The role of newspapers and TV is particularly vital in environmental education. In the late 1960s and early 1970s especially, the media played an crucial role in encouraging citizens' anti-pollution movements and stronger government action. In the late 1980s and 1990s, they also played a vigorous role in drawing people's attention to global environmental issues and encouraging government to take action to control the use of CFCs and stabilise emissions of greenhouse gases. Their role in disseminating environmental information to the public has become increasingly important.

3.4 Businesses

The year 1990 should be remembered as the year when the environmental consciousness of the Japanese business firms was dramatically raised. Many firms, not only in manufacturing but also in various service and commercial sectors, created new divisions to examine the linkage between their activities and the global environment, and to direct their actions towards protection of the global environment. "Corporate citizenship", or the idea that business firms should be like a good citizen, living in harmony with society, became popular. The word "eco-business" was created and earth-friendly goods have been promoted.

There are several business associations which have a strong influence on national environmental policy making. They include *Keidanren*, *Keizai Doyukai*, and the Japanese Chamber of Commerce. They have standing committees which discuss the implications for their positions of environmental issues. They have prepared various forms of environmental reports and guidelines. *Keidanren* (Federation of Economic Organisations) issued its "Global Environment Charter" in May 1991, setting forth guidelines for environmentally sound corporate activity. In October 1991, *Keizai Doyukai*, a society of influential business leaders in Japan, issued "Recommendations on Measures to Halt Global Warming – Tasks to be Undertaken Now for the Sake of Future Generations". They include not only recommendations addressed to private corporations but also messages to government and to the general public. These actions from business have maintained close ties to international business initiatives such as the eco-audit programme of the International Chamber of Commerce (ICC) and the "Changing Course" report of the Business Council for Sustainable Development (BCSD).

3.5 Others

Semi-governmental organisations such as the Japanese Environment Corporation (JEC), the Japanese Development Bank and the Smaller Business Finance Corporation have the special function of providing financial loans to private institutions

Table 2. Survey on the Environmental Management Practices of Japanese Business Firms.

	Percentile of firms giving a positive response (%)	*Number of firms giving a positive response*
Establishment of management guidelines or principles for environmental protection	45.7	414
Setting of environmental objectives or targets	39.4	357
Preparation of environmental action programmes	33.6	304
Appointment of a board member with responsibility for environmental matters	57.8	524
Establishment of special divisions to deal with environmental matters	58.1	526
Periodic reporting to the board member responsible for environmental matters	45.2	241
Preparation of environmental education programmes within the organisation	25.9	255
Conduct of periodic inspections and/or audits on environmental performance	27.9	253
Preparation of environmental reports	27.0	245

Source: Survey conducted by Environment Agency in 1994 (Number of respondents: 906).

which undertake environmental projects. In addition, JEC carries out construction projects such as collective waste water treatment systems for small firms, and green belts to counter air pollution. It also has a special fund (Global Environmental Fund) to subsidise the activities of environmental NGOs.

With respect to international cooperation, the Japanese International Cooperation Agency (JICA) holds a number of environmental training courses, to which it invites trainees from developing countries. It also promotes technology transfer projects and sends experts to developing countries. The Overseas Economic Cooperation Fund (OECF) provides loans and grants for environmental projects in developing countries.

4 Capacity-Building

4.1 General Trends

The basic framework for environmental policy and administration in Japan was established in the 1970s. The Environment Agency's position is relatively weak, although it represents very broad and diverse public interests in health and quality of life issues. In contrast, MITI often represents the strong economic interest of industries. The Environment Agency can play a leading role when its policies have solid support from citizens and the media.

In the 1960s and 1970s, some local governments eagerly promoted environmental measures based on their own initiatives. But the overwhelming control exerted from the centre gradually prevailed, and local initiatives tended to be discouraged. Decentralisation policies will have a positive impact upon local initiatives, and will facilitate the public participation that has been neglected in the framework of national environmental policy.

So far, what has been termed the "political, administrative and business complex" has dominated the country. Consumers have little freedom of choice, and cannot satisfy their own desires. Many ordinary Japanese feel a lack of participation in their country's prosperity and political decision-making. Public participation in environmental policy-making is central to its development, but Japan's domestic environmental legal structure has not emphasised the role of citizens and public review. It lacks a system that allows for citizen review of government decisions and citizens' access to government information.

4.2 Resources

Manpower. The Environment Agency is a rather small institution, having only 1000 employees. However, it should be noted that there are 12,500 full-time officials who are in charge of environmental management and administration including pollution control and nature conservation in the prefectures and municipalities. The breakdown of the figures as of 1993 is shown in Table 3. More than 70% of the officials engaged in pollution control in prefectures and designated municipalities have a scientific or technical background.

Government budget The central government budget for environmental protection for 1993 reached a total of 1,715 billion Yen. The Environment Agency's budget, however, was very small, some 64 billion Yen, or a mere 3.7% of the total budget. Ministries and agencies have a number of financial assistance programmes for environmental projects, e.g., subsidies to local governments for construction of sewage and waste incineration plants, and subsidies to private institutions for R&D in energy conservation technology. In addition, the environmental budgets of the 47 prefectures and 12 designated cities amounted to 540 billion Yen for the FY1993, showing a 240% increase over the 224 billion Yen for the FY1985.

Pollution Control Programmes. For areas with significant environmental pollution, the Prime Minister asks the governors of the affected prefectures to prepare pollution control programmes. The programmes must be endorsed by the Prime Minister. According to the programmes, the related prefectures can enjoy favourable financial terms, such as an increased rate of subsidy from central government.

Table 3. Number of Personnel Engaged in Environmental Administration in Japan.

Agencies	*Number of Personnel*
National Government/Environment Agency	993
Local Government	
- Pollution control in prefectures and designated municipalities	6,373
- Nature conservation in prefectures and designated municipalities	1,210
- Pollution control in other municipalities	4,907
Total:	13,483

Private sector investment: In 1975, total expenditure by the private sector on investment in pollution control plants and equipment for air, water and noise control reached a peak of 950 billion yen, or 17.6% of the total investment of that year. Since then it has remained at the comparatively low level of 300 to 400 billion yen, or 3 to 4% of total private sector investment. Total investment costs for pollution control over the decade from 1965 to 1975 amounted to some 6 trillion yen.

Table 4. Japanese People's Perceptions of the Causes of Environmental Problems.

Q: What do you think is the main cause of environmental problems?	*Reply*
A1: Our daily life is the direct cause of the problems	31.5%
A2: Industrial activities are the main cause, but our daily life is partly responsible for the problems	50.8%
A3: Industrial activities are the cause of most of the problems	10.0%

Source: Public Opinion Poll on Environmental Protection, conducted by the Prime Minister's Office, 1995 (Total sample size: N=2,021)

Market size for environmental technology. In the 1980's, annual sales of pollution control plants and equipment or supply of such plants and equipment to both private and public sectors remained almost constant at 600-680 billion yen. The breakdown of the sales figures shows that a constant annual expenditure of 110-160 billion yen was allocated for air pollution control plants and equipment such as precipitators and flue-gas desulphurisation and denitrification devices. There are no published data available for the operation and running costs of such plants and equipment, but they can generally be assumed to be around 30% of total investment costs. The above data indicate a rough estimation of the cost required to regain and maintain clean air and water in the country. The cost was high, and it may have had a severe impact on certain economic sectors or industries, but its macroeconomic effect was not significant.

Consumers' Initiatives. In order to achieve "sustainable consumption", the role of consumers rather than producers has come to be emphasised. Recent opinion polls in Japan clearly indicate that consumer behaviour and values have changed. A greater proportion of people rate environmental protection more highly than convenience. A significant proportion (about 35%, according to one poll) were willing to pay more for environmental protection through increased commodity prices. There have also been a number of consumer movements seeking new, more environmentally benign lifestyles which harmonise with objectives such as energy saving, reduction of household waste, and recycling of materials.

Table 5. Size of Japanese Environmental NGOs in Terms of Annual Budget.

Size/Annual Budget	*Number of NGOs*
More than 100 million yen	3
50-100 million yen	3
10-50 million yen	7
1-10 million yen	17
0.5-1 million yen	11
Less than 0.5 million yen	44
TOTAL:	86

Source: Nissay Basic Research Institute, 1994

Environmental NGOs. The Environment Agency conducted a survey by questionnaire of environmental NGOs in Japan (August-September 1992). The questionnaire was sent to a total of 661 NGOs, and responses were received from 386. With respect to the geographical area of their activities, 233 identified themselves as "local", 73 as "national", and 75 as "international". Nearly half of the respondent NGOs do not have a single paid full-time member of staff. More than half of the local NGOs operate on an annual budget of less than one million yen (11 to 12 thousand US$). Asked about the kinds of assistance they would like to receive from national or local governments, the largest number of requests made (42%) was for financial assistance, followed by environmental information and data services, government contracts for studies, and facilitating networks between environmental NGOs. To sum up, most of the Japanese environmental NGOs are small in terms of budget and staff, although the importance of their activities is very much recognised.

4.3 Coordination Between Relevant Actors

Coordination between government ministries and agencies. Administration in environmental protection is shared between different ministries and agencies. One major responsibility of the Environment Agency is to provide coordination between them. But this is not always done effectively; the environmental performance reviews of Japan conducted by the OECD (1994) concluded that "agencies and ministries seem to act independently in a spirit of competition rather than cooperating fully". It is said that Japanese government officials identify themselves with certain ministries or the agencies by which they are employed, and they work for the interests of their ministries or agencies rather than of the public. There is a similar problem with the relationship between environment and development authorities in prefectures and municipalities, but the local situation is much better because they work under the political leadership of governors or mayors.

Basic Environmental Plan. In accordance with the new Basic Law, the government prepared the Basic Environmental Plan, setting out the basic directions of national environmental policy. The Plan laid down the following long-term objectives:

(i) "Sound Material Cycle" – A socio-economic system fostering an environmentally-sound material cycle
(ii) "Harmonious Coexistence" – Ensuring harmonious coexistence between humankind and nature
(iii) "Participation" – Achievement of participation by all sectors of society, each shouldering a fair amount of the burden
(iv) "International Activities" – Promotion of international activities

The Plan, however, does contains neither clear policy goals nor quantitative targets. The Basic Law also encourages local governments to prepare their own environmental basic plans.

Cooperation between businesses, citizen, and administration. The environmental awareness of both citizens and businesses has increased dramatically recently, due to the influx of information on the threatened life support systems of planet Earth. Many firms have established special divisions to deal with global environmental issues. On the other hand, consumers' behaviour and values have also changed. The results of opinion polls reveal that a greater proportion of people rank environmental protection above economic growth, and many of them would be willing to pay for a safer environment through increased prices. Cooperation between government and business, as well as between government and citizens, has improved rapidly, especially at local and communal levels. However, businesses and citizens are still caught up in the traditional picture of producer *vs.* consumer, and full cooperation between them has not been developed far. The local environmental authorities could play the role of a catalyst in promoting cooperation.

Roles of action plans and programmes. A number of prefectures and municipalities have already prepared or will soon draft a basic environmental plan similar to the Basic Plan from central government. They have also prepared local action plans or programmes for slowing global warming, traffic pollution control plans, and "local agenda 21", in accordance with the Agenda 21 approved at the Rio Conference. The Environment Agency provided technical guidelines to help local governments work on these plans or programmes. In preparing such plans or programmes, local authorities have consultations with citizens and businesses, in order to build a consensus on the measures to be put in the programme. Businesses are cautious about, and take up guarded positions on central government initiatives, but they are generally more supportive of and cooperative with the actions of local government.

Table 6. Number of Environmental Plans and Programmes Prepared by Japanese Local Governments (as of October 1995).

Type of Plan	*Number of Plans Adopted*
Basic Environmental Plans, Programmes, or Guidelines	22
Local Agenda 21	6
Plans for Stabilising the Global Warming	18

4.4 Development of the Policy Paradigm

Globalisation of the problem. From the 1980s onwards, environmental issues came to be discussed in the broader context of global economic and environmental interdependence. Global environmental risks such as climatic change emerged as critical problems that could affect the future of humankind. As we approach the next century, the paradigm of environmental policies is undergoing rapid and drastic change. At national level, conflicts between economic and environmental policies still exist, but they have been gradually mitigated, as environmental policies have proved beneficial in improving people's quality of life, while their macroeconomic impacts have proved marginal. Now, fresh consideration is required of equity between industrialised and developing countries. A sense of the country's international responsibilities and role in the protection of the global environment has come to be widely shared by politicians, industrial leaders and the general public.

Table 7. Japanese People's Concern for Different Environmental Issues.

Matters of Concern	*Male*	*Female*	*Total*
Problems caused by our daily life, such as water pollution by household waste water, and increasing domestic waste	58.1	69.8	64.4
Global problems such as climate change, ozone layer depletion, and the destruction of tropical rain forests	62.1	50.0	55.6
Air and water pollution caused by industrial facilities	44.6	35.4	39.7
Environmental contamination by industrial and agricultural chemicals	31.6	37.1	34.5
Destruction of our local natural environment	22.7	24.8	23.8
Depletion of material and energy resources	25.0	17.7	21.0
Deterioration of the great natural wealth in national parks	11.9	12.9	12.4

Source: Public Opinion Poll on Environmental Protection, conducted by the Prime Minister's Office, 1993 (Total sample size: N=3,754)

Need for revolutionary change in our society. In the old paradigm of environmental policies, the major policy targets were local problems arising from industrial activities. Polluters and victims were clearly identifiable, and it was enough to monitor and control the emission of pollutants from factories and plants. End-of-pipe control technologies were used effectively to reduce pollutant emissions. But today, we are confronted by more complicated problems which are much more closely linked to our mode of civilisation, culture, technology and social values than the traditional problems; their solution requires revolutionary restructuring of our society with respect to our lifestyles, technology, consumers behaviour and attitudes, business attitudes and management practices, etc.

From "command-and-control" to a mix of "soft-instruments". With the changing environmental objectives, fresh consideration of the policy instruments to be adopted is also necessary. The traditional dichotomy of regulation vs. economic instruments should be re-examined. Whether the instruments be regulatory or economic, they would have limited applicability to management of the global environment. A variety of new "soft-instruments", such as administrative guidance, planning, information and education could be used in addition to the traditional instruments. Furthermore, legislation such as requirements on packaging and labelling and eco-audits would facilitate compliance with standards.

Changing political conditions. For a several decades, the so called "political, administrative and business triangle" has dominated Japanese politics. This triangle formed around public works projects and created a strong pro-development power block, opposing EIA legislation and public participation in environmental decision-making. Japan, however, has entered a period of rapid political change since the establishment of the new coalition cabinet in August 1993. Its new political moves are likely to be towards deregulation, consumer-oriented policies, and decentralisation, all closely related to international economic issues, and will have an effect on the new directions of environmental policy in the coming decade.

International transfer of Japanese experience. Today, Japanese people pay less attention to local pollution issues, but experiences in this area are attracting the growing attention of developing and newly industrialising countries, where environmental degradation similar to what took place in Japan may be repeated. There is a growing need to transfer Japanese technology and expertise to developing countries, although the number of experts who can work in this field is rather limited. Many local governments have initiated international (or "inter-provincial" or "inter-municipal") environmental cooperation with foreign local governments. In particular, cooperation with China, Korea and the Southeast Asian countries has become vitally important, due to Japan's geographical location and economic relationships to these countries.

Environmental aid to developing countries. Japan's ODA (official development assistance) has been increasing, and Japan was the world's biggest donor country in 1989 and 1991. The United Nations has called upon all countries to achieve a 0.7% ratio of overseas aid to GNP, and Japan has pledged to meet this target, although it has not specified any time frame. Moreover, at the end of the UNCED, Japan pledged that it would expand its bilateral and multilateral ODA in the field of the environment to around 900 billion to one trillion yen (10 to 11 billion US dollars) during the five years starting from FY 1992, and that the funds would be made available in addition to conventional assistance.

Stabilisation of global warming. In most countries, a large proportion (40-70 percent) of total CO_2 is emitted from industrial facilities including electric power plants and factories. In Japan, it is roughly estimated that the dominant part (80-90%) of CO_2 emissions from industrial facilities is accounted for by about 3,000 plants and factories. Thus, if it is administratively feasible, regulation of and guidance to these larger plants and factories would prove extremely effective, encouraging them to take stronger action on energy conservation and reduction of CO_2 emissions. MITI's energy conservation programme has been implemented on the basis of such an idea. Furthermore, there has been discussion of the use of economic instruments such as a carbon tax.

4.5 Main Restrictions

Limitations of administrative guidance. Administrative guidance and industrial policy played a positive role in transforming Japan into an economic giant. It was an appropriate policy when Japanese industries were in their infancy. Government intervention and guidance in industrial activities successively stimulated competition among industries in the technological challenge of controlling industrial pollution. Today, however, there is a growing view that government should minimise its intervention in private business activities. Administrative guidance has the advantage that it can be used flexibly, while avoiding rigid application of the law. It sometimes stimulates businesses to take stronger action than that required by law. It can provide semi-official guidelines to businesses even when relevant legislation does not exist. It may, however, opt to rely on the efforts of specific sectors or firms with greater technical potential and financial capacity. On the other hand, it may protect smaller firms and weak industrial sectors, thereby causing economic inefficiency. Moreover, it might prove inefficient due to bureaucratic sectoral policies and conflicts between different ministries and departments.

Necessity of "social breakthroughs". For global environmental problems such as climatic change, collective and cooperative action by government and industry will still be effective. However, the traditional Japanese approach, which has relied heavily on technological countermeasures, might have its limitations. It is difficult to apply "command and control" policies to regulating energy use and changing high-consumption lifestyles. Instead, greater attention should be paid to "social breakthroughs" such as the reform of economic and social systems, including the behavioural patterns of businesses and individuals. In this, citizens' participation and the role of non-governmental organisations, as well as environmental education and information, will be essential. Greater use of market-based economic instruments, such as environmental charges and taxes, should also be considered. Such ideas have already been taken into account in the new Environmental Basic Law, but they have not yet been implemented.

Limitations of environmental impact assessment. In Japan, environmental impact assessment (EIA) has not been legally formalised. With the enactment of the new Basic Law, environmental assessment legislation will be a central policy issue in Japan. Nonetheless, EIA has been conducted under individual laws, administrative guidance from ministries and agencies, and local government ordinances and guidelines. Furthermore, the Cabinet approved the decision on EIA in 1984 as a uniform rule for large-scale projects in which national ministries or agencies are involved. The current practice of EIA in Japan has several limitations with respect to the narrow definition of target projects, limited scope of the study items, and insufficient public involvement. There also are conflicts between the national schemes based on the Cabinet Decision and those prescribed by local government ordinances and guidelines.

Insufficient public involvement in public works projects It is planned to increase domestic demand in Japan in order to weaken export pressure. A huge amount of investment in public works is scheduled. It may remarkably improve Japan's social infrastructure, both in quantity and quality, especially those related to urban environmental amenities and facilities such as sewage systems. At the same time, it may risk to causing serious damages to the natural environment in Japan. Public involvement in the public works projects from their planning stage should be strengthened. The public review process in EIAs should be improved in order to ensure the better integration of public opinion into the project design.

International aspects. In today's globalised economy, a great problem of environmental policy is that it may serve as a tariff-free trade barrier which obstructs free trade and cause distortions in the international allocation of resources. On the other hand, with the globalisation of the economy, there has been a movement within Japanese industry to relocate production bases to other countries. This movement has been triggered by the significant difference in factor costs between Japan and developing countries. Although the movement was not necessarily caused by the difference in environmental regulation and standards, it might well result in "international relocation of polluting industries", and "carbon leakage" from industrialised countries like Japan to developing world. However, trade liberalisation and the protection of the environment should not be considered as mutually conflicting objectives. They call for greater integration of environmental and trade policies at the national level as well as for parallel efforts to promote international cooperation on the basis of multilateral regulations and agreements.

5 Policy Evaluation

Success and limitations of Japan's environmental policy. Japan's environmental policy has focused on pollution control and energy efficiency, and it was in these areas that Japan has been successful. Moreover, most of the Japanese environmental achievements have been local, and they were made possible by the efforts of manufacturing sectors and the use of end-of-pipe control technologies. In large metropolitan areas like Tokyo and Osaka, there are still environmental problems caused by increasing automobile traffic and people's affluent lifestyles.

Scope of environmental policy. Until recently, Japan did not pay much attention to resource recovery and material recycling. Regulating industrial activities alone cannot provide a practical solution to these problems. Moreover, environmental policy in Japan has focused relatively little attention on conservation of natural resources, including plants and animals. Japan is one of the largest users of natural resources, and it depends heavily upon these resources, imported from other countries. Japan is often criticised for its extensive use of tropical timbers and its attitudes to nature conservation.

Technology development stimulated by regulation. As far as industrial pollution control in Japan was concerned, government intervention and guidance on industrial activities successively stimulated competition among industries in the technological challenge presented by the environment. Most of the environmental technologies such as flue-gas desulphurisation and denitrification, computerised monitoring techniques and automobile exhaust gas control technologies were developed through a combination of appropriate government control and competitive market conditions.

Need for consumer-oriented policy and assistance to NGOs. Interventionist policy may no longer be effective in achieving new policy goals, such as the reform of people's affluent lifestyles and business attitudes, and stabilisation of carbon dioxide emissions. Wider use of economic incentives and educational tools should be encouraged. The fundamental limitation of traditional administrative guidance is that its structures are producer-oriented and it can not take into account consumers' behaviour and actions. It is not practically feasible to exert administrative guidance on unspecified individuals. Some new mechanisms such as local authorities' initiatives to encourage the participation of citizens and a variety of NGOs must be put into force. Financial and other assistance to NGOs is also vital.

Wider participation of citizens and businesses. Preparation of action plans and programmes at local level will be an effective tool in facilitating consultation and consensus-building among the diverse entities concerned. What is likely to emerge is a greater effort by citizens and industry to participate in the process of planning and drafting action plans and programmes and working toward improved consultation and communication.

Japanese model as lesson for newly industrialising countries. The Japanese environmental experience is attracting the growing attention of developing and newly industrialising countries. Japanese achievements suggest the importance of an effective combination of market mechanisms and government initiatives. Especially in the newly industrialising countries in Asia, there is a clear need for appropriate government guidance and incentives to encourage environmentally sound business activities. In this respect, the Japanese model might be useful to them, despite its limitations.

References

Environment Agency, 1995: Quality of the Environment in Japan, 1993. Government of Japan.

Imura, H., 1994a: Japan's Environmental Balancing Act, Asian Survey 34, 4, 355-368.

Imura, H., 1994b: Non-Economic Instruments: Regulation and Standards, Administrative Guidance, and Action Plans and Programmes, Proc. Tsukuba Workshop of IPCC WGIII, 248-266.

Imura, H., T. Futawatari and R. Fujikura, 1995: Economic Development, Energy, and Environment in East Asia: A Comparative Study of Japan, South Korea, and China, Journal on Global Environmental Engineering, 21-42.

Kato, K., 1992: Forging Partnerships Among NGOs, Businesses, and National and Local Governments, Environmental Study Quarterly No. 92, 55-65.

Matsui, S., H. Imura and M. Tanaka, 1993: Environmental Pollution Control – Papers Presented at the UNU International Symposium on Eco-Restructuring, 5-7 July 1993. The United Nations University.

GREAT BRITAIN

Albert Weale

Introduction

The term "environmental policy" in the UK covers a wide field, ranging from pollution control covering air, waters, waste management, nuclear emissions, and noise, through land use planning, housing, building conservation and urban regeneration to the conservation and protection of flora and fauna. Because of limitations of space, the principal focus of this chapter will be upon institutions and policies related to pollution control. This is an analytical convenience rather than a description of how policies and environmental effects work. After all, acidifying emissions can adversely affect the stonework of cathedrals, and good land use planning is essential in the design of successful waste disposal strategies. However, this focus will also mean that some important aspects of British environmental policy - including land use planning and flora and fauna protection - will be ignored in what follows.

1 Main Environmental Problems and Environmental Quality Changes since 1970

Main Problems. As a medium sized, post-industrial society, Great Britain has a familiar range of pollution problems: acidification, urban air pollution, photochemical smog, soil degradation and a legacy of contaminated land, freshwater and coastal pollution and problems of noise and dirt. It is also a contributor to the familiar global problems, including greenhouse gas emissions and ozone depletion. Its geographical situation means that, unlike the Netherlands for example, it does not import pollution via rivers that originate in other countries. Its river emissions, however, flow into a series of seas bordered by neighbours who allege problems from UK sources. As with other countries in north west Europe, it has both an import and an export trade in air pollution, with more being exported than imported. It is claimed that it also has some specialist engineering capacity to cope with the disposal of hazardous chemical and nuclear waste.

Main Changes. In the field of air pollution, time-series data on sulphur dioxide emissions between 1970 and 1992 show a substantial drop, with the 1992 output being only some 54% of the 1970 output. In 1992 electricity-generating stations accounted for 69% of total emissions, and other industrial uses accounted for another 21% (Digest of Environmental Protection and Water Statistics, 1994, Table 2.2a). The UK reached its 1998 target for sulphur dioxide emissions under the EC's large Combustion Plant Directive in 1994 (ENDS Report, 1995a: 11).

Nitrogen oxides by contrast show the opposite trend. Over the twenty year period 1970 to 1990 there is an increase of some 19% in emissions, although there is a slight downward trend in the period 1990 to 1992. The principal cause of this increase is to be found in the increase in road traffic from 1970 onward (Digest of Environmental and Water Statistics, 1994, Table 2.9a).

Of the greenhouse gases, emissions of carbon dioxide show a decline in emissions, some of which is due to declining emissions from power stations. The rate of decline slows down considerably after 1975 and there is virtually no improvement from the mid-1980s onwards (Digest of Environmental and Water Statistics 1994, Table 2.9a).

Turning to water pollution, the long-standing practice in the UK has been to assign stretches of river to one of five categories: Good 1A; Good 1B; Fair; Poor; and Bad. Between 1980 to 1990, the most striking feature is the decline in the length of river classified as falling in both of the "good" categories from 69% to 63%. Although it is thought that some part of this decline, particularly in the south of England, may be attributable to changes in survey methods, it is equally clear that the trends are also picking up some failures of regulation, especially of the water utilities before they were privatised, when their capital spending was restricted by Treasury controls. Indeed, it has been alleged by someone inside the water policy community that one of the arguments for water privatisation was the high capital cost of cleaning up emissions under EC directives that would otherwise fall on the public purse (Kinnersley, 1994).

Coastal water quality until 1994 also showed a rather flat trend, with only some 80% of bathing beaches meeting the coliform standard under the EC's Bathing Water Directive. Some caution must be exercised in looking at these figures, particularly as there was a marked improvement in 1995, with some 89% of beaches meeting the standard (ENDS Report, 1995b). However, this improvement has in part been attributed to unusual weather conditions. It is also well known that the UK authorities have been rather stringent in their designation of areas of traditional bathing (in the first year fewer beaches were designated than in Luxembourg!). With these qualifications in mind, the generally slow moving picture appears to reflect the failure to invest in wastewater pollution control equipment prior to 1989.

Figures on waste management are less freely available than for other aspects of pollution. However, one example where figures are available concerns the proportion of recycled paper in annual consumption, where in the period 1980 to 1992 consumption only went up from 30% to 34% (Digest of Environmental Protection and Water Statistics, 1994, Table 7.17).

Clearly not all of the above changes and fluctuations can be attributed to the effects of environmental policy. Economic and industrial policies, influencing the volume and composition of economic activity are important, as are autonomous changes in technology, fluctuations from non-anthropogenic sources, climatic variations and alterations in behaviour patterns. However, apart from the case of sulphur dioxide, the general picture from these few, but important, indicators is of a static, or if anything worsening trend.

2 Development and Main Characteristics of Environmental Policy.

Institutionalisation. Environmental policy has had a long history in Britain. Nineteenth century legislation established the world's first national pollution agency, the Alkali Inspectorate, in 1865 and public health legislation in the nineteenth century gave local authorities powers to control pollution and redevelop slums (Ashby and Anderson, 1981). Despite this relatively long history of efforts to control pollution and improve the environment, the modern institutional development of environmental policy institutions began, as with many other countries, in the late 1960s and early 1970s.

The establishment of an environment ministry was one of Mr Heath's first acts when he became Conservative Prime Minister in 1970. The ministry brought together sections of the bureaucracy that had previously existed in different departments and combined them with the ministry for transport, thus creating one of Mr Heath's "super-ministries". The idea was to create an organisation that was capable of taking a comprehensive look at policy problems (similar organisational innovations occurred in the field of social policy). The incoming Labour government in 1974 split the transport function from that of the environment, and the present structure of the environment ministry stems from that period (Weale, O'Riordan and Kramme, 1991).

Although transport was detached from the environment ministry, the residual environment department was still large covering not only matters of environmental pollution, but also such functions as land-use planning, building and nature conservation and the management of government property. Moreover, the Department of the Environment (DoE) is also responsible for the control of local government, and this includes local government finance. Since the subject of local government finance became politically controversial in the UK during the late 1980s, the span of the DoE's work had important implications for what it was able to do.

Main Laws and Regulations. The institutionalisation of environmental policy was accompanied by some new legislation. The most important change was the 1974 Control of Pollution Act (COPA), which mainly represented a codification of existing practice, although it did introduce a new regime for waste disposal.

Since COPA there have been two principal pieces of legislation. The first is the 1990 Environmental Protection Act, which now provides the statutory framework for most forms of pollution control. The principle informing this act is that of Integrated Pollution Control (IPC). The chief idea of IPC is to minimise pollution to air, water and land, treating the environment as a whole. The intention is thus to avoid worsening a water pollution problem, for example, as a result of dealing with an air pollution problem. In order to achieve this objective, the 1990 Act provides for a single written authorisation for prescribed processes governing all emissions into the environment (Environmental Protection Act, 1990, Part I).

The act also enables the Secretary of State to specify emission limit values for the prescribed processes or particular substances and environmental quality standards for all environmental media. Under this provision of the act public authorities may specify air quality standards, rather than simply emission standards as had happened under previous legislation. Water quality standards, by contrast, had been part of the custom and practice of water pollution authorities since the 1950s, but they had only been put on a statutory basis under legislation passed in 1989, although at the time of writing (December 1995) no actual standards had been put into force. The 1990 Act also controlled especially dangerous substances emitted into water (the "Red List" substances) under IPC.

The 1990 legislation also introduced significant changes to waste management policy. The main principle of this part of the legislation was to establish the legal basis upon which it would be possible to separate the functions performed by local authorities in regulating waste management from the functions of collection and disposal. The act also imposed a duty of care upon waste operators to act responsibly within the regulations (Environmental Protection Act, 1990, Part II).

The second main piece of legislation is the 1995 Environment Act, the principal effect of which was to establish the Environment Agency for England and Wales and the Scottish Environment Protection Agency for Scotland. These will be the bodies that in future will be responsible for operating the IPC regime. The 1995 act also created the powers for local authorities to establish air quality management areas and laid upon the national government responsibility for a national waste management strategy.

Main Instruments. Within current legislation the principal policy instrument is regulation, in the sense of the laying down of standards of environmental performance that products and processes should meet. Primary legislation sets the administrative framework for regulation, and standards are developed by the appropriate public authorities. Standards are set through a process of public consultation in which there are extensive negotiations between the regulated and government. The Environmental Protection Act also gave new powers to local authorities to control air pollution for processes that were not prescribed under the legislation.

Economic instruments only have a minor place in the control of emissions. The regulatory authorities levy cost-recovery charges upon those whom they license, but there are no water pollution charges set in a Pigovian way to provide an incentive to reduce emissions and there is no provision for tradeable permits under an air pollution "bubble".

One major exception to this rule is that a system of landfill charges is due to come into effect in 1996. The revenues from the tax will be recycled into a reduction of national insurance contributions for employers, which is the principal payroll tax in the UK. The government has also imposed Value Added Tax on domestic fuel use, and some supporters of the measure have argued that the policy was motivated by environmental concerns. Others (perhaps rather cynically) suggested that the policy had more to do with raising government revenue. The full rate of tax was due to be imposed in two stages, but the second was frustrated by a backbench revolt among Conservative MPs, worried at its electoral implications.

Policy Style. The regulatory strategy is traditionally conceived by commentators as one of cooperation rather than confrontation with industries to be regulated (e.g. Vogel, 1986). Such a cooperative style of regulation has been traced back to Angus Smith, the first chief air pollution inspector. Thus, emphasis has usually been placed within the British regulatory system upon voluntary compliance where it can be secured.

Such a system was in place until 1985 in respect of pesticide residues, according to which a safety precautions scheme was formally agreed between government and the agrochemicals industry. However, under the 1985 Food and Environment Protection Act, the voluntary scheme was replaced by a statutory one. Other examples, in recent years, might lead to a modification of this traditional view. Thus, when it was first established in 1989 the National Rivers Authority (see below) sought to flex its muscles with a number of high profile prosecutions, and when Mr Frank Feates was Director of Her Majesty's Inspectorate of Pollution (HMIP) he spoke in terms that suggested that the inspectorate was intending to be more forceful in its requirement for new technologies. However, it is also clear that there are strong pressures restricting the development of more forceful regulation.

Issues of policy style cannot be detached from the neo-liberal political strategy of Conservative governments since 1979. There are a number of elements to this. Part of the strategy has involved giving substantial policy concessions to industry in order to boost the UK's economic performance. Hence, at various times, there have been persistent attempts to pursue an agenda of deregulation, and under the 1995 Environment Act the Environment Protection Agency will be required to conduct cost-benefit analyses of proposed measures, a requirement that was criticised by a number of proponents of environmental protection during the passage of the legislation. Similarly, the privatisation of the utilities was premised on a promise of high profits to investors, and regulatory impacts that threaten to reduce profits will meet with substantial, and influential, political resistance. Finally, the neo-liberal agenda has been suspicious of the role of government as a provider of services, holding that private enterprise can be expected to be more efficient, and such sentiments lay behind the changes in 1990 that turned local authorities into regulators of waste services rather than providers.

3 The Main Actors

3.1 Environmental Policy Institutions

The present range of responsibilities of the DoE are set out in *Table 1*. The political head of the department is the Secretary of State for the Environment, who sits in the Cabinet. Although in recent years the position has been held by prominent or upwardly mobile politicians, the position is not among the most senior in the Cabinet: the Chancellor of the Exchequer and the Secretary of State for Trade and Industry should be regarded as stronger. Relative to agriculture and transport, the position is possibly a little more fluid. Since there have been no coalition governments in the UK since 1945, the Secretary of State for the Environment always shares the same party affiliation as other members of the Cabinet. He (the office has never been held by a woman) is assisted by ministers of state responsible for sections of the department's work, and pollution control is identified as one of these components for which a minister will take responsibility. Underneath the ministers are junior ministers known as Parliamentary Under Secretaries. The administrative head is a civil servant, known as the Permanent Secretary, assisted by Deputy and Under Secretaries. Civil servants are by convention career, non-political appointments, although the Secretary of State can appoint temporary advisers.

Table 1. Main Responsibilities of the Department of the Environment.

Main Responsibilities of the Department of the Environment
- Planning and Development Control - Rural Affairs, including countryside and wildlife protection - Water, including drinking water, water resources and marine pollution - Government property holdings - Building construction and building regulations - Pollution control standards, including air quality, toxic substances, global atmosphere, waste control, international measures and data and analysis - Housing and architecture, including public housing - Local authority finance and government

Source: Civil Service Yearbook (London: HMSO)

From 1 April 1996 primary operational responsibility for environmental protection will rest with the Environment Agency (in Scotland the Scottish Environment Protection Agency). This will comprise a board of between eight and fifteen members appointed by the government to implement policy. In England and Wales the new agency will bring together two bodies: Her Majesty's Inspectorate of Pollution (HMIP) and the National Rivers Authority (NRA), as well as some inspectors from local authorities concerned with waste management.

HMIP had been created on 1 April 1987 and brought together specialised inspectorates within the Department of the Environment responsible for air pollution (which by a quirk of administrative history had been the responsibility of the Department of Employment since 1974), radiochemicals, hazardous waste and water quality. HMIP had administrative responsibility for the setting of standards in the fields for which it is responsible. Traditionally HMIP and its precursor, the Alkali Inspectorate, has been small. Around its creation it had some 230-250 inspectors. In 1993/94 it had about 450, but it was due to reduce these numbers, under public sector so called "efficiency savings", to some 434. The consequence of these small numbers is that site visits are relatively infrequent, even to complex plants. The single largest category of visit in recent years have comprised pre-licensing discussions with operators.

By contrast with HMIP the NRA was a non-departmental public body, meaning that it operates at arms-length from the ministry. This body was created in 1989 in the wake of the privatisation of the previously public Regional Water Authorities. The NRA had its own board, made up of part-time members, as well as a full-time Director-General. At its inception it had some 7,000 staff, and so was significantly larger than HMIP.

In bringing together HMIP and the NRA into one body, it is hoped that the organisational structure will be created to operate integrated pollution control. This should be helped with a transfer of some functions from local waste regulatory authorities (see below).

Another non-departmental public body with some environmental responsibility is the Health and Safety Commission, the prime responsibility of which is worker health and safety, but with obligations in respect of residual emissions affecting the public. It works through an executive, and has responsibility for such matters as the control of pesticides, and the labelling of hazardous chemicals.

There are also other ministries that have some responsibility in matters of pollution control. The Ministry of Agriculture, Fisheries and Food has pollution control responsibilities in respect of some marine pollution and the Department of Transport has responsibility for vehicle emissions control.

Local authorities also have important pollution control functions. In Britain there are two broad sets of local authority: those responsible for large urban areas and those responsible for the rest of the country. In the large urban areas metropolitan districts have a full-range of local authority functions, although no one authority covers more than a fraction of the conurbation in which it is located (for example, London has thirty-two districts, with no government for London as a whole). In the non-metropolitan areas there are typically two layers of government (although the structure is currently under review): county councils have been responsible for waste disposal policy and district councils responsible for some aspects of air pollution policy, with both operating within the framework of national legislation and standards.

Local authorities are creatures of statute in the UK and therefore do not enjoy constitutional autonomy. During the 1980s Conservative governments under Mrs Thatcher sought both to limit their powers and control their spending. It should also be noted, conversely, that arrangements differ in Scotland and Northern Ireland. Thus, in Scotland water supply has not been privatised. These differences reflect the political and constitutional barriers to changes in Conservative government policy in the 1980s.

Parliamentary committees have also be important influences upon policy, and it is possible to trace the origin of various policy initiatives to them. Three committees in particular have been important: the House of Lords Select Committee on Science and Technology; the House of Lords Select Committee on the European Communities; and the House of Commons Select Committee on the Environment.

There are a number of other important bodies in the policy making process. The Royal Commission on Environmental Pollution (RCEP) was established in 1970. It is an independent standing body the function of which is to provide scientific and policy advice on matters concerning the pollution of the environment. It is thus the functional equivalent of the German Advisory Council on Environmental Questions or the US Council on Environmental Quality. It has always taken a wide view of its responsibilities, and has produced eighteen reports to date, covering not only problems of pollution (air and water pollution, waste management and lead in the environment) but also questions of administrative organisation (it was behind the establishment of HMIP) and policy principles and approaches.

In addition to the RCEP there are also advisory bodies responsible for specific policy matters. Examples of such bodies include the Expert Panel on Air Quality Standards and the Quality of Urban Air Review Group. These bodies usually have specific terms of reference to advise on the setting of standards. Two particularly important advisory groups are the Government Panel on Sustainable Development and the Advisory Committee on Business and the Environment. Both have wide-ranging remits to advise on the development of policies for sustainable development, and in its first report the former advocated a substantial shift in taxation away from labour and towards taxes on pollution and resource use.

3.2 Green Organisations

Environmental pressure groups in the UK go back to the mid-1880s, when groups like the Coal Smoke Abatement Society and the Royal Society for the Protection of Birds were formed. The 1920s and 1930s saw the formation of other groups, including the Pure Rivers Society, and in the 1950s and 1960s there mushroomed other groups including Friends of the Earth (in general, see Lowe and Goyder, 1983).

Since the mid-1980s membership of the principal environmental groups has grown considerably (O'Riordan, 1991). Some figures are given in *Table 2*. It will be seen that a number of these organisations are concerned with the protection of countryside, wildlife and historic buildings, rather than with pollution control, though the Royal Society for the Protection of Birds has been concerned with questions of pollution control when they have touched its field of interest. Friends of the Earth, Greenpeace and the National Society for Clean Air have been the principal groups concerned with pollution control.

Table 2. Membership of Environmental Groups (in thousands)

Organisation	*1971*	*1981*	*1992*
Council for the Protection of Rural England	21	29	46
Friends of the Earth	1	29	116
National Trust	278	1,046	2,186
National Trust for Scotland	37	110	237
Ramblers Association	22	37	94
Royal Society for Nature Conservation	64	143	250
Royal Society for the Protection of Birds	98	441	850
WWF	12	60	150

Source: Digest of Environmental Protection and Water Statistics (London: HMSO, 1994) No.16, Table 11.22

One distinction often invoked in connection with the environment movement in the UK is that between insider and outsider groups. Insider groups seek to use the formal and informal methods of consultation to influence public policy, shunning publicity stunts in order to achieve their goals. Outsider groups adopt a less compromising attitude to the public authorities, and seek to achieve their goals through public pressure or direct action. Although the distinction is clear in theory, in practice it may be elided, since some groups, like Friends of the Earth, have moved from being an outsider group in the 1970s to being an insider group. Moreover, one should probably see the environmental movement as a *system* of groups, so that as one part of the system adapts its strategy, other groups move in. For example, recent years have seen the growth of more extremist groups including the so called Donga Tribe, activists who have taken direct action against road building by building tree houses in the line of proposed road developments.

Size is not necessarily a good indicator of effectiveness. One of the more influential groups of the 1980s was the Green Alliance, with a membership of no more than five hundred people, but with the status and contacts to gain access to Westminster and Whitehall. Not only does the Green Alliance supply parliamentary briefings and inspire parliamentary questions, but it is credited for example with prompting a meeting in 1987 in the prime minister's office to address problems in the relation between agriculture and the environment.

There are also groups that seek to promote dialogue between government, industry and the environmental movement. One of the leading groups in this category is Centre for Environment and Economic Development. The London based office of the Institute for European Environmental Policy has a similar clientele. It should also be noted that business consultancies advocating "green capitalism" are vocal in the UK.

An important part of the pressure group process is that pressure to improve the environment comes from groups other than those solely devoted to environmental protection. For example, growing concern over the effects of air pollution on asthma sufferers has in part been prompted by the concerns of medical organisations like the British Medical Association.

The Green Party has been through cycles of strength and weakness. Its most successful electoral campaign was in 1989, when it polled some 15% of the popular vote in the European elections. This figure needs to be seen in context, however. To a large extent it was the beneficiary of the protest vote that otherwise would have gone to the Liberal Party had that party not itself been involved in traumas over its merger with the Social Democratic Party. The electoral system meant that the popular vote was not translated into seats. And subsequent in-fighting in the party, broadly along the lines of the familiar fundi/realos split, led to a dramatic decline in support in the following five years (Evans, 1993). However, the Green Party does have some elected members on local councils, and one Plaid Cymru (Welsh nationalist) MP arguably owes his seat to an election bargain with the local Green Party.

3.3 The Media

Environmental issues are treated both in general current affairs coverage and in specialist reporting. Television, radio and newspapers all carry stories about environmental matters.

Newspapers are traditionally divided into popular ("tabloids") and "serious" or "broadsheet" categories. Among the broadsheet newspapers specialist environmental correspondents began to emerge in the mid-1980s, prior to which only Geoffrey Lean of *The Observer* was an environmental specialist. Although there is movement of individuals from post to post, all five main national newspapers (*Guardian, Times, Telegraph, Financial Times* and *Independent*, as well as *The Scotsman* in Scotland) have specialist environmental correspondents, responsible also for issues of conservation and nature protection as well as pollution control. The Sunday papers will also carry specialist reports. Great Britain is also the home of serious weekly periodicals, which cover environmental issues regularly, especially *New Scientist* and the *Economist*.

Radio and television reveal a similar mixture of coverage in both general current affairs programmes and in specialist programmes. It should also be borne in mind that some attention will also be devoted to the effects of pollution on various species in the many nature programmes that are such a conspicuous feature of British television.

It has been argued by Lean (1991) that there is a bias in news coverage towards events rather than processes. That is to say, that great attention will be paid to an oil tanker spillage in news coverage, but no attention to regular emissions that cumulatively do more damage.

3.4 Other Proponents of Environmental Protection

Among business organisations, the Confederation for British Industry, the main employers organisation, has sought to develop awareness of the commercial importance of environmental quality among its members through an environment department. But it is a body with a varied membership, and it has to balance its environmental awareness with concerns among its members about costs, particularly those arising from regulation. The Chemical Industries Association has also sought to develop codes of conduct for the chemical industry and is a regular participant in policy fora.

The retail sector is a relatively strong part of the UK economy, and it is not surprising that retail organisations have sought to respond to public concerns about the environment, in part to attain a competitive advantage. The foremost example is The Body Shop, which has since the 1970s aimed to be environmentally responsible in the purchasing and marketing of its products.

There are attempts to promote environmental awareness among learned organisations and other public bodies. The Royal Society has been important in promoting the public discussion of risk perception, and the Royal Society of Arts has sought to promote the importance of environmental awareness in industrial design and manufacture. A further way in which technical innovation is encouraged is through the so called "Queen's Award to Industry", a special section of which was established in 1993 to promote environmental technology.

Environmental consultancy has grown greatly in recent years, as has the provision of courses in further and higher education to supply the skills of environmental analysis. Ethical investment trusts have also grown, and take an interest in the environmental aspects of the firms in whom they invest.

3.5 Target Groups

The circumstances among industries traditionally regarded as the main polluters have changed considerably in recent years.

In the early to mid 1980s one of the principal actors was the Central Electricity Generating Board, a nationalised undertaking responsible for electricity generation in England and Wales. It was this body, for example, which was one of the main sources of opposition to greater UK participation in international action to control acid emissions (Boehmer-Christiansen and Skea, 1991). In 1988 the electricity supply industry was split up and privatised, and subsequent policy by the electricity generators has been to replace the burning of coal with the burning of gas. Partly in consequence, the coal industry, formerly a powerful actor, has been severely reduced in size and scope.

The water industry in England and Wales has undergone a similar transformation (see Kinnersley, 1994). Between 1974 and 1989 responsibility for water supply and wastewater treatment, as well as water pollution control, rested with the Regional Water Authorities virtual monopolists in their areas of operation. With the privatisation of the water industry, the roles of "poacher and game-keeper" were separated. In anticipation of the increased costs of installing water treatment facilities arising from EC directives, the water industry was allowed a special factor in the formula used by the price regulator to regulate permissible price increases. It should be noted that the price regulator, a public body known as OFWAT, has been critical of the cost of EC directives, arguing that they are detrimental to the customer and unjustified by evidence of benefit.

Technical changes in the heating of houses, with a more away from the traditional open hearth coal fire towards alternative sources of heating, now mean that individuals and households contribute less to smoke and particulate emissions to air (Scarrow, 1972). However, there are still a number of issues where changes in individual and household behaviour are necessary if problems are to be solved. These include: source separation for waste disposal; the disposal of oils; fertiliser and pesticide use on gardens; and the use of noisy machinery. Most important of all are likely to be changes in transport use, with less reliance on the private car and greater use of public transport, if increases in emissions arising from vehicles are to be reversed.

4 Capacity-Building

4.1 Increase in Number, Strength, Professionalism and Independence of Proponents

The traditional characterisation of British policy making is that it is dominated by small elites, relatively closed from a wide range of influences but operating with close informal networks - "club government" as it is sometimes known (Marquand, 1988: 178). Moreover, the Conservative Party has been in office for all but five of the last twenty-five years, and since in international terms party discipline is high in matters of legislation, there is a limit to which the government has to take account of other views. However, there are several factors that have made for a degree of openness in the making of policy, and one should be careful about too easy a set of generalisations about closed input structures.

Firstly, the technical features of environmental policy mean that scientific advisers often have high status in the policy process. The clearest examples are represented by some of the reports of the RCEP that appear to have played a crucial role in changing government policy, for example in lead in petrol. Moreover, since the RCEP takes evidence from interested parties, including representatives of the environmental movement, it can act as a conduit for wider concerns. Conversely, part of the explanation of the UK government's opposition to measures on acid rain or the dumping of waste in the North Sea has come from scepticism within the scientific advisory system about the effectiveness of the measures proposed.

Secondly, environmental pressure groups have themselves become more technically expert, and they have also become adept at using the opportunities provided by the existence of European legislation and policies to embarrass the government. Moreover, there is evidence that some government departments actually initiate informal consultations with environmental groups, although there is dispute as to how far this is simply a token gesture.

Thirdly, the growing influence of parliamentary select committees in the 1980s made it difficult for the government to keep certain items - in particular acid deposition, bathing beaches and waste management - off the policy agenda. Like the RCEP the committees take both written and oral evidence from interested parties, and they have been able to act as a channel of communication.

Public opinion at large has a diffuse, but significant, influence on thinking, even if it is difficult to detect its influence in policy. The growth in public environmental concern in the late 1980s, and in particular the fifteen per cent of the popular vote secured by the Green Party in the 1989 European elections, meant that some changes of at least vocabulary were necessary from the rampant materialism of the mid-1980s. Moreover, with public concern interpreted through a national mass media, it is difficult for a government to ignore trends in opinion. However, it is probably the case that UK public opinion is probably more moved by issues of land use (land drainage, road building etc.) than by issues of pollution.

Despite these trends towards greater openness, standard-setting for pollution control is still conducted by processes that are relatively closed to the public at large and dependent upon negotiation between industry and the regulators. Thus, Dr David Slater, the present head of HMIP, is quoted as saying in connection with the IPC authorisation of electricity-generating stations: "(...) at the end of the day we are charged with getting these plants under authorisation (...). Most industries in this country do not have the capability to produce what the Environmental Protection Act requires, and to expect it overnight is unrealistic" (ENDS Report, 1993, no. 219, p. 15).

Moreover, openness in government often means openness to conflicting, and not always equally resourced, forces. In addition to the arrangements for environmental regulation of the newly privatised industries, there is also a regulatory structure to control prices. In the case of water the price regulating authority (OFWAT) has been at the forefront of criticisms of the NRA for demanding water quality standards that are, it is alleged, too high. Thus, there is built into the processes for regulating the utility monopolies a conflict of regulatory imperative.

4.2 Increase of General Resources for Proponents of Environmental Policy

The resources for proponents of environmental protection derived in part from diffuse public opinion, in part from institutional sources like the courts and in part from international organisations.

Public Opinion. Public opinion surveys show that although variable over time, there is a strong desire on the part of the public for more effective environmental policies. Surveys also show a willingness to take personal action (for example the use of recycling facilities or the source separation of waste) to improve the environment.

We can also distinguish an attentive public that subscribes to environmental journals and magazines, gives money to environmental organisations and will on occasion take some political action, for example writing to MPs. At a rather higher level of commitment, it is clear that some organisations have been willing to collective action for environmental improvement if only to protect their interests. One clear example is provided by those angling associations that have used for a number of years the resources of the common law to prevent pollution.

The Courts and the Law. The courts have played an important role in the development of environmental policy. Within a common law system there is an opportunity for judges to develop and extend principles of nuisance and damage. Moreover, within statutory law there are difficult questions of jurisprudence that arise over such questions as causation. For example, in a recent case the House of Lords decided that Yorkshire Water had "caused" pollution, in breach of a statutory duty, in the case where an unknown polluter discharged a substance into a river contrary to consents that had been granted. Environmental protection groups are becoming more adept at using the courts to control pollution.

International Influences. Without doubt the principal international influence on the development of pollution control policy has been the European Union (formerly the European Communities). Not only have important pieces of legislation - for example in the field of air pollution control - been undertaken to meet obligations under EC directives, but standards generally have often been tightened as a result of EC decisions.

The image of the UK as the reluctant partner in the development of European environmental legislation, as well as other international environmental regimes, needs careful assessment, however. Opposition to the EC's proposed Large Combustion Plant Directive, disputes over water quality standards, the often unhelpful attitude over the North Sea Conferences, and a contingent commitment to carbon dioxide stabilisation, to name but a few issues, suggest that the UK has not been an international leader in environmental protection. Indeed, the UK is not fully convinced of the need for international action in many areas, as the rhetoric of subsidiarity and the repatriation of environmental competences in the EU goes to show.

Moreover, the image that the British authorities seek to portray of a government that bargains hard over policy formulation but delivers on implementation is not entirely supported by the evidence. It is true that EU directives are often implemented within good time, but this can mean little more than formal legal transposition. There are a series of directives on which implementation has been poor: the sulphur content of fuel oils; bathing waters; the air pollution framework directive; large combustion plants; and environmental impact assessment.

On the other hand, this litany of problems is one-sided. There are issues at the European level at which the UK has sought to take the initiative. Héritier et al.(1994) point to eco-audits and integrated pollution control. Other examples might include set-aside for agriculture and the protection of birds. Moreover, both in the natural sciences and in the social sciences British participation in international efforts is high (compare Haigh, 1989, 1992). The British standard BS 7750 on environmental management has been taken as the basis for its international equivalent ISO 14001. More recently, but perhaps with less just justification, the British government has sought to style itself as taking a stringent stand of questions of global climate change.

4.3 Policy Integration

Policy integration can be looked at from two points of view: internal integration, concerned with the consistency of control in respect of different receiving media; and external integration, concerned with the extent to which environmental considerations are incorporated into policy sectors such as industry, agriculture and transport.

Within thinking on pollution control policy there has been great stress placed on issues of internal integration. The basic principle of IPC might suggest that this is substantially achieved with UK environmental policy, but this view needs to be carefully qualified. Although the 1990 legislation sought to introduce integrated control, it did not establish a single environmental protection agency bringing together the NRA and HMIP. That had to wait until the 1995 act and in the intervening period there were "turf disputes" between HMIP and the NRA over responsibility for regulating the "Red List" of dangerous substances discharged to water. Moreover, the focus of integrated control is the prescribed process, not the operating site. Since an operating site may contain prescribed and non-prescribed processes, full integrated permitting is not yet achieved, since some processes might be regulated by central government and others by the local authority.

External integration raises another set of questions. In 1990 the government published a white paper entitled *This Common Inheritance*, in which it accepted the need to place environmental issues on the agenda of government ministries in general (UK Government, 1990). The document in effect summarised a wide range of prevailing policy goals at the time. In practical terms, the meaning of integration was limited to the establishment of inter-departmental ministerial committees to pursue environmental questions. One was to be chaired by the Prime Minister, another by the Environment Secretary and a third was to be explicitly concerned with energy efficiency in government departments. The white paper also promised annual reports on how far its targets were being met and it also promised that each department's annual report would contain a section devoted to environmental issues.

In the four years since the white paper, it is clear that implementation of its promises has been far from perfect. There are few quantifiable targets in the annual report on the implementation of the white paper promises, and the annual reports from departments are often vague on environmental matters. Because of secrecy surrounding the operation of government, it is difficult to know how often and with what degree of seriousness the inter-ministerial committees are meeting, although it is known that two of them only met four times in their first year of operation, and the third on energy efficiency had not met at all.

In January 1994 the government published its strategy for sustainable development. Again the most conspicuous feature of this document was its emphasis upon changes in the machinery of policy making and the establishment of advisory committees, rather than a commitment to specific policy measures. The strategy established a panel of "five wise persons", the Panel on Sustainable Development, to advise the government on strategic issues in sustainable development, and set up a UK Round Table on Sustainable Development with the intention of opening up new channels of communication.

It is possible to argue that action on sustainable development might begin to take place, despite this stress upon question of policy machinery in many of the formal documents. The most striking case here is the case of transport, where partly for electoral reasons, the government is having to consider its commitment to road-building plans in the south-east where many of its supporters are concentrated. For example, it was reliably reported on 24 March 1995 that the Cabinet was split on plans to widen the M25 motorway that circles London, and that some local Conservative MPs were opposed to the widening scheme (Brown, 1995). It also seems likely that tight constraints on public finance will also lead to reductions in road building.

4.4 Development of Policy Paradigm

Traditionally policy elites in Great Britain have stressed its "proud record" of pollution control, arguing that the British system had developed since the nineteenth century flexible, efficient and effective strategies for reducing pollution (Hajer, 1995: 115). An essential element in the argument was that the task of environmental policy was not to eliminate pollution and environmental risk, but to optimise it, that is to set the benefits of reducing pollution in the context of the costs of reduction and control (see e.g. Ashby and Anderson, 1981: 136).

This approach to pollution control was also associated with a preference for flexibility in the setting of standards, a preference that occasioned some dispute between Britain and its European partners in the 1970s and 1980s over the issue of uniform emission standards versus environmental quality objectives. As Nigel

Haigh (1992) has pointed out, this supposed contrast of policy approaches was always oversimplified, since there were always elements of uniform standard setting in traditional British practice.

It would be an exaggeration to say that a new policy paradigm is emerging to replace the traditional approach. Certainly there is a greater willingness to use statutory standards. There is a notional commitment to the precautionary principle, although it is difficult to find unambiguous instances of its application. There is also a commitment to what may be called process-based approaches, most notably the use of eco-audits by firms, and the application of a performance quality standards approach to firms" internal environmental management in the BS 7750 standard.

Against these trends, one should note however that there is little policy development on what might be regarded as the central tenets of ecological modernisation. Thus, there is: little use of pollution charges in accordance with the polluter pays principle; only haphazard encouragement of research and development for cleaner technologies; virtually no work on reconceptualising measures of economic well-being to net out from national income accounts defensive environmental expenditures; and a poor record on energy conservation. Dean Acheson once said that Britain had lost an empire and not found a role. It is tempting, in the case of pollution control, to say that Britain has lost a tradition and not found a paradigm. Interestingly in this context, the Royal Commission on Environmental Pollution is pursuing a major enquiry into standard-setting for sustainable development, and it is possible that this will be the catalyst that will enable the UK fundamentally to reform its policy thinking.

4.5 Main Restrictions

To what extent does the system of policy formation correspond to a closed system involving government, industry and the environmental movement? There are at least two ways in which the British process does not correspond to this characterisation. Firstly, although the environmental movement has some influence in government, the role of industry is, by common consent, stronger. In part this reflects the priority given by the government to reversing the UK's economic decline (as conventionally understood), thus creating a dependence upon industry. Secondly, other groups also have an influence. Inside the policy making system we should note the influence of technical and scientific experts. Externally, we should note the pressure on the UK coming from the European Union in particular, which has not only brought issues on to the political agenda, but also forwarded some policy developments (notably on acid deposition and water quality) that otherwise would not have taken place. However, the main structural restrictions on the development of environmental policy are probably to be found in the combination of four aspects of political culture and institutions.

Firstly, the electoral system imposes severe limits on the scope of party competition, and on the growth of new parties. The UK's system of plurality ("first past the post") voting means that political parties with a plurality of the popular vote normally have their parliamentary majority magnified to a considerable degree. Combined with the convention of party discipline in the House of Commons, this has meant that the Conservative Party has been able to secure majorities (three of them substantial) in the last four general elections (1979, 1983, 1987 and 1992) with only some 43% of the popular vote. Conversely, the electoral system punishes small parties whose support is geographically dispersed. There is thus no chance of a green party playing a potentially pivotal role in the formation of a governing coalition, and therefore there is no opportunity for the sort of policy bargaining that might bring environmental policy more prominently to the fore.

Secondly, there has been a widespread belief among policy elites of all political persuasions that the chief priority of public policy was to counteract the perceived long-term decline in the international competitiveness of the UK economy. This political culture has at various times fostered a belief in government-led modernisation, but in the 1980s the Conservative Party sought deliberately to reduce government intervention in the economy, and this rendered it insensitive to the line of argument that governments could improve economic performance through policies of ecological modernisation (Weale, 1992, chapter 3). Moreover, within the British political system, dominated by two parties, a cross-cutting issue like the environment does not easily fit into the polarised strategies of political competition that are such a central feature of British politics.

Thirdly, one ministry, the Treasury, is extremely powerful, combining the functions of a finance ministry, a tax ministry and an economic policy ministry. Apart from a brief period in the 1970s it has a reputation for keeping a tight control on public expenditure and so was effective in the 1980s in restricting capital expenditure of the then publicly owned utilities. Moreover, it is traditionally suspicious of capital expenditure that cannot show a tangible return within a suitably discounted period of time. Hence, there has been a systematic bias against investing in projects, like many of those in the field of environmental protection, where the returns are intangible or difficult to quantify. Moreover, during the 1980s it came powerfully to reassert the rather traditional neo-liberal principle of the neutrality of the tax system, thus rendering it difficult to justify a policy of taxes on public "bads" and subsidies on public "goods".

This dilemma is compounded by the relatively limited powers of local government. Local authorities have restricted tax raising powers and they own little by way of productive resources. Thus, even in cities like Leicester, which aspire to high environmental standards, there is no prospect of following the practice of some German cities of using revenues from public utilities, like electricity generation, to subsidise public transport.

5 Evaluation

Effectiveness. As the discussion of trends in environmental quality at the beginning of this paper go to show, British pollution control policy is not especially effective at reducing pollution control problems.

From one point of view, however, it can be argued that UK pollution prevention policy is relatively *cost*-effective: given the priority given to economic growth in the 1980s, the rapid organisational changes that destabilised the workings of the regulatory authorities, the tight constraints on capital expenditure within the public sector and the low level of resources, particularly personnel, available to pollution control agencies, one might have expected a much greater deterioration in signs of environmental quality. On the other hand, improvement in some areas, stability in others and modest deterioration in yet others hardly evidences a record of success.

Efficiency. If by efficiency we mean a low ratio of policy inputs to policy outputs, then by the argument just cited the British system is relatively efficient at achieving the modest goals that it has set itself. Moreover, the British political and administrative system is by international standards relatively uncorrupt, so that few resources are wasted in bribes and regulatory rent-seeking.

However, if by efficiency one means something more ambitious, for example raising the productivity of the population of a country to achieve sustainable development, then judgements would have to be harsher. There is little evidence of the commitment to intelligent, coordinated measures (in technical education, research and development, infrastructural investment, data analysis and the like) to make this goal possible.

Concrete Steps. It is difficult, within the context of UK institutions, to point to concrete steps towards pollution prevention and sustainable development. In part this is due to the character of legislation, which tends to establish a framework within which administrative action then works incrementally. Emission standards are tightened and new substances are controlled, but this may happen some years after the enabling legislation has been passed.

Of the identifiable concrete steps the evaluation of the record is at best patchy. This is most clearly evidenced in the implementation of the measures announced in the environmental white paper, *This Common Inheritance*. The measures were in part disappointing in themselves because they focussed so much on changes in the machinery of government, rather than substantive policy measures, but even within its own terms the implementation of the measures has been disappointing. Many of the inter-departmental committees have only met infrequently, and the promise of a greater use economic instruments has been frustrated by the Treasury.

Optimists can point to other trends. At the time of writing, it looks as though the UK may be at the point of a major reversal of its long held policy of favouring the private car over public transport. There is widespread interest at local level at initiatives to do with sustainable development. And sections of industry are coming to see the economic advantages of high environmental standards.

Whether the balance of political forces, however, will work to overcome the constraints is something that only time will tell.

Acknowledgement

I should like to thank Dr David Lewis and Professor Timothy O'Riordan for comments on an earlier version of this paper.

References

Ashby, E. and M. Anderson, M. 1981: The Politics of Clean Air. Oxford: Clarendon.
Boehmer-Christiansen, S. and J. Skea, 1991: Acid Politics. London, New York: Belhaven Press.
Brown, K., 1995: "Cabinet Split over M25 Widening Plan". Financial Times, Friday March 24, 9.
Digest of Environmental Protection and Water Statistics, 1994. London: HMSO.
ENDS Report, 1995a: "UK well ahead of EC targets for acid gas emissions". No. 248, 11.
ENDS Report, 1995b: "Best ever bathing water results boosted by weather". No. 250, 11.
Evans, G., 1993: "Hard Times for the British Green Party", Environmental Politics 2, 2, 327-33.
Haigh, N., 1989: EEC Environmental Policy and Britain. London: Longman.
Haigh, N., 1992: Manual of Environmental Policy: The EC and Britain. Harlow: Longman.
Hajer, M. A., 1995: The Politics of Environmental Discourse: Ecological Modernization and the Policy Process. Oxford: Clarendon Press.
Héritier, A. et al., 1994: Die Veränderung von Staatlichkeit in Europa. Opladen: Leske + Budrich.
Kinnersley, D., 1994: Coming Clean. Harmondsworth: Penguin.
Lean, G., 1991: "The Role of the Media" in L. Roberts and A. Weale (eds), Innovation and Environmental Risk. London, New York: Belhaven.
Lowe, P. and J. Goyder, 1983: Environmental Groups in Politics. London: George Allen and Unwin.
Marquand, D., 1988: The Unprincipled Society. London: Jonathan Cape Ltd.
O'Riordan, T., 1991: "Stability and Transformation in Environmental Government", Political Quarterly 62, 2, 167-85.
Scarrow, H. A., 1972: "The Impact of British Air Pollution Legislation", British Journal of Political Science 2, 3, 261.
UK Government, 1990: This Common Inheritance. London: HMSO.
Vogel, D., 1986: National Styles of Regulation. Ithaca, London: Cornell University Press.
Weale, A., T. O'Riordan and L. Kramme, 1991: Controlling Pollution in the Round. London: Anglo-German Foundation.
Weale, A., 1992: The New Politics of Pollution. Manchester: Manchester University Press.

THE NETHERLANDS

Hans Th. A. Bressers and Loret A. Plettenburg

1 Introduction

Together with several other northern European countries, the Netherlands is seen as one of the leaders in innovative environmental policy. Unfortunately, the actual policy results do not always justify this reputation. Recent signs suggest however that the Netherlands is beginning to practise what it preaches. One notable development is the incorporation of the findings of the report "Our Common Future" into a National Environmental Policy Plan. Another is the renewal of relationships between environmental authorities and their target groups, resulting in the establishment of covenants and other voluntary agreements. This renewal has also been a source of inspiration for the fifth environmental action programme of the EU.

Over the past twenty years Dutch environmental policy has undergone a radical reassessment: the original and, with hindsight, rather naïve belief that the pollution problem could be resolved within a single decade (Staten Generaal, 1972) has made way for nothing less than the aim of bringing about drastic economic reform in order to achieve global "sustainable development". In both cases, the stated objective far outstrips the government's resources to realise it. The difference is that we are now more aware of this fact.

The next section describes the development and perception of environmental problems. Section 3 looks at various aspects of Dutch environmental policies, while section 4 discusses Dutch capacity to support these policies. By way of conclusion, the results of the policies and future expectations are set out in section 5.

2 Environmental Quality

This section deals first with the most important environmental problems affecting the Netherlands, and then moves on to discuss the various "themes" surrounding Dutch environmental policy. For Dutch environmental policy is not compartmentalised according to affected media (water, soil or air) but is formulated and implemented according to an integrated thematic approach. The use of environmental themes provides clearer insight into the causes and effects of complex, interrelated environmental problems.

2.1 Environmental Problems in the Netherlands

One of the biggest and most typical environmental problems in the Netherlands is the manure issue. Dutch livestock production (pigs and chickens) is extremely intensive, particularly in areas with sandy soil. The result is an excessive availability and use of manure and, since the latter is being progressively phased out by law, a consequently gigantic manure surplus. In 1986 manure production reached its peak, with excretion levels of 713 million kg of nitrogen and 273 million kg of phosphate. This represents more than sixty kilos of pure manure per inhabitant! Ammonia emissions from manure are a prime cause of acidification in certain ecologically vulnerable areas. Excessive use of manure both in the past and present has also affected some ground water reserves and is threatening the quality of many others.

Another serious environmental problem is soil pollution. More than 100,000 sites have already been found to be so severely polluted that soil clean-up operations are required. In a densely populated country like the Netherlands, noise is inevitably a serious nuisance factor, while the processing of large quantities of solid waste also constitutes a major problem. The national production of household refuse is incidentally considerably higher than in the rest of Europe, not only per square kilometre but also per inhabitant. Although water quality in the large rivers has clearly improved, the many years of poor quality have led to severe pollution of the underwater soil in many locations. Acidification (due to road traffic, electricity generation, industry and agriculture) and desiccation (mainly due to water extraction for agriculture) threaten the quality of natural areas. Cities, industrial estates and roads are steadily encroaching upon the countryside with serious disruptive effects. In addition, the Netherlands is directly and indirectly a major source of environmental pollution at European and global level.

2.2 Changing the Burden

Climate change. The theme "climate change" can be assessed according to two indicators, namely one for the sub-theme "climate change due to greenhouse gas emissions" and one for the sub-theme "depletion of the ozone layer".

In 1970 CO_2 emissions amounted to roughly 143 billion kg. Around 1978 these emissions peaked at 197 billion kg. After falling in 1980 to about 150 billion kg, CO_2 emissions started to rise once more, to approximately 184 billion kg in 1992 (RIVM, 1991: 22), only to fall back again in 1993 to the 1990 level of 182 billion kg (OECD, 1995: 72).

The environmental impact of Dutch use of ozone-depleting substances decreased by 68% between 1980 and 1993 (Ministerie van Volkshuisvesting, 1994), partly as a result of more stringent international agreements. In 1992 the total volume of CFCs ran to 5,415 tonnes as opposed to 14,019 tonnes in 1986. The volume of halons amounted to 1,110 tonnes in 1992, against 2,586 tonnes in 1986 (OECD, 1995: 187).

Acidification. The decrease in average deposits of acidifying substances on the Netherlands in the period 1980-1993 (37%) resulted mainly from the reduction in SO_2 emissions in the Netherlands itself and surrounding countries. The emission of SO_2 due to incineration (i.e. excluding process emissions and road traffic) declined from 662 million kg in 1970 to 372 in 1980, and to 135 in 1991. The reduction in the 1970s was caused primarily by the increased use of natural gas, while the decrease in the 1980s was due mainly to the use of low-sulphur coal and flue gas desulphurisation in the power generation sector. Substitutions of natural gas for oil in the industrial sector and from high-sulphur crude oil to residual chemical gas in refineries also made their contributions (OECD, 1995: 57). In 1990 65% of SO_2 emissions were caused by incineration processes (CBS, 1992: 115).

Between 1980 and 1991 NO_x emissions fell by a mere 8%. This decrease was partly attributable to the reduction in NO_x emissions from passenger cars since 1988. It is worth noting, however, that between 1970 and 1980 NO_x emissions from road traffic jumped from 147 to 269 million kg.

Discharges of NH_3 fell by 32% between 1980 and 1993, thanks to the increasing use of low-emission fertiliser and the covered storage of manure from 1992 onwards (Ministerie van Volkshuisvesting, 1994: 73).

The manure problem. Between 1985 and 1990 the infiltration of nitrogen into surface water fell by only 3% between 1985 and 1990 (RIVM, 1993: 88-95). Industrial discharges into surface water were reduced by 25% in this period. This improvement was partially cancelled out by an increase in pollution from agriculture.

The pollution of surface water through phosphor decreased by 20% between 1985 and 1990. Within this period infiltration of phosphor into surface water decreased from 33 million kg to 25 million kg. Here too, the reduction in emissions was to some extent neutralised by increased agricultural pollution.

Annual accumulation of phosphor in Dutch soil (and ground water) has fallen gradually from 166 million kg in 1980 to 88 million kg in 1990. However, total phosphor stocks in the soil are still continuing to grow.

The dispersion of waste materials. In 1993 the dispersion of agricultural pesticides had decreased by 43% relative to 1980, partly thanks to the sharp fall in pesticide sales. The decrease in the spreading of other pesticides between 1984 and 1992 is estimated to be 44%. The risks resulting from the dispersion of priority substances were reduced by about 48% in the period 1985-1992. Lead emissions from road traffic amounted to roughly 2.6 million kg in 1975. At the end of the 1980s, the level had fallen to about 0.39 million kg (Adriaanse, 1990: 264).

Furthermore, industrial discharges of heavy metals into surface waters were reduced from a total of about 2,000 tonnes a year in 1970 to about 200 tonnes a year in 1989. The annual average dispersion of heavy metals in surface waters has decreased sharply during the past twenty years, but still exceeds water quality objectives to a greater or lesser extent at most sampling sites (OECD, 1995: 39-40).

Waste removal. The objective of the Dutch government in this respect is to minimise the quantity of waste that needs to be incinerated or dumped, and to promote waste prevention and the re-use of waste materials.

A total of 5.2 million tonnes of waste was generated in the Netherlands in 1990. In the early 1970s about 3 million tonnes of household refuse was produced. By the early 1990s the figure had risen to 4.5 million tonnes. Hazardous waste volumes increased from 287,000 tonnes in 1986 to 1,193,000 tonnes in 1992. This was partly due to improved waste separation practices and changes in hazardous waste definition, but partly due to economic growth (OECD, 1995: 77).

Relative to 1980, the amount of solid waste in landfills had fallen by 19% in 1993, while the level of household refuse in landfills decreased in the same period also. This can be explained by the increase in separate collection of vegetable, fruit and garden waste. The separate collection of bio-waste for composting started in the late 1980s, became obligatory for municipalities in 1994, and now applies to 61% of households. The amount of bio-waste is rapidly increasing, from 320,000 tonnes in 1991 to 620,000 tonnes in 1992 and is expected to reach 1 to 1.36 million tonnes by 1995. In the case of other waste categories too, separate collection is proving similarly successful. In 1992, more than 75% of glass and more than 65% of paper in household waste was collected separately. Paper/cardboard and glass recycling rates in the Netherlands are among the highest in the OECD. In 1990, out of 813,000 tonnes of hazardous waste (excluding contaminated soil, dredging soil, shipping waste and screened sand), 36% was recycled, 28% incinerated and 33% sent to landfill (OECD, 1995: 84-87).

Disturbance. The disturbance policy is aimed primarily at the reduction and prevention of disturbance in people's immediate living environment as a result of e.g. noise, smells, local air pollution, vibration, and the risk of accidents due to the proximity of industry or the transportation of hazardous substances.

In the period 1980-1993, total nuisance levels rose by 17% (Ministerie van Volkshuisvesting, 1994). The number of people disturbed by unpleasant smells, for instance, increased between 1980 and 1989 from 18% to 22% of the Dutch population (RIVM, 1993: 120). Noise disturbance from road, rail and air traffic also increased slightly between 1985 and 1990 (RIVM, 1991: 406).

Desiccation. Desiccated natural land covers roughly 424,000 ha. Until recently the area of desiccation in the Netherlands had been estimated at only 207,000 ha (RIVM, 1993: 132). In many places the relatively highly lying sandy soil areas have become seriously dried out. Desiccation is a serious problem, particularly in the coastal dunes, where it is due mainly to the extraction of potable ground water. The extraction of ground water for drinking and for industry doubled to roughly 1 billion m^3 per year between 1950 and 1988. According to recent research (Beugelink et al., 1992), this has led to a reduction in humid and wet natural habitats of between 17% and 25% since 1950.

Dissipation. Here the focus is on, among other things, the management of the reserves of natural resources which are necessary for passing on a sustainable environment to future generations. Every year these reserves undergo further depletion. Concrete targets for dissipation are still being developed and a useful indicator is unlikely to be available before 1998.

Overall, both nationally and regionally, biodiversity of fauna and flora in the Netherlands is decreasing. There are no species where the situation is entirely satisfactory (OECD, 1995: 93-97).

Dramatic changes have also occurred in the size of natural areas and forests. Since the 1920s, the total area covered by natural areas and forests has declined dramatically, from 7,900 km^2 in 1920 to 4,500 km^2 in 1988. The vitality of forests has also been decreasing in recent years: in 1992, vitality in 29% was low (compared with 8% in 1984), and 6% showed none at all (against 2% in 1984).

While overall progress is being made in extension the legal protection of natural areas and species, and in the creation of management structures to implement nature conservation policy, these are not sufficient in themselves to safeguard the national resources involved (OECD, 1995: 107).

3 Environmental Policies

3.1 The Start of Institutionalisation

As early as 1810, when the Netherlands was under Napoleonic rule, a decree was issued, introducing a licensing system aimed at controlling the hazards, damage and nuisance caused by industry (De Koning, 1994: 13). In 1962 the term "environment" made its first official appearance in the Netherlands. In that year, the minister of Social Affairs and Public Health set up a Public Health Inspectorate which was responsible for environmental protection. However, it was only in 1971 that environmental protection became institutionalised as a task of national government, in the Directorate General for Environmental Protection (DGEP). This occurred during the term of office of the first Minister for Public Health and Environmental Protection. The DGEP was initially allocated very limited resources. As the workload increased and environmental problems gained in urgency, the political will - notably in the Lower House - to provide the DGEP with further resources and powers grew. The allocation of financial and human resources showed a marked increase in the period from 1972 to 1982. In 1982 the Ministry of Public Health and Environmental Protection was dismantled and its tasks were transferred to the new Ministry for Housing, Spatial Planning and the Environment (VROM).

3.2 Environmental Laws and Regulations

The Nuisance Act, which governed hazards, damage and nuisance caused by installations linked to a specific location, dated from 1896. Its predecessor, the Factory Act, had been in force since 1875.

As environmental awareness became far more widespread in the late 1960s and early 1970s, certain sections of the Nuisance Act were tightened up and incorporated into separate laws and regulations. Supplementing the provisions contained in the original Nuisance Act, these also imposed restrictions on particular products and behaviour. The sectoral laws introduced in the course of the 1970s include laws on surface water pollution (1970), air pollution (1971), chemical waste (1976), waste materials (1977) and noise nuisance (1979).

These laws attracted sharp criticism from various sections of society. Citizens alleged that the public participation and appeal procedures were biased against them. Industry, for its part, claimed that the licensing procedures were far too time-consuming. Furthermore, companies asserted that there was insufficient coordination between the various authorities involved in the license-issuing process (De Koning, 1994: 168-169).

To solve these problems, the Act on the General Provisions for Environmental Protection (Wabm) was introduced. This Act made the various sectoral laws subject to uniform rules for the application and granting of licenses while also providing for uniform participation and appeal procedures.

Nevertheless, it was still felt that environmental legislation lacked coherence, and pressure was exerted to further expand the scope of the new General Provisions Act at the expense of the sectoral laws. This culminated in the incorporation in 1993 of the General Provisions Act into a new Environmental Management Act. The name of the Act was changed deliberately, to indicate that the ambit of the Act had been extended to create an all-embracing general law. The Environmental Management Act also opened up the possibility of granting integral environment licences covering all environmental aspects. One consequence was that, effective from 1 March 1993, five licensing systems from five environmental acts - the Nuisance Act, the Air Pollution Act, the Noise Nuisance Act, the Waste Materials Act and the Chemical Waste Act - were subsumed under the Environmental Management Act. This has made environmental legislation far more transparent and easier to enforce. A separate licence is however still required under the terms of the Surface Water Pollution Act, which falls within the competence of another ministry that was unwilling to relinquish this responsibility.

3.3 Policy Instruments

The central instrument in Dutch environmental policy is a ban on performing any environmentally harmful activities without a licence. The various environmental laws put this ban forward as the foundation stone of environmental policy instruments. For a long time, the licensing systems based on the ban were virtually the only instruments available for manipulating environmental behaviour. Recently, licensing has been replaced by general regulations for a number of sectors characterised by large numbers of small companies.

Financial instruments are less important, with the notable exceptions of water effluent charges and the tax differentiation that was used to stimulate the introduction of cleaner cars. Various subsidy schemes are used, but these have only limited influence (Bressers and Schuddeboom, 1994). Attempts to introduce an energy or carbon tax have failed several times, mostly owing to strong resistance from industry and its representatives in parliament and government, e.g. the Minister for Economic Affairs. An independent committee was set up to calculate economic side effects on the basis of hypothetical scenarios. Some possibilities that had emerged as positive were nevertheless swept under the carpet by the Minister for Economic Affairs (who is also responsible for energy-saving policies) even before the report was completed. When in 1994 the administration of Prime Minister Kok took over, a new opening was found. Plans were made for the introduction of a limited energy tax in 1996. This tax is regarded as a first step, to be followed by additional steps taken in the context of the European Union. In December 1995, however, the Dutch Senate was still discussing the new tax and its introduction was still uncertain. One of the counter-arguments was the fear of a loss of competitiveness in Dutch industries. This fear seems unjustified in the light of recent research, which has shown that the Dutch paper industry had, on balance, suffered no loss of competitiveness as a result of the very high water pollution charges, when compared with the Belgian paper industry in the 1970s and 1980s - a period during which such charges were not present in Belgium.

Since the reformulation of Dutch environmental policy in the first National Environmental Policy Plan, new instruments have appeared or gained dramatically in importance. These are aimed at stimulating society and non-environmental government bodies to assume some responsibility for sustainability and environmental quality. This is done through target-group consultation and covenants, company or personal liability, emissions trading, research and information obligations, regulations requiring companies to employ staff with adequate expertise, the creation of institutional facilities (environmental impact assessment, company environment departments and internal company environmental management systems), and so forth. Many of these instruments also operate indirectly through intermediary organisations, and sometimes even lead to the creation of such intermediary organisations. This approach is becoming far more widespread, not only in the Netherlands but - since the fifth Environment Action Programme of the European Union - also at European level.

3.4 Policy Style

Initially there was a wide discrepancy between the policy style of the policy-makers and that of the policy implementers. The policy-makers, civil servants and administrators of the new Department of the Environment adopted a distant and often rather negative attitude towards the policy target groups, such as the industrial sector. The target groups generally had little influence over environmental policy which, consequently, showed a preference for direct regulation. The enforcement of these regulations, however, was left to other agencies (Bressers, 1993; Bressers, Huitema and Kuks, 1994). The policy implementers, such as the license issuers, were understaffed. In addition, they often lacked motivation and received too little support from the responsible administrators. As a result, they tended to concentrate on responding to complaints rather than pursuing an active and systematic licence-issuing policy. Systematic control was virtually non-existent.

The difficulties in providing concrete evidence of environmental infringements were a further reason for refraining from legal action and, instead, for a policy of "talk, talk and talk again". In other words: although the legislation was strict in theory, in practice it was used at best as an informal bargaining counter in negotiations.

In the first half of the 1980s, the then environment minister Winsemius (see Winsemius, 1986: 61-67) initiated vigorous efforts to persuade the environment policy-makers and target groups to abandon their entrenched positions. As this process gained momentum, special "target group managers" were appointed at the ministry. Their task was not only to act as ambassadors of the environmental policy vis-à-vis the target groups, but also to "educate" their own organisation to have more understanding for the target group's position. Contacts between environmental policy-makers and representatives of the target groups became far more regular. It was only thanks to these initiatives that at the end of the 1980s the authoritarian style of Dutch environmental policy-makers was supplemented with a new approach designed to encourage self-regulation. At the same time, the allocation of additional funds had reinforced the policy implementation powers of environmental authorities at provincial and municipal level, while growing public interest in environmental problems had also resulted in stronger administrative support for implementing the environmental laws. At the present time, a confrontation is looming between the implementers' regulation-oriented policy style and the policy-makers' emphasis on the goodwill of companies.

This problem has been recognised in the past few years, and scope has been created within the law to attach more weight in the licence-issuing process to companies' own environmental plans On the other hand, frameworks have been created to ensure that temporary "toleration" of infringements is now subject to clearer rules. However, there is still a long way to go before the problem of opposing policy styles can be definitively solved.

4 Capacity-Building

4.1 The Main Actors and their Strength

Policy institutions. To obtain a clear understanding of the policy institutions in the Netherlands, a subdivision must be made between the vertical and horizontal distribution of powers among them. Horizontal distribution entails powers being divided between the different departments of higher or lower level authorities. This is discussed at length in section 5 (integrative solutions). Vertical distribution refers to the division of powers between authorities at various government levels.

In the Netherlands, the administrative structure comprises three layers of government, namely (1) municipalities and water boards, (2) provinces and (3) national government. These do not operate in isolation, but complement each other (Nelissen et al., 1987: 241). Initially, environmental regulations were made exclusively at local level. The municipality long remained the only government body involved in the issuing and monitoring of licences (Boersema et al., 1991: 65-66). As a result of the environmental laws introduced in the 1970s, the provinces were also entrusted with responsibility for implementing these laws. For this reason, provinces now also have license-issuing powers, namely in cases involving complex and technically detailed issues and potentially highly-polluting companies. National government concentrates primarily on national legislation and regulations, as well as on national environmental policy planning, including setting targets and norms.

Nonetheless, this by no means implies that national government carries sole responsibility for determining the environmental policy which municipalities and provinces are subsequently obliged to implement. The provinces and most municipalities conduct their own environmental policy planning. In addition, the municipalities and provinces enjoy autonomous status when it comes to environmental policy. The law, for instance, lays down that plans made at the various levels of government are not governed by any hierarchical order. Vertical fine-tuning must take place, principally by means of consultation, agreements and the exchange of information. Where this fails, both national government and the provinces have instruments at their disposal to insist on vertical fine-tuning.

Environmental policy planning has become an interesting Dutch policy institution in itself. The planning system distinguishes itself from similar systems in other countries, which were often introduced after observing the Dutch example, in two characteristics. First of all, it is an attempt to operationalise ideas from the "strategic choice approach" in the academic literature on planning. This means that its main goal is not to stipulate future action, but to improve coherence and to focus on the future quality of motivation and openness in external participation in subsequent decision-making (Coenen, 1992). As a second characteristic, more or less in spite of this general orientation, the national and also many of the provincial and local plans contain precise and quantitative goals, for instance on emissions reduction.

The attention devoted to the environment in the political institutions varies over time. Like the general public, political institutions are inclined to show greater interest in the environment at times of economic prosperity. But once awareness has reached a certain level, it tends to be sustained for a while, even after the economy starts to slow down again (Gijswijt and Van der Vliet, 1993).

Green organisations. The nature conservation movement in the Netherlands dates back to the turn of the century. These pioneers of nature conservation initially focused exclusively on the protection of nature as a recreational resource. Nature conservation organisations are generally stable, have a formalised internal structure and are directed by people with a lot of expertise. Negotiations and consultation based on expertise, formal pressure and lobbying are currently the main elements in the strategy of the conservation movements. Until a few years ago, they did not concern themselves with general political objectives (Leroy and De Geest, 1985: 94-97).

In the period around 1968-1972, a series of environmental incidents prompted growing worries over the environment in the Netherlands. A new environmentalist movement sprang up, rooted partly in the traditional nature conservation organisations. This new movement comprised a conglomerate of roughly 700 to 800 associations, organisations and action groups. It differed from the original nature conservation movement in two respects (Leroy and De Geest, 1985: 41-42). First of all, its interest is in a more general combating of pollution and of the relentless depletion of natural resources. Secondly, it has showed a clear preference for expressive and symbolic actions which were used to mobilise the public against the government's environmental policy. The environmental activists came into far more regular and violent conflict with certain prevailing values and norms. Initially, many environmental groups lacked a formal internal structure. This meant that support had to be drummed up specially for each new action. The resulting expertise and skills in mobilisation and organisation of social protest actions have become important strategic weapons in their hands.

Towards the end of the 1970s, as the popularity of "extra-parliamentary" action was beginning to wane (Nelissen et al., 1987: 231), many environmentalist organisations and action groups changed their strategy, and were now prepared to meet the government around the negotiating table. Their timing could not have been better. For the new environment departments at national, provincial and local levels saw the environmental movement as a potential ally, and were therefore eager to accept it as a negotiating partner. In many cases, in fact, their status was even laid down in law. To perform this role, new skills had to be acquired in the fields of policy processes and environmental policy. As a consequence, the environmental movement, currently comprising about 12 organisations, has developed very much into an institutional and professional actor in the field of environmental policy.

A pivotal role in this respect is played by the Stichting Natuur en Milieu (Foundation for Nature and the Environment). This foundation acts as the think-tank of the Dutch environmental movement and supplies many of its representatives on committees and other consultative structures within and outside government. As noted above, the environment department sees the foundation as an important ally, and has therefore always provided the organisation with generous subsidies to guarantee the quality and strength of its arguments.

All nature conservation and environmental movements are united under the national and regional environmental federations in the LMO (National Environmental Consultation). The secretariat of the LMO, incidentally, is based in the building of Stichting Natuur en Milieu. According to the LMO, the environmental and nature conservation organisations in the Netherlands have jointly roughly 50,000 active members. The number of non-active members is far greater. The Vereniging Milieudefensie (Dutch Friends of the Earth) has a relatively high number of active members. The two nature conservation organisations, in particular, have achieved explosive growth rates in membership over the past years, thanks to large-scale recruitment campaigns. It is perhaps worth noting that both are managed by former environment ministers who were immensely popular in the 1980s.

The table below shows the development of the membership of several large and influential nature conservation and environmental organisations:

Table 1. Membership of Environmental Organisations in the Netherlands.

Nature/Environmental Organisation	*Membership around 1975*	*Membership around 1985*	*Membership in 1995*
Greenpeace	90,000	300,000	600,000
World Nature Fund	45,000	115,000	690,000
Dutch Friends of the Earth	8,000	14,500	35,000
Vereniging Natuurmonumenten (Nature Association)	200,000	235,000	722,000

Target groups. One of the Dutch government's strategies for achieving a more integrated environmental policy is the target group policy (see also the section on policy style). This new policy focus means that, after the national environmental objectives have been formulated, target groups and their representatives now have much greater say in all further stages of the political process. The cooperative - and at times even extremely positive - attitude of many industrial organisations has been a great help in this connection.

In creating a consultative structure between government and industrial organisations acting on behalf of the polluters (i.e. the target groups) an attempt is being made to internalise environmental responsibility in individual companies. One of the main aims of this consultation is to define the tasks for a specific sector of industry within the framework of the overall national environmental objectives. These arrangements are usually laid out in covenants and other forms of guidelines, incorporating targets to be realised by the various sectors of industry within a certain time limit. This approach is not only aimed at achieving more broadly-based support for government policy, but also recognises that the expertise necessary to reduce environmental pollution can be gained in large part from the polluters themselves.

One consequence of target group consultation, therefore, is that an increasing number of companies are setting up their own internal systems to make care for the environment an integral part of their business procedures. The industrial organisations can promote this process by developing model environmental management systems, which companies can use to create their own internal systems and procedures.

Although the industrial organisations are generally playing a positive role, this does not mean that all sectors of Dutch industry are interested in reducing the pollution they cause. Far from it: many individual companies are still standoffish, even obstructive, while employers' and employees' organisations also tend to keep their distance. In 1994, however, the latter surprisingly issued a joint declaration affirming their commitment to a sustainable society and calling upon government to pursue a more vigorous policy. However, as soon as concrete measures are proposed, the employers' organisations are generally the first to raise objections.

Experience over the past few years has shown that target groups consisting of large numbers of small units (such as households, farmers, car drivers, retailers and other small firms) are far more difficult to reach by means of a target group policy (Staten Generaal, 1993).

The media and others. In the late 1960s and early 1970s, alarming scientific and more populist publications dramatically highlighted environmental problems, thus vastly increasing environmental awareness among the Dutch public and making the environment an acute political issue. The mass media were also instrumental in mobilising popular support for the up-and-coming environmental movement.

As in the early 1970s, the media still have the power to influence public opinion by devoting more or less of their time to environmental issues. Research by Gutteling and Caljé (1993) confirms that strong media attention to the environment results in greater concern among the population. Ester (1979: 141) asserts that the more threatening the information is, the more likely the (negative) message is to stick in people's minds.

Gutteling and Caljé showed in the same study (1993: 14-19) that reports on the environment in Dutch newspapers differ widely in terms of the quantity and type of information (pollution, risks, incidents, miscellaneous) published. This results in people's views on the environment being largely determined by the paper they read (Wiegman et al., 1989: 846-852). The reason for focusing on newspapers is that these are still an important source of information in the Netherlands, despite the rise of other important media like TV (Galetzka et al., 1994: 7; Gutteling et al., 1991)

4.2 Resources For Environmental Protection

Institutional and legal. In section 3.3 we discussed the government's most important instruments for pursuing environmental policy. This section looks at the means available to other relevant actors for influencing environmental policy

One such instrument is the EIS (environment impact statement), which came into force in the Netherlands in 1987. The EIS must describe the relevant activity, alternatives for that activity and the possible environmental consequences of these alternatives, while also giving a comparative analysis (Barendse and Udo de Haes, 1989: 127). The inclusion of the most environmentally friendly alternative is required by law. The other alternatives that may be incorporated depend on the outcome of the negotiations between the initiator of the EIS (the private entrepreneur or government body seeking to undertake the activity) and the authority empowered to take a decision on the proposed activity. One noteworthy aspect of the Dutch EIS procedure is the existence of an independent committee of experts which advises the competent authorities on the guidelines that the report must satisfy. This committee also subsequently evaluates the quality of the report in the light of these guidelines (Evaluation Committee, 1990). Moreover, the EIS procedure also provides for extensive participation opportunities for all interested parties.

The Environmental Management Act, which came into effect in March 1993, has widened the opportunities for participation as well as for lodging objections and appeals. First of all, it requires the authorities to draw up a draft decision (concerning e.g. a licence) before the actual decision can become law, thus permitting private individuals and organisations to lodge an objection to the proposed decision at an early stage. If need be, the authorities can also be required to organise a public hearing. Secondly, parties need not be directly involved to be admitted to the participation, objection and appeal procedures. The reasoning here is that the environment is perceived to be an issue of general public concern.

Nevertheless, these extensive procedures have become a bone of contention in the Netherlands.

Information. The National Institute of Public Health and Environmental Protection (RIVM) is a scientific institute belonging to the national government. Its task is to provide government with the information required for its health and environmental policies, in order to promote public health. In this connection, the RIVM keeps records and makes predictions of the current and future quality of the environment on the basis of models. So far the RIVM, working in cooperation with other institutes like the CPB (Central Planning Bureau), has published three national environment surveys, in 1988, 1991 and 1993. These surveys support national environmental policy planning. With the aid of the surveys, forecasts can be made of the environmental quality that can be expected in the future if the existing policy is carried out. Furthermore, additional policy alternatives are presented in the light of the expected effectiveness of the proposed government policy.

Twenty per cent of the research carried out by the RIVM within its wide-ranging terms of reference is undertaken on its own initiative. The remaining 80% consists of commissions from the government (RIVM, 1994: 4).

From 1995 onwards, the RIVM will bring out an annual Environment Report, containing the latest environmental data as well as indications of the expected effects of policy measures in the medium term (several years). An Environment Survey covering a period of at least 10 years is to be published every four years. In this way, the RIVM proposes to give concrete shape to its role as an environmental planning agency. Apart from the important part played by the RIVM, a great deal of policy-supporting research is performed at the universities. A special role is played by the extensive policy research that is conducted here (see section 5).

Money. Initially there were serious problems keeping the expenditure of the Environmental Management Directorate General in check, partly because the organisation did not have its own financial department - this was only set up in 1985. A further problem was that a lot of money was spent on research that was difficult to monitor and control. Finally, many of the employees were young and unaccustomed to working within a tight financial regime. It was only in the course of the 1980s, with the institutionalisation of a financial department, that the accountants started to play an important role (De Koning, 1994: 203).

The expenditure of the Directorate General increased from 289 million guilders in 1976, to 608 million in 1980, 998 million in 1985, 1,269 million in 1990, and totalled 1,583 million in 1993. Even after adjusting the figures to take account of inflation, the increase since 1977 remains substantial. Environmental expenditure is predominantly aimed at water pollution (37%), waste management (23%) and air pollution (17%). The total environmental expenditure of all sectors in the Netherlands amounted to 1.1% of GDP in 1980, 1.3% in 1985, 1.9% in 1990 and 2.7% in 1995 (OECD, 1995: 114-116). The total environmental costs for 1991 are estimated at almost 10 billion guilders. The largest part of the government's environmental expenditure is paid for out of specific duties and charges. Some of these duties have an additional regulatory effect. The water pollution duty is a good example of this.

Municipal and provincial environmental policy used to be financed without any contributions from national government. When the environment started to make increasing demands on the lower-level authorities, national government provided substantial special grants on a temporary basis. These individual transfer payments are to be phased out eventually (Brussaard et al., 1993: 47-48).

Public attitude, trust and prestige. Between 1967 and 1994 public opinion concerning important national problems has fluctuated considerably and the importance attached to environmental issues has not been stable over the years. During most of this period only a small percentage (1-5%) of the electorate has seen the environment as the most important national problem. But this proportion peaked in the early seventies (about 20%) and around 1990 (45%). This instability stands in stark contrast to the problem of unemployment which, except for a time in the early 1970s, has been regarded by the electorate as one of the most important national problems throughout the period 1967-1994, although its "support" has alternated to some extend with that of the environment (c.f. Aarts, 1995: 4).

Over the past decades, interest in environmental problems has risen and fallen in a wave-like pattern. What this seems to suggest is that public interest in environmental problems in the Netherlands is subject to a cyclical change which is relatively unaffected by the gravity of the actual problems. One possible explanation could be the varying intensity in media coverage of environmental problems (see media). Another explanation for the cycle is that the main national problems "the environment" and "unemployment" take turns in commanding the public's attention. It would seem that many people still see the environment as a luxury problem. The fickleness of public attention to the environment, incidentally, discourages politicians from taking up strong standpoints and seeking public support for radical policy measures.

International context. In the Netherlands the international context is generally, and perhaps wrongly, regarded as a poor source of political power. The chapter on international environmental policy in the Dutch environmental programme 1995-1998 (Staten Generaal, 1994) tends to focus on what the Netherlands could do to promote environmental policy in the rest of the world, rather than on how international agreements could be translated into national policy. It is characterised by "a mixture of self-interest and a sense of being corresponsible" (see Kaakebeeke, 1994: 341). One clear instance of this was the Netherlands' active stance at the Rio conference.

Nevertheless, in certain cases the international context of environmental policy can most certainly be seen as a potential source of political power for parties seeking to further the environmental cause. The best example of this is still possibly the way in which the former environment minister cleverly used the report "Our Common Future" to lure the other ministers into accepting certain basic principles which eventually and inevitably resulted in the National Environmental Policy Plan. The likelihood is that the current content of the policy plan would never have been achieved in direct negotiations between the environment minister and the other ministers.

4.3 Policy Integration: Inter-Policy Cooperation as an Objective

It has been argued in the Netherlands that the best environmental policy would be to put an end to all environmentally-unfriendly policies (Drees, 1992). This is a pointed reference to, among other things, the policy aimed at stimulating intensive agriculture and livestock farming, which involves the construction of large infrastructural works to increase the mobility of people and goods, land reorganisation and a lowering of the ground water level to facilitate agricultural mechanisation at the expense of the natural environmental.

A further factor is that many environment-related tasks have been entrusted to other ministries. The Ministry of Economic Affairs, for instance, is responsible for energy-saving. The tasks of the Ministry of Agriculture include nature management, prevention of overfishing and - together with the environment department - the elimination of excessive use of manure and pesticides. The Ministry of Transport and Public Works is responsible for water quality management and also for limiting the growth in (car) mobility. The National Environmental Policy Plan (1989) and its successors were therefore signed by all four ministers involved. The realisation of the National Environmental Policy Plan, incidentally, can be seen as a successful attempt by the Ministry of the Environment to take advantage of the wave of sympathy for and attention to environmental issues in the late 1980s, in order to secure the commitment of the other relevant ministries to environmental objectives. This process was certainly not free of tension. Shortly after the publication of the first National Environmental Policy Plan, the government fell, because the Minister of Transport and Public Works and the Environment Minister lost the backing of their Conservative Liberal Party after seeking to abolish tax relief for commuting in an effort to slow down the growth in mobility. Within the framework of the National Environmental Policy Plan, municipalities and provinces have been entrusted with all sorts of tasks designed to make the environment an integral concern in other policy fields. Their environmental policy plans have still to reveal how this aim is to be given concrete form. In the meantime, this approach has resulted in the creation of "bridgeheads" at other ministries or departments, in the form of individual officials or entire departments which put the environmental point of view within their own organisations.

One particular way of achieving policy integration is the so-called "field-oriented policy". In this connection, government and non-government organisations with importance for the quality of the environment are brought together as far as is possible. The idea is that these groups work together, while retaining their individual powers, to define the problems for the entire field and formulate a joint approach for tackling the problem. It is hoped that this approach will facilitate the recognition of "win-win" situations and prevent conflicting interests from leading the various parties to become entrenched in their respective positions. Experience with this form of network management has so far been mixed.

The environment issue has also had far-reaching consequences for the nature of some policy networks. The agricultural world was traditionally seen as "the green front" in the Netherlands, a close-knit and closed network which not only consisted of ties between farmers, their organisations and the Ministry, but even had its own university and bank. The excessive use of manure and pesticides, in addition to overproduction, has seriously tainted the image of the agricultural sector and impaired the strength of the policy network. As a consequence, the sector is gradually opening up to the outside world. Similarly, in the fields of drinking-water production and water management in general, the era of "pumping and billing" has come to an end. The new (environmental) challenges mean that all organisations, both government and non-governmental, must abandon their strictly technocratic orientation and become more focused on consulting with other relevant parties (Bressers and O'Toole, 1994).

4.4 The Policy Paradigm

The objectives of a policy are linked to a wider pattern of values surrounding the policy field (Fischer, 1985; Hoogerwerf, 1990). The value patterns in respect of the environment which helped to determine the policy objectives can be subdivided into the following "policy paradigms": "human health", "nature" and "sustainability". In addition, "ecological modernisation" can also be regarded as an emerging paradigm.

Human health. The earliest general government memorandum on environmental policy (Staten Generaal, 1972) centred on public health, which was defined as physical, mental and social well-being. It did not go so far as to assign intrinsic value to the well-being of non-human organisms (Bressers, 1979: 80). This limitation was debated in Parliament.

Nature. The debate marked the start of a period in which, at least in official policy documents, the demands made on the quality of the various environmental sectors were based on the "general ecological function". It was added that "additional demands with a view to human uses" will be made wherever necessary (e.g. Ministerie van Verkeer en Waterstaat, 1975). Naturally it is questionable to what extent this far-seeing standpoint, where ecological considerations prevail over anthropocentric concerns, ever had a sufficiently broad base of support. In all likelihood, it is at least partly the expression of a naïve optimism regarding the manageability of environmental problems and the "malleability of society" in general. In the second half of the 1980s, however, it became clear that policy effects were repeatedly falling short of expectations. At the same time, recognised environmental problems turned out to be more serious than initially assumed, and the general public became aware of a growing number of environmental problems. As a result of these developments, the environmental focus shifted away from "the intrinsic value of nature" towards "care for tomorrow".

Sustainability. "Care for Tomorrow", which in Dutch also means "worries about tomorrow", was the title of a ground-breaking collection of studies published by the RIVM (State Institute for Public Health and Environmental Protection) in 1988. In these studies, the gravity of all environmental problems in the Netherlands and the outlook for the future if policy remained unchanged were presented for the first time as an interrelated whole. Almost immediately after the publication of "Our common future", the Dutch government designated "sustainable development" as the general guideline for overall Dutch government policy. It was evident from the RIVM report that this would have to have far-reaching consequences. These were translated into the first National Environmental Policy Plan (NMP), entitled "To Choose or to Lose" (Staten Generaal, 1989).

Subsequently, the concept of sustainability was developed further in the term "ecological user space", which sought to define permissible environmental impact and permissible use of raw materials given the capacity of the global environment and the needs of future generations and other countries. Please note, incidentally, that this still concerns exclusively the needs of human beings.

Ecological modernisation: The definition of sustainability in terms of "ecological user space" was criticised by the Nederlandse Wetenschappelijke Raad voor het Regeringsbeleid (a scientific advisory council for government policy) in 1994. It was pointed out that the interpretation of sustainability involves, implicitly or explicitly, political choices in which different interests have to be weighed against one another. Consequently, various lines of action could be defended from different perspectives. The people who had conceived the NMP were accused of having opted too glibly for a particular far-reaching interpretation, without addressing the economic consequences of the policy until the actual implementation took place. The Ministry of the Environment, incidentally, denies that the economic consequences were ignored when the NMP was drawn up. It is also curious that, according to the analysis in their own report, the objectives of the NMP would eventually have to be tightened up even if the least far-reaching line of action recommended by the scientific council was followed.

This debate does show how, now that the renewed environmental policy has reached the implementation phase, the opponents within society are beginning to stir themselves. The achievement of a synthesis and reconciliation of ecological and economic interests, both at micro-level ("pollution prevention pays") and at macro-level ("ecological modernisation"), is becoming increasingly indispensable in order to arrive at a strong environmental policy.

5 Evaluation

The previous sections illustrated the development of Dutch environmental policy. In this closing section, an attempt is made to assess this policy and its development.

Effectiveness. The Dutch situation regarding environmental policy is exceptional, not so much for the environmental policy itself, but because of the enormous number of studies devoted to its implementation and effects in all its different aspects.

More than one hundred policy evaluation studies have taught us a lot about the actual effects of permits and other policy instruments (cf. Schuddeboom, 1994). Until recently many firms operated without a licence and control visits were often rare.

From 1970 to 1980 organic pollution from industrial effluents nevertheless fell by two-thirds. But research (Bressers, 1983a, 1983b, 1988; Schuurman, 1988) has showed that almost all the credit for this reduction goes to the introduction of charges. The licences introduced at the same time had had little effect.

The results of Dutch environmental policy during the 1970s and 1980s, the period before the reformulation of Dutch policy in the first National Environment Plan 1990-1994, were, all in all, not particularly impressive. The dominant policy instrument - i.e. licensing - was bedevilled by implementation problems in practice, and as a result, largely ineffective. One might sum up the results of Dutch environmental policy in the 1970s and the 1980s in the following sentence: About half of the policy objectives formulated in, and with the available knowledge of the 1970s have been achieved, but the information we have now tells us that we need to achieve objectives that are at least twice as far-reaching.

So it was inevitable that a complete rethink of environmental policy strategy took place. Although it is rather early to assess the results of the new policies, recent goal-attainment in various - though not all - aspects of the environment appears as a fairly positive sign (see section 2).

Efficiency. In the 1980s (1980-1989) total environmental costs rose from less than four billion guilders to 7.3 billion guilders. The direct environmental expenditure of companies ran to 1,757 million guilders in 1989. 800 Million guilders were spent in 1989 on "executive tasks": i.e. drafting laws, norms and regulations and the licence-issuing system. Arentsen, Klok and Maarse (1993: 224) conclude that in the 1980s a significant proportion of the government's environmental expenditure was spent on policy development. They also note a tendency to pass responsibility for part of these "executive tasks" on to companies. In practice this often involves the services of consultants, who help with the preparation of license applications and negotiations with government. To make the system more efficient, general rules have been drawn up for specific sectors so that large numbers of companies no longer need to apply for individual licences.

After the 1980s, there was substantial additional expenditure on the environment. In the second National Environmental Policy Plan (Staten Generaal, 1993), the growth in total environmental costs was estimated (in 1994 prices) at 17.5 billion for 1995 (almost 2.5 times 1989 levels), 22.8 billion in 2000 and 26.5 billion guilders in 2010. For the year 2000 the government expects that 3.25% to 3.5% of GDP will be spent annually on environmental measures.

To evaluate the efficiency of a policy, it is necessary to know its effects. Unfortunately, this knowledge is not always available. In broad terms, it can be said that soon after the National Environmental Policy Plan had resulted in an intensification of policy efforts, attention was devoted to safeguarding efficiency and preventing dissipation of resources in this "booming sector".

On the other hand, it would have been a virtual miracle if such rapid growth had resulted exclusively in efficient measures. While overall support for environmental policy remains as strong as ever, there are also complaints about overly stringent norms (incineration installations, soil clean-ups), lack of flexibility (noise nuisance) and inconsistencies between the target group policy and the issuing of licences.

Integrative solutions. From the late 1970s and early 1980s onwards, people became increasingly aware of the disadvantages and limitations of the sectoral environmental policy pursued up to then (Leroy, 1994: 42). As a consequence, efforts were made in various ways to integrate the policy. Environmental policy was no longer reviewed according to environmental compartments - such as air, water and soil - but according to themes (e.g. acidification, the manure problem and desiccation), geographical areas (e.g. vulnerable sandy areas), material flows (e.g. cadmium) and target groups (e.g. traffic, industrial sectors and agriculture). The new approach was later called the "internal" integration of environmental policy. This internal integration has taken on concrete form in plans, laws and institutions so that, thanks to these efforts, major progress has been achieved.

Since the National Environmental Policy Plan was published, far more attention has been devoted to the "external integration of environmental policy" (interpolicy cooperation - see section 4.3). What this entails is that such matters as transport policy, physical planning, agricultural policy, economic structure policy, water management, building regulations, energy policy and educational policy are fine-tuned with an eye to the national environmental policy (Staten Generaal, 1990). This fine-tuning should take place not only at national level, but also at provincial and local level. Despite excellent initiatives in this respect, this form of integration is still making slow progress, because traditional policy objectives remain an overriding concern. Over the past few years this has become abundantly clear in important government decisions on infrastructure. In these cases economic interests have always, either implicitly or explicitly, taken precedence over environmental considerations.

Redefining the policy mix. The challenge of environmental policy has shifted from winning corporate cooperation to harnessing corporate creativity. Creativity cannot be squeezed out of companies by a regulatory stranglehold, at least not now that so many competing aspects are involved. For this reason the Netherlands is now emphasising consultation between government and target groups while encouraging self-regulation among businesses. Such a policy calls for delicate handling, because consultation can only succeed if realisation of the environmental objectives is ultimately perceived by all participants to be "inevitable", and this perception can only be achieved by means of sufficient social and political pressure. In such a twin-track policy, therefore, it is vital to achieve an optimal fine-tuning of, on the one hand, legislation and enforcement and, on the other, consultation and self-regulation.

Clearly, we believe that consultation and self-regulation form an indispensable track of environmental policy. However, this track soon comes into collision with direct regulations which stipulate requirements concerning e.g. equipment or technology, and even with those that set specific targets. Irritation among the business community provoked by such "twin-track control" can easily undermine precisely the goodwill and commitment which are the key benefits of target-group consultation.

The only way out of this dilemma is to make a distinction, in terms of both the application and the nature of the legal instruments, between positive and active companies, law-abiding but passive companies, and non-cooperative companies. In other words, legal instruments, which will certainly continue to be based on licensing requirements for some time, must be applied more flexibly. Companies with an environmental management system - one which is recognised by the authorities and preferably also certified by an external body - can qualify for an integral licence on the basis of that system and an approved environmental business plan. In such cases, policy enforcement can take place partly through audited progress reports. With other companies, however, government must retain the option of issuing instructions, and even, where necessary, of demanding the use of certain clean-up technologies.

Target-group consultation is playing an increasingly central role in the Netherlands and various other European countries. In this light, it is time to look for ways to institutionalise the consultative process in order to protect the legal and democratic integrity of government.

References

Aarts, C. W. A. M., 1995: Nationale politieke problemen, partijcompetentie en stemgedrag.

Adriaanse, A., 1990: Milieukerngegevens Nederland, Kerngegevens vanuit het milieubeleid. The Hague: Ministerie van Volkshuisvesting, Ruimtelijke Ordening en Milieubeheer, Centrale directie voorlichting en externe betrekkingen.

Arentsen, M. J., J. Th. A. Bressers and P.-J. Klok, 1993: Twintig jaar Nederlands milieubeleid: Van urgentienota naar NMP en verder, in: D. H. van Dijk (ed.), Milieu van jaar tot jaar 1992-1993. The Hague: Delwel, 13-49.

Arentsen, M. J., P.-J. Klok and J. A. M. Maarse, 1993: De uitvoering van beleid, in: J. Th. A. Bressers, J. J. A. Thomassen and F. A. van Vught (eds.), Politicologie en het openbaar bestuur. Alphen aan den Rijn: Samsom HD Tjeenk Willink, 216-229.

Barendse, G. W. J. and H. A. Udo de Haes, 1989: Milieu-effectrapportage: Een planbeoordelingsinstrument, in: P. Glasbergen (ed.), Milieubeleid, Theorie en praktijk, derde, ongewijzigde druk. The Hague: VUGA Uitgeverij B.V., 127-148.

Beugelink, G. P. et al., 1992: Effecten op natuur van grondwaterwinning. Bilthoven: RIVM.

Boersema, J. J., J. W. Copius Peereboom and W. T. de Groot, 1991: Basisboek milieukunde. Meppel: Boom.

Bressers, J. Th. A., 1979: Beleidsanalyse en de ontwikkeling van het waterkwaliteitsbeleid, Bestuurswetenschappen 33, 2, 70-82.

Bressers, H. Th. A., 1983a: The Role of Effluent Charges in Dutch Water Quality Policy, in: P. B. Downing and K. Hanf (eds.), International Comparisons in Implementing Pollution Laws. Boston etc.: Kluwer Nijhoff, 143-168.

Bressers, J. Th. A., 1983b: Beleidseffectiviteit en waterkwaliteitsbeleid. Enschede: University of Twente.

Bressers, J. Th. A., 1987: Milieubeleid op maat, in: T. P. W. M. van der Krogt et al., Big is beautiful? Schaalveranderingen in overheid en samenleving. The Hague: VUGA uitgeverij, 361-380.

Bressers, H. Th. A., 1988: A Comparison of the Effectiveness of Incentives and Directives: The Case of Dutch Water Quality Policy, Policy Studies Review 7, 3, 500-518.

Bressers, J. Th. A., 1993: Beleidsnetwerken en instrumentenkeuze, Beleidswetenschap 7, 4, 309-330.

Bressers, H., D. Huitema and S. M. M. Kuks, 1994: Policy Networks in Dutch Water Policy, in: H. Bressers, L. J. O'Toole and J. Richardson (eds.), Water Policy Networks, Special Issue Environmental Politics 3, 4, 24-52.

Bressers, H. and L. J. O'Toole, 1994: Network and Water Policy, in: H. Bressers, L. J. O'Toole and J. Richardson (eds.), Water Policy Networks, Special Issue Environmental Politics 3, 4, 197-217.

Bressers, H. Th. A. and J. Schuddeboom, 1994: A Survey of Effluent Charges and other Economic Instruments in Dutch Environmental Policy, in: C. Chung (ed.), Applying Economic Instruments to Environmental Policies in OECD and Dynamic Non-Member Economies. Paris: OECD, 153-172.

Brussaard, W. et al., 1993: Milieurecht, derde druk. Zwolle: W.E.J. Tjeenk Willink.

CBS (Centraal Bureau voor de Statistiek), 1990: Environmental Statistics of the Netherlands. The Hague: Staatsuitgeverij.

CBS, 1992: Algemene milieustatistiek 1992. The Hague: Staatsuitgeverij.

Coenen, F. H. J. M., 1992: The Role of Municipal Environmental Policy Planning in Environmental Management, paper IIAS World Congress, July 1992.

Drees, W., 1992: Strategiën voor het NMP-2, Milieu 7, 2.

Ester, P., 1979: Sociale aspecten van het milieuvraagstuk. Assen: van Gorchum.

Evaluation Committee (Environmental Protection Act), 1990: Towards a Better Procedure to Protect the Environment, English summary. The Hague: Staatsuitgeverij.

Fischer, F., 1985: Critical Evaluation of Public Policy: A Methodological Case Study, paper IPSA world congress, Paris.

Galetzka, M., J. Gutteling and O. Wiegman, 1994: The Acid Rain Situation in the Netherlands is Unique, draft. Enschede: University of Twente.

Glasbergen, P. (ed.), 1989: Milieubeleid, Theorie en praktijk, 3e druk. The Hague: VUGA uitgeverij.

Gutteling, J. M. et al., 1991: Veranderingen in de berichtgeving over milieurisico's in vier Nederlandse dagbladen, Massacommunicatie 19, 1, 18-36.

Gutteling, J. M. and J. F. Caljé, 1993: De invloed van het milieu in het nieuws: Mondiale risico's en risico's dichter bij huis, Milieu 8, 1, 14-19.

Gijswijt, A. J. and M. van der Vliet, 1993: Regering en parlement over mondiale milieuproblemen, Milieu 8, 1.

Hoogerwerf, A., 1990: Reconstructing Policy Theory, Evaluation and Program Planning 13, 3, 285-291.

Huberts, L. W. and W. J. van Noort, 1989: Sociale bewegingen in de jaren negentig, Stand van zaken en vooruitblik. Leiden: DSWO Press.

Kaakebeeke, W. J., 1994: Nederland in het internationale milieuveld, in: P. Glasbergen (ed.), Milieubeleid. The Hague: VUGA, 341-358.

Klok, P.-J. and S. M. M. Kuks, 1994: Het doelgroepenbeleid, in: P. Glasbergen (ed.), Milieubeleid, Een beleidswetenschappelijke inleiding, 4e herziene druk. The Hague: VUGA uitgeverij, 79-96.

de Koning, M. E. L., 1994: In dienst van het milieu, Enkele memoires van oud-directeur-generaal Milieubeheer prof. ir. W.C. Reij. Alphen aan den Rijn: Samson HD Tjeenk Willink.

Leroy, P., 1994: De ontwikkeling van het milieubeleid en de milieubeleidstheorie, in: P. Glasbergen (ed.), Milieubeleid. The Hague: VUGA, 35-58.

Leroy, P. and A. De Geest, 1985: Milieubeweging en milieubeleid, Sociale en politieke aspecten van de milieukwestie. Antwerpen enz.: De Nederlandsche Boekhandel.

Ministerie van Verkeer en Waterstaat, 1975: De bestrijding van de verontreiniging van het oppervlaktewater: Indicatief meerjarenprogramma 1975-1979. The Hague: State Press.

Ministerie van Volkshuisvesting, 1994: Ruimtelijke Ordening en Milieubeheer, Milieuprogramma 1995-1998. The Hague: State Press.

Nelissen, N. et al., 1987: De Nederlanders en hun milieu. Zeist: Kerckebosch B.V.

OECD (Organisation for Economic Co-operation and Development), 1995: Environmental Performance Reviews. Netherlands. Paris: OECD.

de Ridder, J. A. and O. Scholten, 1993: Milieugolven in de media: Dagbladen en de beleidsvorming 1965-1990 over het milieu, Beleidswetenschap 7, 3, 331-357.

RIVM (State institute for health and environment), 1988: Zorgen voor morgen. Alphen aan den Rijn: Samsom HD Tjeenk Willink.

RIVM, 1991: Nationale milieuverkenningen 2 1990-2010. Alphen aan den Rijn: Samson HD Tjeenk Willink BV.

RIVM, 1993: Nationale milieuverkenningen 3 1993-2015. Alphen aan den Rijn: Samson HD Tjeenk Willink BV.

RIVM, 1994: Jaarverslag 1993. Bilthoven: RIVM.

Schuddeboom, J., 1994: Milieubeleid in de praktijk, 3rd edition. Alphen aan den Rijn: Samsom HD Tjeenk Willink.

Schuurman, J., 1988: De prijs van water. Arnhem: Gouda Quint.

Staten Generaal (proceedings Dutch parliament), 1964: Memorie van Toelichting bij het wetsvoorstel Wet Verontreiniging Oppervlaktewateren, nr. 7884. The Hague: State Press.

Staten Generaal, 1972: Urgentienota Milieuhygiëne, nr. 11906. The Hague: State Press.

Staten Generaal, 1989: Nationaal Milieubeleidsplan: Kiezen of verliezen, nr. 21137-2. The Hague: State Press.

Staten Generaal, 1990: Nationaal Milieubeleidsplan-Plus, nr. 21137-20. The Hague: State Press.

Staten Generaal, 1993: Nationaal Milieubeleidsplan 2: Milieu als maatstaf, nr. 23560. The Hague: State Press.

Staten Generaal, 1994: Milieuprogramma 1995-1998, nr. 23905. The Hague: State Press.

Tellegen, E./Tommel, D. K. J. 1984: Milieuactie en milieubeleid, in: Boersema, J. J. /Copius Peereboom, J. W./de Groot, W. T., Basisboek milieukunde, Over de analyse en oplossing van milieuproblemen. Meppel: Boom, 65-72.

Wetenschappelijke Raad voor het Regeringsbeleid, 1994: Duurzame risico's: Een blijvend gegeven. The Hague: State Press.

Wiegman, O. et al., 1989: Newspaper Coverage of Hazards and the Reactions of Readers, Journalism Quarterly 66, 846-852.

Winsemius, P., 1986: Gast in eigen huis. Alphen aan den Rijn: Samsom HD Tjeenk Willink.

GERMANY

Martin Jänicke and Helmut Weidner

1 Main Environmental Problems and Improvements Since 1970

Germany's geographical and political situation as well as its economic structure bring with them specific environmental problems and, as a rule, require greater efforts in environmental protection than is the case in many other countries. Population density and the degree of industrialisation are among the highest in Europe. A large proportion Germany's industrial sector is environmentally problematic, its agriculture is highly industrialised, its transport network one of the densest and traffic volume high and increasing. Germany has a long coastline and, as it shares borders with nine other countries, most of its main rivers have their sources in neighbouring countries.

In the 1970s and 1980s air pollution from stationary and mobile sources, the risks from nuclear power plants and the pollution of the aquatic environment by industrial and household effluents played a dominant role in public debate.[1]

Pollutants (such as SO_2, particulates, CO, NO_2, VOC, and lead) from energy production and fuel use led to a deterioration in ambient air quality in urban agglomerations, in particular to incidents of smog (with detrimental effects on public health), and caused widespread damage to plants and trees (Waldsterben). Germany was subject to transfrontier air pollution, mainly from East European countries, but it was also a significant exporter of air pollutants (especially SO_2) itself.

The pollution of rivers and lakes was a great cause for concern in the 1970s, partly because of its high visibility: many water bodies were covered with polluted foam. Some water bodies with a slow circulation faced the threat of being unable to support life, as a result of the high concentrations of phosphorous, nitrate and other oxygen-demanding substances there, others were so highly

1 See for this and the following information about the environmental situation: Umweltbundesamt, 1994; SRU, 1974, 1978, 1987, 1994; BMU, 1987, 1992; OECD, 1993.

contaminated with heavy metals that they had to be withdrawn from use for recreational activities or as sources for drinking water. The heavily polluted Rhine became a symbol of this reckless treatment of natural resources. Already in the 1970s there was growing concern about the improper disposal of domestic, industrial and hazardous waste. During the 1980s there were several industrial accidents (particularly at chemical plants) which raised public awareness of the environmental risks of living in a highly industrialised society.

The following description of environmental quality change focuses on West Germany only (East Germany is described briefly in an annex; figures for the unified country will be given in brackets). Since 1970 remarkable progress has been made in several areas. With respect to air pollutants, emissions in particular of sulphur dioxide, particulates, and several heavy metals have been drastically reduced; emissions of volatile organic compounds (VOC) and methane (CH_4) have declined, but to a lesser degree; total nitrogen oxide emissions have remained almost stable. Total CO_2 emissions were 10% lower in 1993 than in 1973, but still as high as they were in 1983. (In East Germany there was however a reduction of about 50% from 1987 to 1993.)

Table 1. Evolution of Air Pollution Emissions.

Pollutant	*1975*	*1985*	*1992*
CO_2	732	731	740
NO_x	2,511	2,908	2,426
SO_2	3,308	2,369	875
CO	13,994	8,939	6,577
Particulates	809	566	430
VOC	2,696	2,569	2,092
CH_4	5,747	5,326	5,045

Source: Statistisches Bundesamt, 1995, pp. 722 f.

Ambient air quality has generally improved with respect to SO_2, particulates and CO, but an increase in road transport has led to a deterioration (with respect to NO_x) in urban areas. Increased concentrations of ozone have been observed frequently in recent years, especially during the summer months ("summer smog").

As regards water pollution control, the huge investments in municipal and industrial waste water treatment plants have led to improvements in water quality with respect to some parameters and pollutants (e.g., BOD, suspended matter, heavy metals). In some aquatic systems there have been considerable improvements, notably in the Rhine for instance. However, ground water quality has deteriorated in many areas, due mainly to agricultural activity. The generation of municipal waste has been stabilised and in the early 1990s even reduced (contrary to the situation in most industrialised countries). The incineration of waste on the high seas and the discharge of liquid waste into the North Sea was halted completely in 1989, thereby ending simultaneously Germany's last remaining practices of direct dumping and discharge of industrial wastes into the oceans. But in the early 1990s West Germany was still a prominent exporter of toxic waste. And illegal dumping is by far the most frequent offence (and one which is very much on the increase) among nearly 25,000 (32,000 in the unified Germany) criminal violations of environmental law (Umweltbundesamt, 1994).

Improvements in environmental quality have been made easier by economic modernisation and increases in efficiency. Primary energy consumption has remained almost stable since 1973. Consumption of materials in West Germany in 1990 was higher than in 1960, but there was a slight decrease during the 1980s. In contrast to abiotic inputs there was an increased input of biotic materials. In spite of a reduction in manufacturing industrial activity, overall water abstraction has grown (two thirds being used by the electricity generating industry).

Germany has a relatively high proportion of protected land (11.9%), but the proportion of developed land increased steadily to 12.2% in 1989 (an increase of more than 10% within a period of eight years).

The improvements in environmental quality did not lead to a more favourable public opinion on the environmental situation. A public opinion survey carried out in West Germany in 1991 revealed that 48% of the sample judged their environment at the time to be in a "bad" or "very bad" state; in 1992 and 1993, 53% of those surveyed held the same view (Umweltbundesamt, 1994: 77-78; see also ipos, 1994).

Today the following areas should be considered most problematic: air pollution and damage caused by private transport (the growing number of passenger cars, with an increasing mileage, the trend towards bigger-engined cars, the increase in road haulage and road construction); increased illegal dumping and local opposition to conventional waste treatment methods (landfills, incineration); the presence in 1995 of 170,000 (sometimes heavily) contaminated sites, as a result of improper waste disposal and production practices in the past; a general growth of the proportion of developed land; the still serious rate of damage to trees; pollution of soil and water by the agricultural sector (Germany's production of drinking water depends approximately 70% on ground water); increasing volumes of sewage sludge with a high concentration of toxic substances, due to improved techniques of effluent treatment (making its disposal, especially on agricultural land, a controversial issue); and global challenges caused by the so-called greenhouse gases (CO_2, etc.).

2 Development of Environmental Policy and Principal Regulations

2.1 Development

Government regulation to protect humans and their environment against hazardous industrial activities has a long tradition in Germany (Wey, 1982). It is embodied in private and public law, e.g. building, public health and sanitation, and in criminal law. The Water Rights Act and the Factories Act date back to the last century. The various regulations enacted to prevent noxious and offensive emissions from becoming a nuisance or being prejudicial to health and property were systematised and formulated for the first time in the Prussian Industrial Statute of 1845. At that time, the main emphasis was on protecting health and property in the vicinity of the emitting premises.

In the 1960s air pollution laws existed on the state level, but it was not until 1969 (Hartkopf and Bohne, 1983) that environmental policy developed into a national policy area based on a comprehensive concept of environmental protection. In 1969, a centre-left coalition government formed by the SPD and FDP came into power for the first time since the founding of the Federal Republic. Although neither election campaign nor party manifestos had placed much emphasis on the environment, the official Government Declaration to the Federal Parliament in October 1969 made the solving of environmental problems one of the primary tasks of government. According to Müller (1986: 53), "the view cannot be dismissed that, in 1969, the initiators of this declaration were not then fully aware of the political implications and subsequent explosiveness of the environment as a political theme". There was no noticeable public demand for this, nor was pressure being exerted by organised interest groups to which government would have had to respond. Therefore, the initiative which formed this new political area must have come not from forces outside parliament, but rather from within the political system itself. This is what Reich calls an "inside initiative model" (Reich, 1984). It was not until the 1970s that the environmental movement outside parliament became a driving force in environmental policy (Brun, 1978; Poguntke, 1993; Raschke, 1993).

The government's first initiatives, beginning in September 1970 with a quick-start programme (*Sofortprogramm*) for environmental protection and continued in the 1971 environmental programme for the coming years, were orientated initially around international developments, especially around early environmental legislation in the USA (the 1969 National Environmental Protection Act).[2] The initiatives of the Council of Europe concerning water and air pollution (The Water Charta/The 1968 Clean Air Charta, European Nature Protection Year 1970) and, above all, the UN Conference on the Environment in Stockholm in 1972 were of considerable importance.

2 For the evolution and main characteristics of modern environmental policy in Germany see Wey, 1982; Hartkopf and Bohne, 1983; Müller, 1986; Hucke, 1990; Malunat, 1994; Weidner, 1995.

One of the first measures taken by the federal government (October 1969) involved the transfer of responsibilities relating to pollution control (water, air quality, noise abatement and waste disposal) from the Ministry of Health to the Ministry of the Interior, which became the central Ministry for environmental protection at the federal level. In subsequent years, the Minister of the Interior (Hans-Dietrich Genscher) acquired further powers. In 1972, an amendment to the German constitution granted the federal government so-called concurrent legislative power for statutory regulations on waste management, air pollution control, noise abatement, protection from radiation, and on criminal law relating to environmental protection matters. Concurrent legislative power (i.e., power shared between the federal government and the states) means that the federal government has the right to issue detailed regulations, and that in this case, federal law supersedes state law. However, in the areas of water management, regional planning, nature conservation, and landscape preservation, the federal government is only authorised to issue so-called framework laws as a basis for detailed and specific legislation to be drawn up by the states.

Modelled on the U.S. Council of Environmental Quality, the German Council of Environmental Advisers, the *Rat von Sachverständigen für Umweltfragen* (SRU), was set up in 1971 to provide advice on environmental matters. This council, made up of independent social and natural scientists (mainly university professors), published its first report (on the automobile and the environment) in 1972. Its first comprehensive report on the environment came out in 1974. To provide the Federal Ministry of the Interior (and from 1986 the Ministry of the Environment) with technical, scientific and administrative support, the German Federal Environmental Agency, the *Umweltbundesamt*, was established in Berlin in 1974, modelled on the U.S. Environmental Protection Agency, but with less autonomy and no regulatory function.

In the course of deliberations on the environmental programme and other programme statements, three principles were deduced that were to act as central guidelines for environmental policy: the principles of precaution (*Vorsorgeprinzip*), cooperation (*Kooperationsprinzip*) and the Polluter Pays Principle (*Verursacherprinzip*). Precaution only attained its commanding position within the three basic principles in 1976, when the federal environmental programme was updated.

In general, when compared internationally, the federal environment programme was progressive and demanding for its time. The question of how the responsible federal ministry, the Ministry of the Interior, succeeded in obtaining support for it from the entire federal government without major conflict has been answered as follows (Müller, 1989: 5-6):

> "[Strong] political backing came from the Chancellor's Office, (...) [but] reformers also benefited from the general reformist climate associated with the new government. In the face of wholesale change in official policy, the defenders of established policy areas and interests inside the 'state apparatus' adopted a wait-and-see attitude. Represent-

> atives of industry had been able to bring their particular interests to the table during the work carried out by various groups set up to draft the environmental programme, and had therefore joined the decision-making process."

Given these favourable conditions at the outset, several pieces of federal legislation concerned with pollution control were quickly enacted between 1971 and 1974, such as the Air Traffic Noise Act (1971), the Leaded Petrol Act (1972), the Waste Disposal Act (1972), the DDT Act (1972), the Federal Immission Control Act (1974) and the Environmental Statistics Act (1974), as well as a number of regulations, decrees and administrative directives, such as the Technical Instructions on Air Quality Control (*TA Luft*) of 1974. In the following years, after a period of stagnation (from about 1974 to 1978) mainly due to the oil crisis and subsequent world recession, numerous further regulations were passed, covering almost all areas important in environmental protection – e.g. the Nature Conservation Act (1976), the Waste Water Charges Act (1976), the (somewhat watered-down) Chemicals Act (1980), the Ordinance on Large Combustion Plants (1983), the Waste Avoidance and Management Act (1986, fundamentally amended in 1994), the Environmental Impact Assessment Act (1990), the Environmental Liability Act (1990), and the Packaging Ordinance (1991). The comprehensive Closed Substance Cycle and Waste Management Act of 1994 (the "*Kreislaufwirtschaftsgesetz*") has opened up new paths for waste recycling or avoidance and has increased the opportunities for private initiatives in waste management.

In June 1990, the programme for reducing CO_2 emissions, a threat to the global climate (greenhouse effect), was passed. This programme, along with the goal of reducing CO_2 emissions by 25-30% of their 1987 level by the year 2005, has led to detailed measures and voluntary agreements with industry (which are nonetheless judged insufficient to reach this ambitious goal). In 1995 the chemical industry, for example, published advertisements about a reduction in CO_2 emissions by about 30% by the year 2000, based on 1990 emission levels. The organisations of the electricity generating industry declared their intention of reducing CO_2 emissions by 25% by 2015 (Stromthemen 5/1995: 2).

In 1994, environmental protection was written into the constitution as a goal of the state (*Staatsziel*). There is, however, no detailed, all-embracing basic environmental law, although an official expert proposal for the codification of existing environmental law (*Umweltgesetzbuch*) exists at the time of writing (Kloepfer et al., 1990; Jarass et al., 1994). Because of the large number of legal regulations at federal and *Länder* level, of which many are highly detailed, Germany has one of the most complex legal systems for environmental protection in the world.[3]

---

3 For a comprehensive overview and analysis of German environmental laws and regulations see Kloepfer, 1989.

2.2 Basic Instruments

In Germany there is a strong preference for, and continuity in the use of traditional regulatory instruments (Rehbinder, 1992; Wicke, 1989; Rose-Ackermann, 1995). This command-and-control approach consists mainly of licensing and laying down standards. The licensing of large emission sources, which includes public participation, consists of a complex procedure usually accompanied by highly detailed technical prescriptions (based upon the latest available technology). This approach forces the responsible authorities to formulate detailed technical requirements and to prove that these are both in keeping with the state of the art in technology and feasible. The result of this is often a time-consuming and controversial procedure, accompanied by strong administrative efforts at control in the form of monitoring and sanctions. There are also other regulatory instruments available – for example prohibiting specific activities or the generation of certain substances, imposing general restrictions on individual substances in combination with a step-by-step procedure for their further reduction or elimination, and the imposition of emissions standards for special target groups (as in the Ordinance on Large Combustion Plants, for instance) addressed usually at a specific group of larger-scale emitters.

Economic instruments based on the Polluter Pays Principle (PPP), have thus far played a less important role than in Scandinavian countries (OECD 1993, 1995). They are used mainly in the areas of waste water (effluent charges), petrol (tax differentiation in favour of unleaded fuel), and waste management. There is also a complex system of direct and indirect subsidies to promote environmentally sound or helpful activities in the private sector. Public expenditure on these subsidies has increased continuously. But, as economic studies have shown, the efficiency and effectiveness of these instruments is low (see SRU, 1987, 1994). They have also been criticised from a broader ecological perspective as sending the wrong signals. The proposal for the introduction of an energy tax or a CO_2 charge caused an intense debate and was met with strong resistance, although it was favoured by a majority of experts, the large environmental organisations, the Greens and the SDP. In spring 1996 it still had not come into existence.

Most of the environmental improvements have been realised by the dominant command-and-control approach, which is combined often with subsidies, sometimes with charges. This approach has come under criticism (theoretical as well as practical) from numerous experts for its inefficiency and lack of flexibility. This has not been without effect. In recent years the German government has begun a systematic review of this approach (see Weidner, 1995). On the one hand, it has examined possible ways and means of speeding up planning and industrial siting procedures, as a reaction to complaints from industry. On the other hand, it has modernised its instruments by stressing the role of information and negotiation. Guidance through information is no longer restricted to government (e.g. through environmental reports or labelling), but has also been used by private actors such as Greenpeace, e.g. in campaigns against polluting sectors and firms or in support of more progressive firms. Negotiation has led to an increasing number of voluntary agreements between industry and public authorities or environmental organisations (Lautenbach et al., 1992). Recently for instance the German car industry voluntarily agreed to take back outdated models. Alternative forms of dispute resolution (e.g. mediation) have become increasingly important. The new instruments of information and negotiation still need to be improved. There is a lack of standardisation, e.g. in life-cycle assessments (LCAs). Voluntary agreements are often not formulated clearly enough, the obligations of individual members within sectors or organisations and the monitoring mechanisms are often not adequately defined. However, much scientific and political activity regarding both instruments can be seen at present.

The environmental liability law, passed in 1990, was another important step towards generating alternative approaches to environmental regulation. For instance, the principle by which fault or negligence are not considered when determining liability, the partial alleviation of the victim's burden of proof, the extension of liability to cover damages caused even by normal, properly functioning equipment and operations, and the obligation to hold environmental liability insurance are all significant improvements. The Packaging Ordinance of 1991 obliges industry and suppliers to reclaim and recycle virtually all packaging in accordance with phased-in quotas. This has led to the so-called "green dot" (*Grüner Punkt*) programme which, in combination with a licensing fee designed to cover the costs of collection and recycling of packaging materials, make up Germany's so-called "dual system" (*Duales System Deutschland*), a privately run enterprise to co-ordinate waste collection and recycling in Germany.

Special types of planning such as air quality planning (*Luftreinhaltepläne*) or waste management plans exist, but in contrast to many other developed countries (such as the Netherlands, Canada, Austria or Japan) there is no comprehensive national environmental plan as foreseen in Agenda 21.

2.3 Policy Style

Not only individual experts and independent institutions in Germany and abroad have come to the conclusion that German environmental policy is based largely on an inflexible approach and conventional attitudes toward regulation, rooted in old-fashioned policing law (danger avoidance) and "statist" ideology. Increasing problems with this policy style when trying to meet environmental objectives have led to a change in direction, towards consensus building, a broader form of cooperation that includes scientists and members of environmental organisations critical of current environmental policy.

German environmental policy is now characterised by intensive formal and informal cooperation and negotiation between regulator and regulated. The style of policy implementation is undergoing a process of change. There are increasing indications that a broader, more intensive involvement by organised environmental activists (non-governmental environmental organisations) at macro- and meso-level consultations on environmental policy and decision-making is taking place. In the German environmental policy of today we see a combination of hierarchical and cooperative elements, characterised as "negotiation under the shadow of hierarchy" (Scharpf, 1991).

3 Evaluation of Policy

The OECD Report on Environmental Performance in Germany, published in 1993, came to a more or less positive view. It emphasised especially that: "The de-coupling of economic growth from the flow of several major pollutants over the past two decades is indicative of Germany's remarkable achievement in reconciling economic growth and environmental objectives" (OECD, 1993: 205). However, it also points out that major challenges remain, for instance in waste disposal and pollution from agriculture and transport, and that Germany's environmental problems have increased substantially with unification. Regarding Germany's approach to policy-making, the OECD expert group doubted that the environmental achievements had been reached at the lowest possible cost, and it proposed that prioritisation be improved, and that diverse measures linked to the characteristics of specific ecosystem be used. It also recommended extending the use of economic instruments, promoting voluntary environmental agreements in the industrial sector, improving the dissemination of information on environmental developments by the federal government, especially with regard to national and international availability of data (OECD, 1993: 208-209).

In spring 1994 the Council of Environmental Advisers judged the development of environmental policy in Germany more harshly (SRU, 1994). As far as air quality control is concerned, the report identifies further success in reducing emissions, although the weak measures undertaken in the transport sector are criticised. The basis for a comprehensive nature conservation programme not confined to selected, individual areas is still a long way off. It is pointed out that damage to forests is "still as much a cause for concern" as it has been in the past, and that the objective of soil conservation is being jeopardised by what are known as "streamlining laws" (*Beschleunigungsgesetze*, which restrict public participation in the interests of speed); in water protection there has been steady improvement in the quality of moving bodies of water; the situation as regards data on the environment still requires fundamental improvement; waste management policy is now judged positively (as compared to the SRU report of 1987), but the limitations on public participation in this field gives rise to scepticism notwithstanding; the necessity of better supervision to prevent the illicit export of waste is underlined. In respect of hazardous substances and health risks, the report acknowledges that the measures taken, including legislative ones, have led to improvements, but that there are still obvious deficits to be made good, for instance in effective product labelling to indicate risks to health. Regulations on environmental impact assessments are judged to be inadequate. The Environmental Liability Act is greeted as a step in the right direction, but considered too narrow in scope; the report suggests that "a collective fund with state participation be set up to deal with environmental damage which cannot be attributed to an individual polluter" (SRU, 1994: 32).

The following two statements sum up the SRU's general conclusions on the present state of environmental policy in Germany and its future development:

> "The environmental policy of the last two governments was unable to satisfy fully the demand for an environmental policy programme that is cohesive and transcends individual sectors; most regulations and measures are still concerned with individual environmental media. A number of more comprehensive schemes have, however, been initiated" (SRU, 1994: 32).

> "Without doubt there has been some progress (...) in individual areas of environmental policy. The development of a fundamentally new programme for ecological modernisation has not, however, been successful. In principle we have a fairly active, but still fundamentally technocratic, and to some extent merely symbolic policy of small steps" (SRU, 1994: 177).

Compared internationally, Germany was no frontrunner in the new policy area. It started in the early 1970s – later than the US, Sweden, Japan or the UK (a frontrunner in the 1960s) – and in these early years it adopted many measures from other countries. Compared to the US and Japan, measures against air pollution from cars or power stations arrived relatively late. But in the 1980s this situation changed, and Germany became an influential player in the international environmental policy arena. This bears not only on what is now a leading position in the world market for pollution control technology, but also on environmental achievements relative to other industrialised countries.

Compared with other European countries, the use of emission-reducing technologies is extremely widespread within Germany. Capacity for reducing emissions of sulphur dioxide and nitrogen oxide, and the market share of lead-free petrol are larger than in most other European countries. Germany's maximum permissible emission levels for air pollutants are generally among the strictest on the continent. This is also true for the permissible levels of sulphur in domestic heating fuel and diesel oil, as well as the emission standards for waste incineration plants. The picture is similarly positive in respect of sewage purification: in West Germany over 90% of the country's households are connected to the sewage mains, and 90% of the sewage is purified biologically. The maximum permissible level for dioxin emissions from waste incineration plants has set standards in environmental technology throughout the world. Germany was among the first countries in the world to stop the production and use of chlorinated fluorocarbons (CFCs). Approximately 3,800 environmentally sound products manufactured in Germany carry the "blue angel" label (introduced in 1978), and other countries, including the European Community, have adopted this market incentive to consumers and producers. Germany's state research and development (R&D) funding for renewable energies such as solar and wind energy (now a boom industry) ranks highly. The same is true for expenditure on environmental protection, as mandated primarily by environmental regulations aimed at industry and public budgets. The Packaging Ordinance of 1991 had some influence on solid waste policy in other industrial countries. The absence of a speed limit on motorways, or its continuing position as the leading exporter of toxic waste, on the other hand, show that Germany is far from being an absolute model for other countries.

The German government is nevertheless playing an active role in formulating international agreements. This is true not only in the context of the European Union – although there are important exceptions here (cf. Héritier et al., 1994) – but in the larger international framework as well. This became especially noticeable at the UNCED Conference in Rio in June 1992 and at the follow-up Conference in Berlin (Spring 1995), where Chancellor Kohl stressed the ambitious goal of a 25% reduction in CO_2 emissions by the year 2005 (now relative to 1990 levels). Further examples include the implementation of the agreement on long-range air pollution (the Helsinki and Sofia protocols), the Montreal protocol, establishing measures to reduce ozone depletion, as well as collaboration to protect the North and Baltic Seas.

4 Capacity of Proponents of Environmental Policy

4.1 Public Institutions

Germany has a broad network of well-established environmental policy institutions. At federal and state (*Länder*) level there are specific authorities responsible for the formulation and implementation of environmental policy. Local authorities usually have environmental policy departments. At federal and state level, again, there are larger agencies responsible for environmental research, planning and development.

The main actor at federal level is the Ministry for the Environment, Nature Conservation and Nuclear Safety, which was founded relatively late on, after the Chernobyl accident in 1986 (see Pehle, 1988). Below the Ministry for the Environment there are several government institutions, in particular the Federal Environmental Agency, the Federal Office for Protection from Radiation, and the Federal Research Centre for Nature Conservation and Countryside Ecology. Parallel to the Ministry for the Environment are various other federal ministries with specific responsibilities in environmental policy-making or other environment-related areas, notably the Ministry of the Economy, the Ministry of Transport, the Ministry of Agriculture and the Ministry for Education, Science, Research and Technology. Meanwhile, most ministries, even the Ministry of Foreign Affairs and the Ministry of Defence, have a special environmental protection department.

The main function of the Council of Environmental Advisers is to advise government when this is specifically requested, although it sometimes deals with issues on its own initiative. In 1986, an environmental committee was established within the German federal parliament. There are other (sometimes higher-level and more influential) scientific advisory institutions (Parliamentary Commissions of Enquiry, the German Advisory Council on Global Change) as well, who can also exert a strong influence on public and scientific debate.

As is provided in the constitutional division of powers, central government is primarily responsible for policy formulation and international cooperation, while implementation and enforcement are tasks for the *Länder* and local authorities. As almost all environmental laws and ordinances must pass through the *Bundesrat* – the upper house of parliament, made up of delegates from the state governments, which also has the power to introduce its own bills – this body is also involved in environmental policy-making.

At present all 16 German states have established a ministry mainly or exclusively responsible for environmental matters. According to the federal constitution (Article 28, *Grundgesetz*), local authorities have a right to self-administration, putting them in a relatively strong position. Their environmental tasks include some mandatory and some voluntary duties which they perform "in their own right", as well as others delegated to them by state or federal government. Important actors in the field of water or energy supply are the municipally owned utilities. Some German cities play an active role in international networks for environment and climate protection (ICLEI and Climate Alliance).

The environmental institutions of the European Union have for a long time been seen as a restrictive factor as far as Germany is concerned. This has changed somewhat in recent years. The directives on drinking water, environmental impact assessments, information rights and environmental audits have had a positive influence on innovation in this country (Héritier et al., 1994).

The courts had at times (especially in the 1970s) a significant influence on programme formulation and environmental policy implementation in Germany. However, unlike in other key political areas of government policy (e.g., abortion), the German Federal Constitutional Court has not intervened on a massive scale with adjudications on environmental matters, although some of its decisions have helped to make the concept of precaution more viable. Up to now, criminal law and the courts have played only a minor role in environmental matters. A few spectacular decisions by the Federal Supreme Court, in particular on questions of liability, are expected to increase industry's risk-awareness with respect to both products and activities.

The administrative courts exercise comprehensive judicial control over administrative activities which involve the violation of individual rights (Article 19 of the federal constitution). These courts have an important role to play in environmental policy and conflicts, because they are often mobilised by third parties before, during and after planning or licensing, with the aim of bringing about a tightening or weakening of environmental regulations, putting a stop to projects or shutting down plants. The lower administrative courts have tended to rule in favour of environmental concerns; however the higher courts have overruled a number of these decisions and have in general tried to prevent the lower administrative courts from narrowing too far government's scope for discretion. But all in all, the courts have not played a major role in actively shaping environmental policy.

4.2 Environmental Organisations

The level of organisation among environmental interest groups has increased steadily (Leonhard, 1986). Environmental associations and nature protection societies currently have more than four million members (Hey and Brendle, 1994). They have been able to create a broad, relatively close-knit communication network at local, national and international levels. The major environmental and nature conservation organisations include BUND (the German association for the conservation and environmental protection) with 215,300 members, the World-Wide Fund for Nature (WWF) with 103,600 members, NABU (the German conservation association) with about 196,700 members and Greenpeace. Greenpeace Germany – by far the largest single source of finance for Greenpeace International – claims 530,000 financial supporters in 1994 and an income of DM 71.2 million (DM 67.8 million in 1993). The umbrella organisation DNR (the German Conservation Circle) has 3.3 million members (Hey and Brendle, 1994: 140), but its 90 member organisations include also interests such as animal protection, local history, and fishing.

The Green Party (*Die Grünen*, and since unification *Bündnis 90/Die Grünen*) is now firmly established within the party system. In 1979 they were represented for the first time in a state parliament. In the federal elections of 1983, the Greens achieved 5.6% of the vote and entered the *Bundestag* (the national parliament). In the October 1994 elections, the Greens became the third largest party (with 7.3% of the popular vote) in the *Bundestag*. In eight states they are or have been coalition partners with the Social Democratic Party. Now there are also some (albeit primarily gestural or tactical) discussions about a possible coalition with the Christian Democratic Union – an indicator of the Green Party's degree of integration within the party system.

4.3 Economic Proponents of Environmental Protection

The main winner in this new political field is eco-industry. With expenditure on the environment of DM 45 billion by government (DM 25 billion) and industry in 1992 (DM 5.3 billion in 1971), West Germany spends one of the highest proportions of GNP within the OECD (1.6%).[4] According to the research institute DIW, about 550,000 jobs, nearly 2% of total employment, depend directly or indirectly on this area of industry (about 130,000 in East Germany). Its rate of growth is above average and its position in the world market is dominant. Recently there

4 Between 1980 and 1992, state and industry together spent DM 180 billion on pollution control equipment (Stromthemen 11/1995).

has also been a sharp rise in the growth of a new "environmental market", which has established itself alongside the traditional market for end-of-pipe technology. It too is expected to grow in the coming years, partly as a result of the introduction of new policy instruments such as environmental auditing.

Since about 1988, "green" enterprises and environmentally committed business associations – B.A.U.M (with about 500 members), Future (about 200 members) and *Unternehmensgrün* – have become more and more important. This new "green business sector" has become the motor of ecological modernisation by way of "integrated" cleaner technologies and a generator of environmentally friendly demand.

This change within the economic sphere cannot be explained by broad public opinion (which has exhibited traces of "environment-fatigue" in the last years). Its main driving force seems to be the ecological awareness of consumers which (according to opinion polls) has increased steadily in the period from 1975 to 1994.

5 The Changing Constellation of Actors

The development of German environmental policy may best be described by the change in the constellation of actors and in the dominant environment-political paradigm.

After a period of "environmental ignorance", the new SPD/FDP government introduced in 1969 a new environmental policy without there being public pressure to do so. Since air and water quality were seen as the main problems, dilution of pollution was considered an appropriate solution. High chimney policy dominated up to 1983 – contrary to the technologically oriented philosophy of the first government programme for action on the environment. Waste collection was another battlefield in this strategy against the visible symptoms of pollution.

The oil crisis caused a backlash, which characterises the second stage (1974-82). This backlash mobilised an active citizens' movement, which had its roots in the anti-nuclear movement of the early 1970s. Now it strongly opposed motorway construction, the airport expansion in Frankfurt am Main and any kind of air and water pollution from larger plants, especially fossil fuel burners.

In the third stage (1983-87) it was again government – now a centre-right coalition government led by Helmut Kohl – that became the driving force, using a combination of command-and-control instruments and subsidies. The main improvement was the introduction of clean-up technology for power stations and cars. The entry of the Greens into the *Bundestag*, thereby changing the agenda of the German party system, was significant. This also improved the informational infrastructure of the environmental movements, which now became more centralised and professional. In addition, *Waldsterben* and heavy smog periods in some parts of the country led to a change in the media's attitude. Environmental issues became an increasingly acceptable topic for publishers of the mass media. This also led to some direct criticism of industrial polluters. The Kohl govern

Figure 1. Changes of Paradigm and Actor Configuration in Germany.

Stage 1 (1969-74): "Dilution":

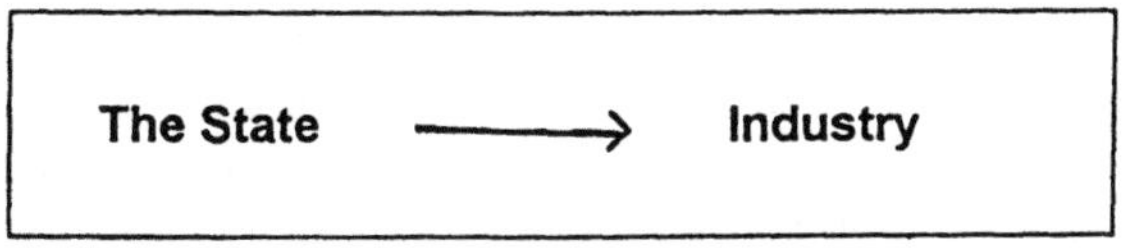

Stage 2 (1974-82): "Dilution plus end-of-pipe treatment":

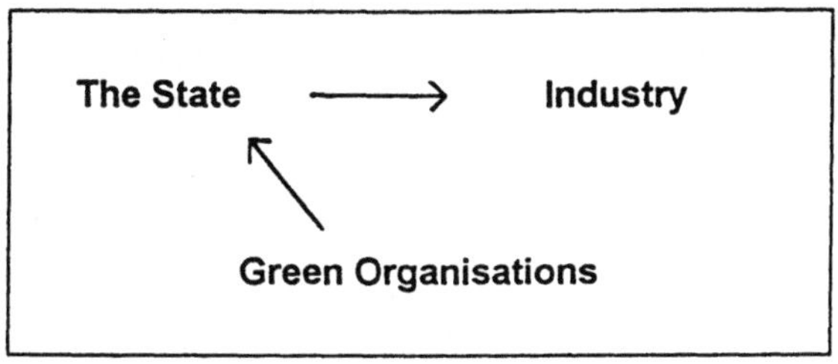

Stage 3 (1983-87): "Intense end-of-pipe treatment"

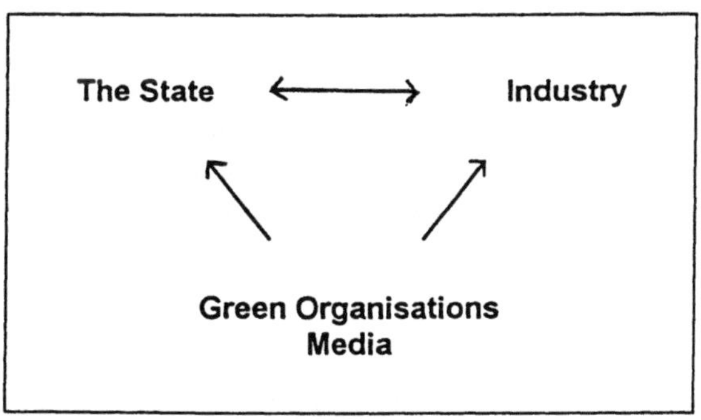

Stage 4 (1988-94): "Ecological modernisation":

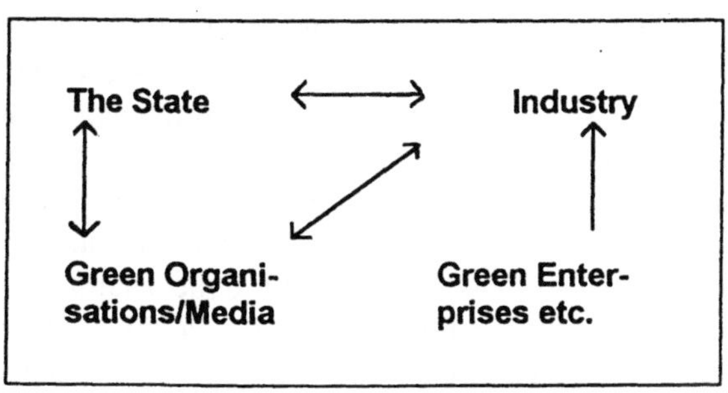

ment, on the other hand, had a closer relationship to industry than the Social Democrats before him. This facilitated close cooperation also in the field of environmental politics.

The crucial fourth (1988-94) stage of ecological modernisation followed the Chernobyl accident and some major pollution incidents in the Rhine. It is also associated with the then minister of the environment, Klaus Töpfer. Now, a trilateral cooperation between the environmental administration, industry and – increasingly – the environmental organisations came into operation. Bilateral dialogue between environmental organisations and enterprises (now threatened by direct attacks) gained great importance. Large enterprises in the chemical industry, for example, regularly have such contacts. The large warehouse company *Hertie* has improved the environmental soundness of about 3,500 of its products in cooperation with the environmental organisation BUND.

The significant new actors were pioneer "green" enterprises and their organisations, influencing the production chain from retailer to supplier in the basic industries (e.g. the producers of chlorine). The motor of change was now not so much government regulation, but a broad coalition of modernisers stimulating the self-interest of enterprises in respect of costs (energy, materials, waste, transport etc.), and new "green markets". The environmental awareness of consumers and the threat of government intervention, combined with negotiation and consultation, were important background facilitators in this process. The entrance of the Green Party into coalition governments at state level was also an influential factor, tightening for instance restrictions on the nuclear power industry (since the early 1990s there has been a de facto moratorium on new installations).

A fifth stage has begun with the federal elections in 1994 and the change in the ministry of environment. It is a stage characterised by heavy economic problems and a more cautious approach to environmental problems. At present it appears that it could become a stage of stagnation.

6 Capacity-Building

Capacity development in environmental policy may be defined as involving: the number, strength, competence and constellation of governmental and non-governmental proponents of environmental interests; their institutional opportunity structure, and the availability of information and public awareness.

As shown above (section 4), the number, influence and competence of governmental or societal actors has grown steadily in Germany. Today all relevant institutions have special departments for environmental protection, from the Ministry of Defence to the large industrial organisations. There is a new added capacity for some intra-market self-regulation, stimulated by economic actors.

The institutional structure of the political system and its proportional voting system (restricted only by a 5% hurdle) worked to the benefit of representatives of environmental interests, giving the Green Party a fair chance early on. The federal structure may also have exerted some positive influence, because some states – especially those with red-green coalition governments – could assume the role of frontrunners. Some innovative influence could also be observed in the relatively strong position of local communities and their utilities (*Stadtwerke*). Green citizens' action groups have occupied a relatively strong position at local level.

But the mix of competences under the federal system sometimes leads to impasses and fragmentation. Centralised collection of and access to environmental data, for example, are not always easy (see Zieschank, 1992).

Participation rights were institutionalised early on in areas such as urban planning, nuclear energy and ambient air quality control. They have been actively used by citizens' action groups. After the unification of Germany some participation rights have been restricted (mainly in East Germany) by the above mentioned "streamlining laws", which were extended even further in 1996.

In the 1980s the integration of environmental interests into parliamentary institutions became an important issue. In general, the degree of interdepartmental integration is now relatively high. The integrative capacity of the economic system regarding environmental interests is also highly developed. Since 1993, "consensus talks" on energy policy have taken place with some integration of environmental interests. The Green Party was however excluded from the discussions. The policy networks in transport, construction, and agriculture in particular are still closed and homogeneous.

Informational capacity, i.e. the capacity to produce, distribute and utilise environmental knowledge has grown fast in the 1980s and is today remarkably large.

This is true for the research infrastructure of the environmental organisations. The first independent research institute dedicated to environmental research, the *ÖKO-Institut Freiburg*, was founded in 1977. Today the umbrella organisation for similar ecological research institutes in Germany, AGÖF, has about 80 member institutes (Hey and Brendle, 1994: 133). After the Chernobyl accident (1986) the *ÖKO-Institut* was asked for the first time by the federal Ministry for Economic Affairs for an expert opinion (in addition to the results of an analysis by an established research institute) regarding the phasing out of nuclear energy. At the time, this drew heavy criticism from within government, but today, such complementary reports from independent environmental research institutes have become relatively standard practice in Germany. Now even the established economic research institutes usually have an environmental department.

The universities were for a long time an unimportant if not negative factor as regards the environment (working mainly for the "other side", as it were). Today there are special curricula for environmental protection in all German universities. Large federal institutes for research into nuclear energy have now been to some extent "converted" to environmental research.

In addition, the base of environmental organisations in the mass media is strong and has grown fast up to now. The "green" national daily *Die Tageszeitung* was founded in 1979. Today it sells about 60,000 copies. Environmental magazines, which hardly existed in the 1970s, had by as early as 1989 a total circulation of about 4 million. As many as 16 such magazines had a circulation of more than 100,000 each (Vasata, 1991: 81) (This number does not include the many specialist magazines for environmental experts.).

In the 1970s, environmental organisations and citizen's action groups received some support in their public relations work from state-owned television and the liberal media, for example *Der Spiegel* or the *Frankfurter Rundschau*. But by the late 1980s even the conservative press and media outlets with close links to economic interests, long-time opponents of environmentalist thinking, had changed their attitude (Voß, 1995: 127). The mass media generally display a positive attitude towards environmental issues and demands. Newspapers and television have become more and more actively involved in pollution issues and, in several cases, they have exerted strong influence on the activities of industry and government.

The parliamentary position of the Green party provided an important informational infrastructure for the ecology movement. A similar effect was caused by parliamentary commissions of enquiry, where "green" research institutes such as the *Öko-Institut* after 1980 took on the role of expert witnesses and improved their reputation (and financial resources). The more public role of scientific "outsiders" and a steady increase in public funding for research led to a certain "greening" of the research community. In the late 1980s a growing demand for environmental knowledge could be seen within the economic system. As a result of rising environmental awareness, there was increasing environmental reporting on industry. Now it was enterprise which discovered its own potential for ecological modernisation. This has stimulated a boom in environmental consulting and further education in recent years.

7 Overcoming Restrictions?

Up to now, however, institutional capacity for environmental action in Germany has been deficient in many areas. A striking example is to be found in transport policy. Even the introduction of a speed limit on motorways has proved too demanding for the existing capacity, although there has been a strong tendency amongst politicians (and pressure from the EU) to take some action in this direction. A soil conservation law has met with strong resistance (a new government proposal has recently been published), as have restrictions on land use for development. The introduction of an energy tax has been repeatedly frustrated – although 67% of managers are in favour, provided wage costs are reduced at the same time (Wirtschaftswoche, 15 June, 1995: 21). The energy sector (in combination with the energy-intensive basic industries), the automotive and the construction industry still have a strong and restrictive influence on political institutions.

This raises the question of whether the institutions have sufficient capacity to realise the goal of a long-term stabilisation of the environmental situation (sustainable development). Capacity must be related to functional requirements. Moderate goals, such as the dilution of air and water pollution, could be achieved with a low capacity for environmental policy. In Germany this was even possible without public pressure and with a small, additive administration. The end-of-pipe approach was implemented against a background of strong public pressure, changes in the political constellation, the development of a highly advanced monitoring technology and the establishment of a special eco-industry in tandem with a specialised eco-bureaucracy.

But as mentioned above, the capacity of environmental policy has proved too small to implement a strategy for sustainable development, which includes such difficult tasks as the stabilisation of land use, soil and ground water conservation, or systematic reduction of material inputs into the production process – goals stressed in the latest reports of the German Council of Environmental Advisers (SRU, 1994, 1996) as well as in a recent private study on "Sustainable Germany" (BUND and MISEREOR, 1996) and broadly discussed in a special parliamentary commission (Enquete Kommission "Schutz des Menschen und der Umwelt", 1994).

For a truly "reductive" strategy regarding resource input in production, the coalition of ("additive") eco-bureaucracy and eco-industry cannot be seen as a driving force. It could even be seen as a restrictive factor in such a strategy. Here it is the capacities of ecological modernisation within the economic system and the role of the green business sector, with their effects on the "dirty" product chain, that have become more important.

A strategy of sustainable development requires not only a broad network of actors, with pioneer enterprises as frontrunners, acting under public and government pressure (this actor constellation was established in the fourth stage of development described above). The next necessary stage in capacity development should be a new strategic role for government, supported by institutionalised co-operative planning, open (pluralist) policy networks in areas such as transport, energy or construction policy, and a reformed tax system, changing the economic framework of resource and land use.

The German government, and in particular the Ministry of the Environment – at present under heavy pressure from the labour market and from traditional basic industries – is still far from establishing the green economic policy which might lead to such far-reaching goals. But things have gained their own dynamic, as can be seen from solid waste management, where heavy competition for recyclable is now in evidence. It may not therefore be too unrealistic to hope that – under additional heavy pressure from environmental organisations including the Green Party – Germany will remain an innovative player in the global arena, although the backlash after the 1994 general election cannot be ignored.

Annex: East Germany

The case of East Germany deserves a brief mention here, principally – and often paradoxically – as a contrast to West German environmental policy.

The first element of this paradox is the early institutionalisation of environmental policy. The socialist GDR was the first country in the world to introduce – in 1968 – an article on environmental protection into the constitution. As early as 1970, a comprehensive environmental protection law was passed – the second in Europe, after Sweden's. A Ministry for Environment and Water Management was also created in 1971 (15 years before the Federal Republic's!).

The second element of this paradox is the poor result of this early institutionalisation. Although a slight reduction of SO_2 emissions could be observed in the early 1970s, the heavy reliance on lignite since 1979 made the GDR the industrialised country with the highest per capita emission of that pollutant, thirty times higher than that of Japan (Mez et al., 1991). In other fields the environmental performance of the GDR was equally poor. Out of 16.6 million inhabitants, 9.6 million had to drink – periodically or continuously – poor quality water. The number of areas suspected to be contaminated (69,693) was extremely high (Umweltbundesamt, 1994: 210). The use of agricultural pesticides was roughly twice as intensive as in West Germany and other countries. The proportion of protected land was, however, relatively high.

The third element is that some of its demonstrable environmental achievements were due not to environmental policy, but to economic pressures:

1. Problems with Soviet oil supply after 1979 led to a radical (c. 25%) reduction in road transport.
2. Problems with agriculture led to a harshly restrictive strategy (including a charge for land use) on the transformation of agrarian into developed land. The effect of this was an actual stabilisation in the total area of developed land from 1970 onwards, in stark contrast to the rapid increase in West Germany (Wirtschaft und Statistik, 7/1991: 434).
3. Finally, shortages of raw material supplies in the former GDR stimulated remarkable activity in recycling. The "*SERO-System*" collected solid waste and paid for it; although the payment was small, it ensured that the system was supplied with well-sorted materials for re-use in industry. In a sense, the old GDR introduced a kind of "ecology of economic scarcity" (which ended in 1990).

Linked to the third, a fourth element of the paradox in the former GDR was the relatively early debate on both the economic and ecological advantages of a highly efficient material and energy use (Roos and Streibel, 1979). The economic results of this debate remained nonetheless unspectacular. The systemic restrictions on economic innovation also hindered this form of ecological modernisation.

The paradox may well be completed by the fact that, after the political upheavals in late 1989, East Germany not only underwent an extraordinarily rapid transformation – through the imposition of West German environmental law and a massive influx of public money – into a region with West European environmental standards (with the exception of the extreme soil pollution); its economic structure now tends to be more strongly environmentally oriented than the West

German. The improvements go beyond the effects of radical de-industrialisation: The general modernisation of industry, using the latest technology, has also led to some improvements in the environment. CO_2 emissions, for example, were about 50% lower in 1994 compared with 1987, and this reduction continued even against a background of rapid economic growth after 1992 (Umwelt, 5/1995: 185). The radical decline of the most dirty basic industries was connected with a strong upswing in the service sector and some of the cleaner branches of manufacturing industry; whether the formerly strong and ecologically very harmful basic chemicals industry will recover remains to be seen.

Radical de-industrialisation has, on the other hand, also created high unemployment, which has caused a rapid decline in environmental consciousness. Improving the very poor environmental situation in the former GDR is nevertheless also seen as an opportunity for creating new jobs.

As regards environmental capacity, the principal lesson to be learnt from the former GDR may well be that the mere existence of government institutions does not automatically generate sufficient capacity in environmental politics, unless it is also supported by a strong environmental movement, a critical media and an innovative economy.

References

BMU (Bundesministerium für Umwelt, Naturschutz und Reaktorsicherheit), 1987: Umweltpolitik. Bilanz des Umweltministers für Umwelt, Naturschutz und Reaktorsicherheit. Bonn: BMU.

BMU (Bundesministerium für Umwelt, Naturschutz und Reaktorsicherheit), 1992: Environmental Protection in Germany. Bonn: Academica Verlag.

Brun, R. (ed.), 1993: Der grüne Protest. Herausforderung durch die Umweltparteien. Frankfurt/Main: Fischer.

BUND and MISEREOR, 1996: Zukunftsfähiges Deutschland. Ein Beitrag zu einer global nachhaltigen Entwicklung. Basel: Birkhäuser.

Enquete-Kommission "Schutz des Menschen und der Umwelt" des Deutschen Bundestages (ed.), 1994: Die Industriegesellschaft gestalten. Perspektiven für einen nachhaltigen Umgang mit Stoff- und Materialströmen. Bonn: Economica.

Hartkopf, G. and E. Bohne, 1983: Umweltpolitik I: Grundlagen, Analysen und Perspektiven. Opladen: Westdeutscher Verlag.

Héritier, A., S. Mingers, C. Knill and M. Becka, 1994: Die Veränderung von Staatlichkeit in Europa. Opladen: Westdeutscher Verlag.

Hey, C. and U. Brendle, 1994: Umweltverbände und EG – Strategien, Polische Kulturen und Organisationsformen. Opladen: Westdeutscher Verlag.

Hucke, J., 1990: Die Entwicklung eines neuen Politikfeldes, in: K. v. Beyme and M.G. Schmidt (eds.), Politik in der Bundesrepublik Deutschland. Opladen: Westdeutscher Verlag, 382-398

ipos, 1994: Einstellungen zu Fragen des Umweltschutzes 1993, ipos-Schriften 876/875. Mannheim: Institut für praxisorientierte Sozialforschung (ipos).

Jarass, H. D., M. Kloepfer, P. Kunig, H.-J. Papier, F.-J. Peine, E. Rehbinder, J. Salzwedel, and E. Schmidt-Aßmann, 1994: Umweltgesetzbuch – Besonderer Teil. Berlin: Erich Schmidt.

Kloepfer, M., 1989: Umweltrecht. Munich: C.H. Beck.

Kloepfer, M., E. Rehbinder, E. Schmidt-Aßmann, and P. Kunig, 1990: Umweltgesetzbuch – Allgemeiner Teil. Berlin: Erich Schmidt.

Lautenbach, S., U. Steger, and P. Weihrauch, 1992: Freiwillige Kooperationslösungen im Umweltschutz. Ergebnisse eines Gutachtens und Workshops. Cologne: Bundesverband der Deutschen Industrie, e.V.

Leonhard, M., 1986: Umweltverbände. Opladen: Westdeutscher Verlag.

Malunat, B. M., 1994: Die Umweltpolitik der Bundesrepublik Deutschland, Aus Politik und Zeitgeschichte B 29, 15-28.

Mez, L., M. Jänicke, and J. Pöschk, 1991: Die Energiesituation in der vormaligen DDR. Berlin: edition sigma.

Müller, E., 1986: Innenwelt der Umweltpolitik. Sozial-liberale Umweltpolitik – (Ohn)Macht durch Organisation? Opladen: Westdeutscher Verlag.

Müller, E., 1989: Sozial-liberale Umweltpolitik. Von der Karriere eines neuen Politikbereichs, Aus Politik und Zeitgeschichte, B 47-48, 3-15.

OECD, 1993: Environmental Performance Reviews: Germany. Paris: OECD.

OECD, 1995: Environmental Taxes in OECD Countries. Paris: OECD.

Pehle, H., 1988: Das Bundesumweltministerium: Neue Chancen für den Umweltschutz? Verwaltungsarchiv 79, 2, 184-209.

Poguntke, T., 1992: Between Ideology and Empirical research: The Literature on the German Green Party, European Journal of Political Research, 21, 4, 337-356.

Poguntke, T., 1993: Alternative Politics. The German Green Party. Edinburgh: Edinburgh University Press.

Raschke, J., 1993: Die Grünen. Sie wurden, was sie sind. Cologne: Bund Verlag.

Rehbinder, E., 1992: Rethinking Environmental Policy, in: G. Smith, W. Paterson, and S. Padjet (eds.), Developments of German Politics. London: McMillan, 227-246.

Reich, M. R., 1984: Mobilizing for Environmental Policy in Italy and Japan, Comparative Politics, 64, 4, 379-401.

Roos, H. and G. Streibel (eds.), 1979: Umweltgestaltung und Ökonomie der Naturressourcen. Berlin: Verlag Die Wirtschaft.

Rose-Ackermann, S., 1995: Umweltrecht und -politik in den Vereinigten Staaten und der Bundesrepublik Deutschland. Baden-Baden: Nomos Verlagsgesellschaft

Roth, R. and D. Rucht (eds.), 1991: Neue soziale Bewegungen in der Bundesrepublik Deutschland. 2nd revised edition. Bonn: Bundeszentrale für politische Bildung.

Scharpf, F. W., 1991: Die Handlungsfähigkeit des Staates am Ende des 20. Jahrhunderts, Politische Vierteljahresschrift, Vol. 32, No 4, 621-34.

SRU (Rat von Sachverständigen für Umweltfragen), 1974: Umweltgutachten 1974. Stuttgart/Mainz: Kohlhammer.

SRU (Rat von Sachverständigen für Umweltfragen), 1978: Umweltgutachten 1978. Stuttgart/Mainz: Kohlhammer.

SRU (Rat von Sachverständigen für Umweltfragen), 1987: Umweltgutachten 1987. Stuttgart/Mainz: Kohlhammer.

SRU (Rat von Sachverständigen für Umweltfragen), 1994: Umweltgutachten 1994. Für eine dauerhaft-umweltgerechte Entwicklung. Stuttgart: Metzler-Poeschel.

SRU (Rat von Sachverständigen für Umweltfragen), 1996: Umweltgutachten 1996. Stuttgart: Metzler-Poeschel.

Umweltbundesamt, 1994: Daten zur Umwelt 1992/93. Berlin: Erich Schmidt Verlag.

Vasata, V., 1991: Medienverantwortlichkeit im Umweltschutz, in: OFW (ed.), Umweltmanagement im Spannungsfeld zwischen Ökologie und Ökonomie, Wiesbaden: Gabler, 73-96.

Voß, G., 1995: Umweltschutz in den Printmedien, in: G. de Haan (ed.): Umweltbewußtsein und Massenmedien. Berlin: Akademie Verlag, 123-129.

Weidner, H., 1995: 25 Years of Modern Environmental Policy in Germany. Treading a well-worn Path to the Top of the International Field, WZB-discussion papers FS II 95-301. Berlin: Wissenschaftszentrum Berlin.

Wey, K-G., 1982: Umweltpolitik in Deutschland. Kurze Geschichte des Umweltschutzes in Deutschland seit 1900. Opladen: Westdeutscher Verlag.

Wicke, L., 1989: Umweltökonomie. Eine praxisorientierte Einführung (2nd edition). Munich: Franz Vahlen.

Zieschank, R., 1992: Umweltbeobachtung und Umweltberichterstattung in der Bundesrepublik Deutschland, in: H. Weidner, R. Zieschank and P. Knoepfel (eds.), Umwelt-Information. Berlin: edition sigma.

DENMARK

Mikael Skou Andersen

1 Environmental Problem Pressure

With more than 500 islands, Denmark has a coastline of about 7,000 km and its environmental policy has, compared with other countries, focused to a considerable extent on the aquatic environment. The many bays and fjords have been vulnerable to municipal and industrial discharges, as well as to agricultural run-off. The establishment of a water quality policy was thus one of the most important results of the work of the Pollution Council in the late 1960s (Forureningsrådet, 1971), and it remains an important concern.

Denmark is only moderately affected by transfrontier pollution. Thanks to the lime layers settled underground, acidification of lakes or forests has more or less been neutralised, and in fact, Denmark is itself a net exporter of air pollution, due to its reliance on coal for its energy supply. With respect to water quality, the interior Danish sea waters (Belts and Kattegat) have been affected predominantly by point and non-point sources from Denmark itself. Water exchanges only slowly between the Baltic Sea and the interior Danish waters, and any influx of transfrontier marine pollution comes mainly from the North Sea, although scientific doubts about this remain. Run-off of nutrients from Danish agriculture accounts for two thirds of the total run-off from the nordic countries (Bernes, 1993).

In 1993 the independent Danish Economic Council published an account of 20 years of environmental policy (Det Økonomiske Råd, 1993). With regard to water, the balance showed that the content of nitrate in ground water had increased, while average oxygen content in Kattegat had deteriorated by about 25% and fish resources had decreased considerably relative to 1970. Only the number of beaches with unacceptable bathing water had been reduced substantially. With regard to air pollution, SO_2-emissions have declined by about 50% from 1972 to 1992, while NO_x emissions have increased by 50% and CO_2-emissions have been stable since the early 1970s. The Economic Council concluded, that "The environmental policy pursued in Denmark over the last 20 years has not resulted in substantial improvements of the marine environment or of ambient air quality. However, due to economic growth, and the derived increase of the use of resources, this does not imply that environmental policy has been useless". The Council stressed, however, the ineffectiveness of the predominantly regulatory approach, and recommended the use of market instruments.

The findings of the Economic Council confirmed other studies and reports published in the same period, which were critical of the Danish approach to environmental policy (DMU, 1993; Andersen, 1991). Still, success in environmental policy is a relative phenomenon, and although major improvements had not been achieved, it is worthy of note that CO_2-emissions have been stable over this 20 year period, despite increases in transport, production and energy use.

In the 1970s the environmental debate quickly turned into a national debate on energy policy and, in particular, the role of nuclear power. Without its own energy sources, Denmark was reliant exclusively on imported oil and coal when the energy crisis broke in 1973. Even so, nuclear power was not introduced, and a policy of energy saving, efficiency measures and supply diversification, including the use of wind and solar energy, was decided upon. The availability of oil and gas in the North Sea also affected policy options. As a result, energy requirement per capita has delinked itself from the OECD average since 1970, and is now about 25% below the OECD average. In the other Nordic countries, with similar GNP per capita, energy requirement is well above the OECD average (Jahn, 1995). The main reason for the Danish record is the efficiency of energy supply systems (combined power and heating production) as well as the insulation and efficiency measures in households. Danish industry, not subject to energy taxation before 1992, remains relatively inefficient as compared to other countries (Worrell, 1994). In households, energy use for room heating declined by 45% per cubic metre from 1972 to 1989 – and by about 25% in absolute terms (Energiministeriet, 1990).

2 Development and Main Characteristics

2.1 Institutions

The basic environmental institutions were established in the early 1970s. The first Minister of the Environment was appointed in 1971. The Ministry of Pollution Control was established in 1971, and is often claimed to be the first of its kind, although both France and the Netherlands established a Ministry in the same year. In 1973 the Ministry was renamed the Ministry of the Environment, and its operational wing, the National Agency for Environmental Protection (*Miljøstyrelsen*), was established in the same year. The first framework law for environmental protection was passed in 1973, and came into force in 1974. Only the introduction of environmental statistics gathering followed very late, in 1990.

The policy approach chosen by Denmark was in many ways exemplary. Fragmented regulations were replaced by a general Environmental Protection Act, which ensured an integrated approach. For instance, the licensing procedure for polluting industries has integrated air, water, waste, etc. since 1974. A separate Ministry of the Environment, with substantial resources, was institutionalised very early on. In addition, a separate Minister for the Environment was appointed from the very beginning, and his position within the government has been relatively significant. Recently, the Ministry has merged with the smaller Energy Ministry, to form an influential Ministry of Environment and Energy.

The problems relate to local administrative structures. In 1968 a reform of local authorities had been completed, which gave the municipalities greater powers and established 14 regional counties. Most of the officials in the Ministry of the Environment came from the Ministry of the Interior, and brought with them a way of thinking about the relationship between national and local authorities, which made it natural to entrust considerable powers to local authorities when drafting the Environmental Protection Act. The municipalities were cast as the main administrative unit responsible for the implementation of environmental policy. This choice was hardly discussed in the political arena (Andersen, 1989). While local authorities in many countries are responsible for implementation of environmental laws, the Danish municipalities are conspicuous for the extent of their autonomy. They raise their own income taxes, and nowhere in the EU do local authorities have such a large and direct share of taxation. The combination of environmental responsibility with exceptionally strong interests in local tax revenues has proven to be troublesome (cf. Andersen, 1994b).

2.2 Major Laws

After a regulatory reform in 1991, in which about 22 different environmental laws were merged together, there are now 4 main environmental laws: The Environmental Protection Act (the framework law), The Nature Conservation Law, The Planning Law and the Waste Deposit Law. While the Environmental Protection Law deals with the basic problems of air and water pollution and with waste management, the Nature Conservation Law regulates the conservation and preservation of sites and landscapes. In addition to these, the Planning law sets out the fundamental principles of physical planning and the Waste Deposit Law concerns soil pollution.

2.3 Policy Instruments

While the use of policy instruments was originally very much oriented towards traditional regulatory instruments, there has in the last decade been a gradual shift towards more anticipatory instruments. Several economic instruments have been introduced, although often with a mixed incentive and revenue purpose, for instance a CO_2-tax, a CFC-tax and a national waste tax. Voluntary agreements are also being used, but on a more modest level than in the Netherlands, as there is a great deal of scepticism in the Environmental Ministry towards this instrument. Recently a bill making "eco-assessments" compulsory has been proposed by the Minister of the Environment, to promote eco-auditing and life cycle analysis.

2.4 Policy Style

The Danish model emphasises public control and responsibility at local level. In the traditional fields of waste management and waste water treatment the municipalities bear sole responsibility, and privatisation or out-sourcing has not been common. Municipalities and counties are responsible for the implementation of environmental policies, and they have substantial discretionary powers in law. In particular they are responsible for the inspection of local industries and farmers, and for monitoring compliance with environmental regulations.

The roots of decentralisation are to be found in the constitutional conflict in the late 19th century between Copenhagen's ruling bourgeoisie and farmers, who were struggling for independence and local self-governance (Østergård, 1990). In the 1890s Copenhagen's ruling bourgeoisie put the parliament out of action through rule by decree, until the farmers seized power in both houses of parliament in 1901. Since then the profound scepticism among farmers with regard to central rule in Copenhagen has paved the way for a marked devolution of competences to local authorities in most public services (Mørch, 1982). The poet, priest and educator, N.F.S. Grundtvig, was a main inspiration to the farmers movement, and stands as the ideological founder of decentralisation (Thodberg and Thyssen, 1983). Another source of inspiration came from German town governments. At a time when the working class began to outnumber farmers in the 1930's, and the social democrats came into power, the preference for local self-governance had become institutionalised, and the tradition for delegation and consultation was continued.

3 The Main Actors

3.1 Environmental Policy Institutions

An interesting trait of Danish environmental policy is the country's lack of independent advisory institutions on general environmental policy issues, such as the German "Rat von Sachverständigen für Umweltfragen" or the British "Royal Commission on Environmental Quality". The Danish Pollution Council existed only from 1969-1971 to report on the state of the art in science, and thus prepare the ground for an environmental policy. In recent years the influential Economic Council has begun to report on the state of the environment and to publish policy recommendations in this field, but their work is mainly directed towards economic policies.

Still, Denmark has an important and unique environmental institution, the Environmental Board of Appeal, which also is known as the "high court" of the environment (Basse, 1987). The existence of this board reflects the Danish tradition for administrative appeal, rather than for the use of courts and civil actions. Decisions taken by environmental authorities, for instance regarding local environmental permits, may finally be appealed against before the Environmental Board of Appeal. The use of administrative appeal procedures incurs lower procedural costs, and the appealing party does not have to pay for a lawyer, since the Board must inform itself about the case. Plaintiffs may bring complaints if they have a direct and judicial interest in a case. Environmental organisations are now also recognised as plaintiffs. The Board of Appeal may overrule local decisions if binding standards or norms have been violated. They often help to define principles for the administration at lower levels of government, but if a decision has been taken within the limits of local discretionary competence, it is difficult for the Board to overrule it. Only issues of principal interest can be appealed against before the Board, and its rulings are final. Appeal cases must first be settled by the Environmental Protection Agency, but the Board of Appeal has often overruled decisions taken by the Environmental Protection Agency. In the early 1980s especially it served also to put the brakes on a somewhat zealous Environmental Protection Agency (Basse, 1987).

The Environmental Board of Appeal is institutionally a mixture of legal body and corporatist interest mediation body. Headed by a judge, the Board consists of experts nominated by interest organisations (industrial, agricultural and other) and by the Environmental Protection Agency. The environmental interest organisations are not represented on the Board and the establishment of the Appeal Board in 1974 was also a result of pressure from industry. In practice, the Board functions as a forum for consultation with interest organisations during the implementation process (Andersen, 1989).

3.2 Green Organisations

The most important environmental or green organisations, in terms of membership, activism and influence, are the Danish Society for the Conservation of Nature, Worldwide Fund for Nature (WFN), Greenpeace Denmark, NOAH (Danish section of Friends of the Earth) and The Anglers' Association, but a host of green organisations are active.

The Danish Society for the Conservation of Nature (*Danmarks Naturfredningsforening*) is the most important and influential green organisation. With a membership of about 250,000 it has become the largest environmental organisation in Denmark. It was established in 1911, and joined local citizens with concerned scientists in an influential alliance (Rehling, 1994). In 1937 its influence culminated when it was entrusted with the right to initiate and implement proposals for the preservation of sites and landscapes. This delegation of implementation responsibility to a private organisation was conspicious, but not unusual within the decentralised and corporatist Danish policy style as it had developed during the first half of the 20th century. For many years the Society for the Conservation of Nature remained a careful and venerable organisation, which came to regard itself almost as a public institution.

During the 1980s, the Society for the Conservation of Nature developed into a more activist environmental pressure group (Svold, 1989). The old generation was turned out, and a new more environmentalist director, David Rehling, was appointed. Probably the most controversial event in modern Danish environmental history was when Rehling, in a televised discussion with the Fishery Minister after eutrophication had resulted in dead fish and lobsters in the middle of Kattegat in 1986, pressed the government to accept the Nature Society's costly demands for water pollution control. The Nature Society demanded a 50% reduction in discharges and run-off, to be achieved within 3 years. Within 4 weeks these far-reaching targets, which substituted emissions standards for water quality standards, had been approved by the parliament and the government. Eventually, Rehling himself wrote the draft of the motion that was finally approved in parliament (ATV, 1990; Schrøder, 1990; Røjel, 1990; Dubgaard, 1991; Andersen and Hansen, 1991).

The Society for the Conservation of Nature has, since 1911, retained the two pillars of its membership: ordinary members and concerned scientists. While ordinary members are involved only in local committees, the national advisory committees that play a key role in the organisation's policy-making are filled by scientists. Despite its impressive membership figures the Society remains thus a somewhat elitist organisation, with relatively strong networks to the scientific community.

Membership figures and influence of other environmental organisations are somewhat less impressive, but all environmental organisations, have, as the Society for the Conservation of Nature, been granted the right to appeal against local environmental licenses and permits. WFN Denmark has 27.000 members and a similar number of regular donors. Greenpeace Denmark has about 18.000 members, but its membership figure has declined by more than 50% in the last 5 years. It has had difficulties with assuring broad public support after criticism was voiced against the organisation of its support for dubious whaling campaigns. Greenpeace is seen as less "serious" than the Society for the Conservation of Nature, but has actually been quite influential in lobbying the Parliament's Environment Committee and the Ministry of the Environment in matters regarding international conventions. In this way a division of labour has evolved, where the Society for the Conservation of Nature focus more on national issues and Greenpeace more on international issues.

NOAH, the grassroots organisation which was established after a famous event at the University of Copenhagen in 1968, has been weakened substantially over the last years. Membership figures are unknown, but are probably less than 1000. It refused to accept the right of appeal, claiming that it was a no more than a symbolic gesture, offering no real opportunities for environmentalists to influence pollution control. The Anglers' Association is small as well, but has been very influential in its own field, protection of water courses, thanks to continuous lobbying of national and local authorities.

3.3 The Media

Environmental issues have been an important news item for a surprisingly long period. The issue attention cycle has not experienced a downturn; on the contrary, recent years may even have witnessed an upturn in interest in environmental issues, in ecological farming and green products for instance. According to a recent survey by the Institute for Market Trend Analysis, 84% of Danish newspaper journalists regard environmental issues as being of high or very high political salience. Among journalists specialising in business and trade affairs, environmental issues rank as the fifth most important news item after "corruption", "dismissals", "management affairs" and "mergers" (Politiken, 26 November 1995). In the leading Danish business newspaper, *Børsen*, environmental news has become one of the top priorities, and it is difficult to follow the Danish debate on environmental policy without reading this newspaper. The media plays an important political role in bringing forward new scientific evidence and also in its frequent efforts to create bandwagons for environmental issues. The unstable parliamentary situation in Denmark provides part of the explanation for what makes possible conflicts on environmental issues interesting from the media's perspective. Situations where the government could be faced with defeat on environmental issues have been especially interesting for the media.

There has been little research into the role of the media in the political process, but research on the role of the media in the decision-making process leading up to the Plan for the Aquatic Environment revealed that the TV-reporters had seen their own role as that of environmental advocates (Andersen and Hansen, 1991). Following the dead lobsters in Kattegat, the national TV-station began to campaign on the deteriorating aquatic environment. For several weeks the news focused daily on marine pollution, and willingly broadcasted relatively one-sided statements from marine experts. Reporters perceived that the public wanted better environmental protection, but that they needed to "help" policy-makers take initiatives. Speaking relatively openly about their own role, one TV-reporter even claimed: "Rehling, he was invented by us" (Andersen and Hansen, 1991: 81).

3.4 Target Groups

The most important organisations representing target groups are the Federation of Danish Industries (*Dansk Industri*) and the Agricultural Council (*Landbrugsrådet*).

The Federation of Danish Industries is normally the most important partner for the environmental authorities, when new legislative proposals have to be negotiated. The federation is an all-embracing umbrella organisation, with a well-staffed environmental office. Both in 1973, when the framework act for environmental protection was passed, and in 1986, when the Plan for the Aquatic

Environment was negotiated, the Federation decided to go along with the proposed measures, although it also asked for and obtained some concessions. The Federation or some of its sub-associations are involved every time new administrative decrees are tabled, and it is quite influential in that part of the implementation stage where actual guidelines are drawn up. Basically, the Federation does not question the need for environmental legislation. Its policy seems to have been to go along and seek influence where it could not prevent unwelcome measures being taken, but it has recently become more vociferous on the issue of the CO_2-tax.

The Agricultural Council has historically been more adverse to the idea of pollution control. The Agricultural Council is an umbrella organisation for three independent organisations of farmers; *Erhvervsjordbruget* (the large landholders), *Familielandbruget* (the small landholders) and *Landboforeningerne* (the alliance of local farmers' associations). In 1973 the Council required economic compensation for pollution control requirements, and although the Pollution Council had indicated the problems of large livestock farms and manure disposal, the new framework law did not eventually apply to "Greenland, the Faroe islands and agriculture". The Agricultural Council resisted environmental regulations vigorously for about 10 years, and as late as 1985 its Chairman declared that the very idea that agriculture polluted stemmed from "a Marxist mindset". Today, agriculture is subject to probably the most detailed environmental regulations of all trades and industries. To avoid the fertiliser tax advocated by environmentalists, the Council has gradually had to accept a more and more finely-meshed net of command-and-control regulations. Its most recent policy seems to be to make the best of things by advocating new subsidies for environmentally friendly farming. In particular the Small Landholders' Organisation is more receptive to the idea of restrictions on agricultural practices, and the Council has not always been able to act unanimously. Once one of the strongest and most powerful organisations in Denmark, environmental issues have been part of the reason for its gradual decline. The organisations seem still to possess a kind of veto right on crucial issues, although a pesticide tax was recently passed despite their objections.

4 Capacity-Building

4.1 The Green Majority in Parliament

To understand the development of environmental policy in Denmark, it is necessary to consider the particular parliamentary situation that prevailed in most of the 1980s, and which allowed the Environment Committee (*Miljø- og planlægningsudvalget*) of the Danish Parliament (*Folketing*) to play an important, but unpredictable, proponent role.

Denmark has a parliamentary tradition of minority governments, i.e. governments which do not control a majority of seats in parliament (Damgaard, 1992).

When Poul Schlüter's conservative government, consisting of four centre and right parties, seized power in 1982 it was, like its predecessors, a minority government. Its economic and financial policies were based on a close alliance with the small Social Liberal Party (*De Radikale*). However, in environmental questions the Social Liberals had difficulties in supporting the government, and the Socialdemocrats and two left parties were able to form the so-called *green majority*. Poul Schlüter's main concern was economic policy, as well as holding on to power after several decades of social democratic rule. His government had to accept that in environmental, and even in security and some foreign policy matters, alternative majorities would often decide policies.[1] In this way he introduced a completely new constitutional practice, which was for a government to accept being outvoted without resigning.

In most of the period from 1982-1993 the *green majority* was relatively influential. The Minister of the Environment was often questioned and required to present plans to the Environment Committee, in which the leftwing Socialist Peoples Party in particular played an active role in promoting new policies and initiatives. The Environment Committee did not work out an all embracing strategy; its members brought smaller and larger issues to the Committee on an ad-hoc basis, and there was little systematic use of the power which the Committee possessed. The Energy Committee was also active.

From 1982-1988 the Minister of the Environment was from the small Christian Peoples Party. The Christian Peoples Party was actually eager to present itself as a green party, alert to the interests of future generations, and as their Minister, Christian Christensen, held his party's the only portfolio in the fragile minority government, it was difficult for his coalition partners to turn his proposals and ideas down. In fact he willingly accepted many of the ideas and proposals coming from the Environment Committee, such as the demand for a veto on the EC small car exhaust directive in 1985.

Many of the important environmental policy initiatives, such as the Plan for the Aquatic Environment (1987) were drawn up during his period as Minister – despite opposition from within government, in particular from the Minister for Agriculture. While the green majority had set the reduction targets for agricultural run-off, the government succeeded in convincing the Social Liberals to support its choice of policy instruments, excluding a controversial fertiliser tax. Without too much vision himself, Christensen was accused of being a victim of zealous officials in his Ministry. Nevertheless, since 1988 Danish environmental policy has by and large revolved around how to implement the ambitious plans and decisions made during his period as minister.

In 1988 Poul Schlüter persuaded the Social Liberals to join his new minority coalition government, and Lone Dybkjær came to hold the Environment portfolio. As one of the architects of environmental policy in the Environment Committee her effort was to bring some order to the many new but diffuse regu-

1 In security policy the alternative majority gave birth to Denmark's policy of NATO "footnotes" in the cruise missile issue.

lations, and to construct a compromise between the governments' general emphasis on deregulation and the environmentalist position. This proved difficult and, for different reasons, the Social Liberals quickly left the government again. The green majority was re-established, but the years from 1990-1993 were calmer, reflecting the fact that issues other than the environment dominated the political agenda. Still, the green majority managed to introduce a CO_2 tax against the will of the government. In 1993 the social democrats returned to power.

Because of this particular situation in parliament, the existence of the green majority, environmental issues often came high on the political agenda. The members of the Environment Committee were particularly eager to demonstrate that they would impose more ambitious targets or deadlines than would the government. The industrial and economic interest organisations began to hesitate in their negotiations with the government, because they no longer knew what would happen when parliament began to intervene. In order to reach a compromise, they would usually have to make concessions to the government, but since the government did not hold a majority, they did not know whether they would be asked to make new concessions to parliament. In the 1980s the corporatist system lost its hold on policy-making, and not only within environmental policy. In terms of the fable of the hare and the tortoise, Denmark might be said to have acted more as the hare in environmental policy, bounding along at impressive speed, with rather ambitious policy goals. The political process was at times more similar to the competitive environmental policy-making of the United States in the early 1970s, than to the traditional pattern of Scandinavian corporatist structures (cf. Lundqvist, 1980).

A Green party was founded in 1983, but it never managed to gain the 2% of the votes necessary to get candidates elected to the *Folketing*. The Green party achieved its best results in the local elections in 1985. In the most recent elections in 1994 the Green party did not even put up candidates. The reason for the lack of success for the Greens is bound up with the "greening" of the traditional parties and with the historical development of the left. Other parties, notably the Social Liberals and the Peoples Socialist Party, already had a well-developed environmentalist profile, and the new Green party appeared more or less dilettante in comparison. But even the centre and right parties chose to go along with the greening of the voters in the 1980s, and rather than face defeat, they often tried to keep pace with the green majority and introduce their own, environmentally friendly proposals. There is no doubt that electoral competition helped environmental policy gain an extraordinarily strong momentum in Denmark, and made a green party to some extent superfluous.

The Social Democratic Party, which with about 35% of votes remains the largest political party in Denmark, has gradually sharpened up its environmental policy. In the first years of the green majority it was quite evident that the social democrats often went along with more radical proposals than they would otherwise have done in order to annoy the conservative government. While interests from trade unions, regional electricity utilities and transport lobbies now and then make their influence clear, the party has become increasingly enthusiastic about environmental policy.

Since 1993, when the Social Democrats returned to power (in coalition with Social Liberals, Centre Democrats and Christian Peoples Party), the former green majority in a sense gained control of the environmental policy institutions. While the conservative-led government had been reluctant to assure implementation of the often ambitious policy goals, and did little to clarify the technical problems, the new government was eager to revitalise most of the plans and initiatives previously approved by parliament. The new Environment Minister, Svend Auken, declared that "green issues would be the thread running through the government's policy", and the government introduced a number of important environmental policy initiatives, such as a CO_2-tax on industrial emissions and a plan for job creation in the environmental sector.

4.2 Openness of Political Input Structures

Despite the prioritisation of many environmental issues, the basic corporatist arrangements for environmental policy-making do remain in place. The standard operating procedure for new proposals remains to try and iron out differences in preparatory committees, where organised interests are consulted. Still, in comparison, the Danish political system appears to have fostered a relatively open attitude to policy ideas coming from outside the established circles. Apart from the existence of the green majority, there are a number of factors which have led to a relatively high degree of openness.

First, the rights of access to information mean that local citizens can ask for access to local environmental authority files. In the early 1980s, when the environmental debate slowly began to re-emerge onto the scene, two biologists – Larsen and Christensen – used the relatively good opportunities in Denmark for gaining access to public files, and began to monitor whether different industries were complying with the conditions of their discharge permits, and to what extent local authorities were enforcing their own requirements. For most of the larger and smaller companies they investigated Larsen and Christensen found serious breaches of the terms of environmental permits. They attracted a great deal of public attention by reporting all breaches to the police, who were inexperienced in dealing with pollution crime (Larsen and Christensen, 1985). Subsequently, their campaign was influential in making parliament improve the mechanisms available in the environmental framework law for monitoring and assuring compliance.

Secondly, the technical and scientific character of environmental policy mean that the research agencies of the Environmental Ministry are influential in bringing new scientific evidence into the debate. In important agricultural matters especially, their findings have sometimes been challenged by research agencies from the Agricultural Ministry, and some of the political debate has been short-circuited by public dispute between environmental and agricultural experts – often on complicated issues, such as the turnover of nitrogen in different layers of marine waters. The government initiated an international peer review of all environmental research institutions, it removed them at the same time from the National Agency of Environmental Protection and established a relatively independent National Agency of Environmental Research, and it finally pooled 300 million krones in a national environmental research programme to promote cooperation between the different sectoral research institutes. As a result, the scientific community has had better opportunities for developing a common understanding, and public dispute between experts has become a rarer phenomenon. While the intention may have been to limit the access of scientific expertise to the debate, the institutionalisation of research communities has been significant in giving more weight to their statements.

Thirdly, an effort has been made to consult more with the environmental organisations during the policy-making process and to increase funding for grassroots initiatives. Although the environmental organisations are not represented in the Environmental Board of Appeal, their rights of appeal were enlarged and formalised in the 1981 and 1991 amendments to the framework law. They are also more systematically included in committees preparing legislation etc., but they remain substantially less influential than economic interests. The Ministry of the Environment has also established a Green Fund to support various kinds of local citizen and grassroots initiatives, including environmental lobbying in Brussels.

Fourthly, in specific sectoral issues, independent advisory institutes have been set up, often on the initiative of the former green majority, in order to provide policy-makers in parliament with an independent source of information and investigation. The Council on Renewable Energy Sources has been established to provide recommendations on the situation of renewable fuels, while the Transport Council has been established to carry out research on more sustainable transport policies. In the later case in particular, the establishment of the Council was seen as a necessary move against an unwilling Transport Ministry.

4.3 Policy Integration

When the World Commission released its report in 1987 it acquired a significant role in the Danish debate on environmental policy (Verdenskommissionen, 1988). It provided the starting signal for attempts to integrate environmental policies into other sectoral policies. However, the first step towards integration underlined the ambiguity of the process.

The Ministry of Agriculture, rather than the Ministry of the Environment, was given responsibility for implementation of the Plan for the Aquatic Environment. The Ministry of Agriculture took this opportunity to water down many of the technical details in the plan (cf. Andersen, 1994a), and it began to flesh out its own plan, the Plan for Sustainable Agriculture. Despite this difficult birth for the integrative approach, it seems as if the transfer of implementation responsibility has been important in changing the way of thinking in the Agricultural Ministry. Changes in the CAP also, the increased competition on food products after the opening of Eastern Europe, and consumers' demand for ecological products have caused the Agricultural Ministry to distance itself from the agricultural organisations who complain about these developments (Daugbjerg, 1995).

In a follow-up to the Brundtland report, the Energy Ministry and the Transport Ministry were required to draw up plans for sustainable development and CO_2-reductions. While the Energy Ministry presented a rather impressive plan, Energy-2000, with detailed prescriptions for a 20% CO_2-reduction, the Transport Ministry aimed only at stabilisation and provided few details. Energy-2000 builds on the tradition of national energy plans, but while previous plans mainly concerned supply diversification and supply security, Energy-2000 is an ambitious attempt to integrate environmental targets with energy planning. The aim is to reduce CO_2-emissions by 20% and reduce energy consumption by 15% by the year 2005 (relative to 1988 levels). Energy-2000 is a catalogue of policy initiatives which will make these targets attainable, and an important instrument is a CO_2-tax as well as reform of existing energy taxation, making these reflect more accurately the actual energy use and environmental burden. Other measures are improved standards for appliances, extension of wind energy and other renewable resources, as well as a substitution for coal of natural gas, and improving efficiency in the remaining coal-based power plants.

The Transport Ministry has been more concerned with its traditional task of expanding the motorway network and with the new fixed links across the Great Belt and Øresund. In the latter cases it has done its best to pursue secretive decision-making procedures, and to play down the importance of environmental impact assessments. Only after the change of government did it present an update of its 1988 Transport Plan, Transport 2005.

Denmark has many sectoral "action plans" rather than an all-embracing National Environmental Policy Plan (the Dutch model). These action plans are often quite explicit with regard to the measures that have to be taken, and are as such more immediately operative than general plans that have to through a subsequent operationalisation phase. An attempt to establish a more comprehensive environmental planning model was made with Denmark's Nature and Environment Policy Report released in 1995 (Miljøministeriet, 1995). The report itself is a parallel to the annual reports of the key economic ministries. Observers raised the question of whether the Environment Ministry was now an intersectoral "Super-ministry" of the same rank. There is no doubt that the Environment Ministry in Denmark is a relatively important Ministry, but a closer look at the Nature and Environment Report reveals that most of the sector-specific policies remain in the hands of the sectoral Ministries. The most innovative step in the report was the introduction of the concept of *environmental utilisation space* in the Danish policy. This concept refers to the amount of resources that can be consumed at a sustainable level per capita, and its implications are far-reaching (Opschoor and Weterings, 1994). At a more political level, its significance lies in its focusing on the consumption patterns of the individual, and it may thus signal a new post-Brundtland phase in environmental policy-making, where the conflictual relationship between environmental and economic growth will come under scrutiny.

4.4 Policy Paradigm

The Brundtland report's emphasis on the need for cleaner technologies and the recommendations for the use of economic instruments has come to play an important role in Denmark. Simultaneously with the release of the Brundtland report in 1987, the environmental protection agency embarked on a new policy based on cleaner technologies. The aim was, by means of subsidised pilot projects, to develop specific cleaner technologies for selected industrial processes, and by way of example to show how pollution control at its source could be compatible with economically profitable solutions for the individual business (Miljøstyrelsen, 1986). In contrast to the former and more legalistic going-by-the-book strategy, the cleaner technology programme put officials of the environmental agency in a practical dialogue with industries about the development of cleaner technologies.

In 1988 the Ministry of the Environment published an influential discussion paper, entitled "simple and effective", which indirectly suggested a political trade-off between environmental policy and deregulation, emphasising the combination of economic effectiveness, regulatory reform and improved environmental protection (Miljøministeriet, 1988). Its practical focus was on cleaner technologies and the use of economic instruments, as means of simplifying and deregulating environmental policy, while at the same time improving performance. The discussion paper prepared the reform of the environmental laws (cf. above) and in the revised Environmental Protection Act, the use and development of cleaner technologies, as opposed to end-of-pipe measures, was accorded a key role. Cleaner technologies will in future form the basis of the permit procedure (Miljøministeriet, 1989).

In 1987-1988 a number of economic instruments were introduced, such as a national waste tax, a CFC-tax and packaging taxes. They provided revenue for increasing the budget of the Ministry of the Environment, during a period of restricted public spending, the waste tax financing more than half of the Ministry's budget. Previous reservations about the use of economic instruments were abandoned, and in the following years there was more and more emphasis on the role of economic instruments in a more preventive environmental policy. In 1993 a tax reform included a substantial package of green taxes, shifting the burden of taxation from income to other sources of revenue, including pollution taxation. Income taxes were reduced, and green taxes provided about 25% of the top-up revenue (Andersen, 1994c).

While there was initially a large measure of consensus on the use of economic instruments for environmental policy, the change of government in January 1993 slowly introduced more serious political cleavages in the debate on economic instruments. While the conservatives and liberals had been in government they had prepared and made several policy recommendations for the use of green taxes, but finding themselves in opposition they gradually became more hostile to the idea of green taxes. In fact green taxes and energy taxes are important for the state budget, providing about 8-10% of the total revenue (Andersen, 1994c).

The principle of integrating economy with ecology gained support in the early 1990s. Due to the key positions occupied by professional economists in most ministries and interest organisations this was important for the consensus that emerged. The Ministry of Taxation as well as the Ministry of Finance released comprehensive reports on economic instruments and the integration of environmental concerns into economic policies (Skatteministeriet, 1992; Finansministeriet, 1994). The difficult employment situation also provided a stimulus for rethinking environmental policy in terms of the know-how, exports and jobs that the environmental sector could produce. A report published by the Ministry of Trade and Industry estimated that the energy and environment sectors account for 70,000 jobs and 22 billion krones of annual exports (Erhvervsfremmestyrelsen, 1994).

All in all, the conventional view of a conflict between environmental policy and economic policies has been given the lie. Not least the wind mill experience (the export of windmills now amounts to more than two billion Danish krones annually) has shown how being at the forefront of a green technology may generate unique expertise and subsequent export revenue. Perhaps this new line of thinking was most clearly expressed by Environment Minister, Svend Auken, when he spoke about Denmark as a laboratory environment for a new more environmentally friendly lifestyle:

> "Denmark can be made an experimentarium for a new life style, which combines modern industrial production and welfare with ecological responsibility. It requires a high priority for research and technological innovation, while new models of governance and organisation, which pass on and develop the cooperative traditions as well as the traditions of public participation and discussion, can perhaps be developed" (Miljøministeriet, 1995).

4.5 Main Limitations

In the fable of the hare and the tortoise, the main limitation of the hare's approach is the question of whether it will tire before it reaches the finishing line. The crucial question in Danish environmental policy is whether implementation by local authorities will succeed. According to conventional implementation theory, decision-making processes that are conflictual may spill over into conflictual implementation processes. The policy targets have proven difficult to achieve, also because they have often been set without due consideration of the technical options available.

As mentioned above, local authorities play a key role in implementation, and in Denmark they are formally more independent than local authorities in most other EU member states. In most fields of pollution control there are serious problems with implementation at the local level. While the larger municipalities manage fairly well, many medium-sized and smaller municipalities (with less than 10,000 inhabitants) lack the resources and political will to enforce the requirements made by national legislation. Danish municipalities have on average three people employed to take care of the environment; in about 50 municipalities, however there is less than one person employed (Miljøstyrelsen, 1995). These people are responsible for local environmental administration, including compliance control. Because of insufficient resources there are normally 500 firms for each environmental inspector, so several years may pass between each control. Local authorities were recently allowed to charge an administrative fee for environmental inspection, but the increased revenue has not resulted in more staff. One municipality was reported to earn 11 times as much from the new fee system as it spent on compliance control (Rohmann, 1995: 24). Evidence of shortfalls has led the government to announce new measures for improving compliance at local level, but given Denmark's historical preference for decentralisation, it seems unlikely that the municipalities will be required to hand over responsibility to the state level.

5 Policy Evaluation

To conclude this case study on capacity-building it seems appropriate to reflect on the role of the green majority in Danish environmental policy-making. Why did such a coalition emerge in Denmark, but not in other Nordic countries? How environmentalist were the policies of the green majority, and how decisive were they?

There are three basic factors which explain how such a coalition could appear. In principle, the absence of majority governments and the historically weaker position of the social democratic party (which in contrast to Sweden and Norway never gained a majority of seats in the parliament) is explained by the complementary greater significance of rural interests in Denmark. Thus, the absence of a social democratic hegemony and the presence of more libertarian preferences in all political parties created a comparatively more open political culture. However, openness was promoted not by means of consensus-seeking, but thanks to the constant tensions and conflicts between different parties and interests. Secondly, in the 1970s a popular movement against the use of nuclear energy and for the use of renewable energies had much success, and showed the political parties that the electorate was becoming greener. Thus, several of the parties began to develop their environmental policies quite early on, and a certain electoral competition in greenness began. Finally, Schlüter's quiet change of constitutional practices and his relative ignorance of environmental policy issues provided the short-term explanation for the appearance of the green majority.

The existence of the green majority by no means implied that environmentalists were in control of the policy-making process, nor that Denmark's environmental policy was always to the fore, when compared to that of its neighbours. Environmental demands had to be balanced with their implications for the economy, or for specific affected sectors. Compromises with the government were often preferable and achievable. Danish environmentalists claim dissatisfaction with Danish policies, and international comparisons do reveal that Denmark's environmental policy is stronger in some fields than in others. However, since Denmark's environmental problems are largely domestic, it has not been possible to "externalise" the solutions to the same extent as has been seen in Sweden or – in particular – Norway.

References

Andersen, M. S., 1989: Miljøbeskyttelse – et implementeringsproblem, Politica, 21:3, 312-328.

Andersen, M. S., 1991: Styring med miljøafgifter, Samfundsøkonomen nr. 2, 5-13.

Andersen, M. S., 1994a: EFs landbrugsreform og miljøet – et harmoniproblem, in: J. Deigård (red.), Den internationale udvikling og dansk landbrug. Rapport nr. 78. Statens Jordbrugsøkonomiske Institut, København, 89-119.

Andersen, M. S., 1994b: Governance by green taxes: making pollution prevention pay. Manchester, New York: Manchester University Press.

Andersen, M. S., 1994c: The green tax reform in Denmark: Shifting the focus of tax liability, Journal of Environmental Liability, 2:2, 47-50.

Andersen, M. S. and M. W. Hansen, 1991: Vandmiljøplanen: Fra forhandling til symbol. Harlev J.: Niche.

ATV (Akademiet for de Tekniske Videnskaber), 1990: Vandmiljøplanens tilblivelse og iværksættelse. Vedbæk: ATV.

Basse, E. M., 1987: Miljøankenævnet. København: GAD.

Bernes, C., 1993: The Nordic Environment: Present State, Trends and Threats, Nord 1993:12. Copenhagen: Nordic Council of Ministers.

Damgaard, E. (ed.), 1992: Parliamentary change in the nordic countries. Oslo: Scandinavian University Press.

Daugbjerg, Carsten, 1995: Policy Networks and Policy Choices in Changing Environments: Environmental Policy in Swedish and Danish Agriculture, Paper Presented to the Political Studies Association Annual Conference, York University, 18-20 April. Aarhus: Department of Political Science.

Det Økonomiske Råd, 1993: Dansk Økonomi. København.

DMU (Danmarks Miljøundersøgelser), 1993: Miljø og samfund – en status over udviklingen i miljøtilstanden i Danmark. Faglig rapport nr. 93. København: DMU.

Dubgaard, A., 1991: The Danish Nitrate Policy in the 1980s. Rapport nr. 59. København: Statens Jordbrugsøkonomiske Institut.

Energiministeriet, 1990: Redegørelse om energiafgifter. København.

Erhvervsfremmestyrelsen, 1994: Miljø/Energi – en erhvervsøkonomiske analyse. København.

Finansministeriet, 1994: Betænkning om grønne afgifter og erhvervene. København.

Forureningsrådet, 1971: Vand: En redegørelse fra målsætningsudvalget og hovedvandudvalget. København: Statens Trykningskontor.

Jahn, D., 1995: The social paradigms of environmental performance: The Scandinavian countries in an international perspective, paper prepared for the workshop on New Nordic Memberstates and the impact on EC environmental policy, Sandbjerg Manor, April 6-8.
Larsen, V. J. and P. Christensen, 1985: Håndbog for kratluskere: hvordan man afslører en miljøforbryder. Harlev J.: Niche.
Lundqvist, L. J., 1980: The Hare and the Tortoise: Clean Air Policies in the United States and Sweden. Ann Arbor: University of Michigan Press.
Miljøministeriet, 1988: Enkelt og Effektivt. København.
Miljøministeriet, 1989: Renere teknologi – Konferencegrundlag. København.
Miljøministeriet, 1995: Natur- og Miljøpolitisk redegørelse. København. (English version: Denmarks Nature and Environment Policy).
Miljøstyrelsen, 1986: Udviklingsprogram for renere teknologi. København: Miljøstyrelsen.
Miljøstyrelsen, 1995: Miljøtilsyn 1993. Orientering fra Miljøstyrelsen nr. 3. København.
Mørch, S., 1982: Den ny Danmarkshistorie. København.: Gyldendal.
Opschoor, H. and R. Weterings, 1994: Environmental utilisation space: an introduction, Netherlands Journal of Environmental Sciences, Vol. 9, 198-205.
Østergård, U., 1990: Peasants and Danes: Danish National Identity and Political Culture. Working paper no. 75. Aarhus: Centre for Cultural Research.
Rehling, D., 1994: Indlæg ved Nordisk Parkkongres. Vejle 25. August, unpublished manuscript.
Rohmann, A. F., 1995: Miljø til discountpris, Teknologidebat no.3, 24-26.
Røjel, J., 1990: Fra Anarki til Hysteri. København: Samleren.
Schrøder, H., 1990: Et økologisk råd. København: Rhodos.
Skatteministeriet, 1992: Skattepolitisk redegørelse. København.
Svold, C., 1989: Danmarks Naturfredningsforening – fra pæn forening til agressiv miljøorganisation. Århus: PLS.
Thodberg, C. T. and A. P. Thyssen (eds.), 1983: N.F.S. Grundtvig. Tradition and Renewal. Copenhagen: The Danish Institute.
Verdenskommissionen, 1988: Vores fælles fremtid. København: FN-forbundet og Mellemfolkeligt Samvirke.
Worrell, E., 1994: Potentials for improved use of industrial energy and materials. Utrecht: Universiteit Utrecht.

SWITZERLAND

Peter Knoepfel

1 The Main Environmental Problems and Improvements since 1970

Significant differences can be observed in the degree of success with which the eight problem areas of classical environmental policy have been dealt over the past 25 years. Least progress has been achieved in the areas of noise, nature and landscape, whereas the most significant improvements can be observed in the areas of waste and the prevention of major incidents. The levels of success attained in the other four environmental policy problem areas (water, air, soil and substances) lie between these two extremes. The situation in the individual areas can be summarised as follows[1]:

- With over 95% of households and industries connected to central sewage treatment plants, Switzerland should be at the top in the European and, indeed, world-wide league table for *water protection*. However, approximately 30% of these treatment plants are currently in need of upgrading; the fourth treatment stage which is required for phosphate precipitation has only been implemented in 25 of approximately 1,000 plants. A significant improvement has been observed in the quality of surface waters, in particular with respect to the problems of organic pollution, over recent years. However, trends show that new pollutants are on the increase. Moreover, the ground water which provides 80% of Switzerland's drinking water is under threat from increasing levels of nitrates, chlorides and chlorinated hydrocarbons. Definitive ground water protection zones have been established at planning level for only 50% of the ground water sources. The level of pressure exerted by the general public on the authorities in the area of water protection policy can still be described as high. Systematic efforts for the revitalisation of streams and rivers and the redevelopment of dried-up river beds (behind reservoirs in mountain areas) have only recently been initiated.

1 Cf. summarises: Knoepfel 1994; Knoepfel and Wälti (1996) (UNIVOX survey results).

- Whilst significant improvements have been achieved over the past 25 years in *air quality* with respect to SO_2 and heavy metal contamination, No_x and ozone limit values continue to be massively exceeded in agglomeration areas. Whereas levels appear to be sinking in the case of No_x, ozone levels appear to be stable. "There is still significant room for action in the area of acid and nitrogen pollution in sensitive ecosystems" (BUWAL, 1993: 237). The upgrading of existing plants from the point of view of air pollution and the implementation of the more stringent limit values for emissions in industry, which were defined in 1992, has for the most part been completed. Gaps still remain, however, in the small business sector. As opposed to this, despite the establishment of clean-air measurement plans in all cantons for areas with high levels of air pollution, little success has been registered in the area of traffic. A considerable amount of pressure continues to be exerted by the public on the authorities to do more to improve air quality.
- Significant successes have been recorded in *waste policy* over the past five years: the rise in the levels of recycling (paper, aluminium cans, PET bottles, tin cans), the reduction in packaging volumes, the reduction of mercury content in batteries and the increase in composting all demonstrate the remarkable progress which has been made in this area. Most waste incineration plants have been cleaned up in the interest of preventing air pollution. Sufficient disposal capacity has been created, and excess capacity exists in some regions. The overall situation with regard to waste policy can be defined as very positive.
- In contrast, *noise protection policy* continues to be defined as inadequate. 30% of the population is exposed to excessive noise from traffic. Noise levels exceeding 70 decibels have been recorded on almost 1,000 km of local roads. 4% of the population is exposed to excessive noise levels from railway traffic. The level of noise pollution has, however, remained virtually stable over the past three years. Development measures are lacking above all on cantonal and local authority roads. According to BUWAL, very few traffic restrictions or speed reductions have been imposed. The federal authorities are also extremely lazy (BUWAL, 1993: 254) in their efforts to establish measures for the reduction of noise pollution caused by railway systems and air traffic. It would appear, however, that the general public has become accustomed to the present situation. Demands that the authorities be seen to make greater efforts in the area of noise protection are few and far between.
- Despite the relatively high degree of pressure from the general public, success in the area of *soil protection* has been limited. Soil monitoring, clean-up of contaminated sites and soil protection measures in agriculture are seen as particularly inadequate. The annual loss of arable land remains high at 24 km^2. Indicative limit values have been exceeded at 17 of 102 monitoring sites (above all for copper, cadmium and lead) and increasing levels of soil compacting and erosion have also been recorded. The clean-up of contaminated sites (approximately 10,000 suspected contaminated sites) is not on the agenda in Switzerland until "the onset of a new environmental discipline"

(BUWAL, 1993: 155). The amount of land allocated to housing and roads and hence the "sealing" of land which prevents drainage is still on the increase. Local authority regional planning has not yet succeeded in reversing the trend here.

- The degree of success attained in the area of *chemicals policy* can be defined as average to high. Although it is impossible for the authorities to exercise complete control over small business activity, thanks to the adoption of increasingly comprehensive eco-evaluations and improved eco-toxicological testing procedures, there has been a significant improvement in the levels of self-monitoring carried out by companies. A total of 380 new substances were registered with the corresponding tests with BUWAL from 1986 to mid 1993.
- Significant success has been registered for policy on the *prevention of major accidents*, the implementation of which has been stepped up as a result of the chemicals spillage in Basle in November 1987. The cantons and sectors of industry have developed various aids for the implementation of policy in this area and a risk register is also being compiled. Examples of the success in the area include the reduction in potential danger from chemicals and a new safety philosophy among companies. Deficits still probably exist, however, in the small business sector.
- The very limited success in the area of *nature and landscape protection policy* is regrettable, particularly in view of the high levels of expenditure and effort which have been invested in terms of both costs and manpower. The extension of housing zones and public infrastructure building, the disappearance of streams, the reduction in fruit growing areas and the annual increase in the road and transport networks (800 km per year) speak volumes. The number of animal and plant species which are endangered or already obsolete is frightening. "Despite considerable efforts, the condition of nature and landscape has deteriorated further" (BUWAL, 1993: 277). However, "in some individual areas (...) the negative changes" observed in the 1980s as compared with the 1970s have "diminished somewhat". "The overall level of intervention is however still extremely high. The state with its so-called investments is itself contributing to the destruction of the landscape" (BUWAL, 1993: 276).

A comparison of the efforts involved and outcome of administrative activities in the eight areas which constitute Swiss environmental policy gives the results shown in *Table 1*. In this table, one of the four extremities is clearly occupied, two others are at least partly occupied by corresponding findings.

- *Low effort/low results*. Everyone who is concerned with the implementation of environmental policy in Switzerland knows that, for whatever reason, noise and soil protection are neglected with respect to the allocation of personnel in most cantonal authorities. It is hardly surprising, therefore, that measures in these two areas have shown a very low degree of success despite the fact that in the case of one of the areas (noise), a thoroughly practicable regulation has

Table 1. Level of Effort and Results Obtained in the Eight Areas of Swiss Environmental Policy.

<table>
<tr><th>effort \ results</th><th>high</th><th>average</th><th>low</th></tr>
<tr><td>high</td><td></td><td colspan="2">water</td></tr>
<tr><td>average</td><td>waste
prevention of major accidents</td><td>air</td><td>nature and landscape</td></tr>
<tr><td>low</td><td colspan="2">materials</td><td>noice
soil</td></tr>
</table>

existed for some time now (seven years[2]). According to information from the UNIVOX survey, the pressure on the authorities to take action in the area of noise protection has remained consistently low since 1987, however soil protection is now (1993) top of the list of all environmental areas in this respect. The public is clearly demanding that more be done here.

- *High effort/low results.* This permutation, which is of interest for the research of deficits in policy implementation, is not filled by any of the eight areas of environmental policy. However, the area of water protection is today quite close to this extreme position. This finding would probably be more extreme if the level of financial expenditure were also to be taken into account. The deficits result on the one hand in the inadequacies in performance of the installed treatment plants, and on the other hand – particularly – in the (continued) excessive use of farmhouse manure and pesticides in agriculture. It is of little use to the condition of ground and drinking water if household and industrial sewage is carefully "cleaned and polished" by the local authorities while farm-

2 The Noise Protection Ordinance of 15 December 1986 (RS. 814.41) has been in force since 1 April 1987.

ers continue to spread massive volumes of nitrates and phosphorous in the open ground.

- *Low effort/high results.* According to our findings this, the most interesting position from the perspective of policy implementation, is at least partly occupied by environmental chemicals policy. These findings should be an indication of the efficacy of the distinctive principle of cooperation applied in this area and the system of self-monitoring which is based on this principle. The effort on the part of the authorities can also be minimised here through the consistent application of the polluter pays principle just as it was probably maximised in the case of water protection through the community pays principle applied hitherto in the area of water protection" (Knoepfel, 1994: 219 f.).

2 The Development of Environmental Policy and the Main Regulations

2.1 Development

The oldest piece of environmental legislation in Switzerland is the Forest Law (*Forstgesetzgebung*) which came into force in 1902. It was passed as a result of the large number of avalanches in the mountain valleys which were caused by excessive tree-felling. The law stipulates the maintenance of the entire Swiss shelterwood area and imposes an obligation that compensation be paid for shelter areas which have been felled. This principle is highly progressive even in today's terms as it is presumably at the core of what can be seen as *sustainable development*. The Federal Act on the Protection of Nature and Habitats originates from the 1960s (1966). This act defines the activities of the nature and landscape protection authorities with respect to the protection of the native animal and plant worlds. It was originally directed at nature reserve policy. Since 1987 it has provided – extended – habitat protection in the planning and construction of state-owned buildings and in the granting of licenses or allocation of state contributions. Landscapes of national importance which are worthy of protection have been recorded in a national inventory on the basis of this legislation. These areas are not to be reduced if possible; projects carried out within the protective boundary have to be passed by a special national commission. Since 1966, local authorities and environmental organisations have the right to take legal action in the case of such projects. The ratification of a popular initiative on the protection of moorlands (6 December 1987)[3] resulted in the introduction of an absolute ban on construction within and modification of such landscapes. The same applies for biotopes recorded in national inventories which must be included by local authorities in their land-use plans.

During the 1970s, a systematic policy for water protection emerged at national level when the Water Protection Act of 1955 was reinforced by effective subsidy

3 57.8% yes votes (17 cantons and 6 half-cantons); poll participation: 47.7%.

mechanisms (1972). In 1975, for the first time in the history of Switzerland, a systematic policy was introduced for the limitation of emissions which prescribes the mandatory raising of emission limit values in the event of critical immission levels being exceeded (Decree on the Discharge of Waste Water). Following the Stockholm environmental conference of 1972, clean-air policy in the area of domestic heating systems was introduced at urban and subsequently at cantonal level. Moreover, the cantonal authorities began to carry out (on the basis of the labour protection legislation of 1961) upgrading and redevelopment projects in industrial and small business concerns where the general public had complained about excessive noise and air pollution emissions. However, within the European context, chemicals, air and noise protection legislation were introduced at a very late stage in Switzerland; the Swiss Environmental Protection Act (*Umweltschutzgesetz*) did not come into force until 1985 (Environmental Protection Act of 1983). Largely divorced from corresponding efforts by the state under previously enacted land use legislation (1979), the most important decrees for the implementation of the act were passed during the second half of the 1980s and began to be implemented by the cantons towards the end of the decade. These decrees are based on standard emission limits which are negotiated with the relevant sectors and which were revised at more stringent levels in the early 1990s. Additional emission limits are defined for specific pollution situations under the relevant decrees.

After 1992, the issue of water protection in agriculture, which had been neglected in the 1980s, was finally addressed with the introduction of restrictions on livestock numbers and regulations on the use of fertilisers. It was during this time that the cantons began to address the extremely difficult and previously largely unsuccessful implementation of clean-air measurement plans for areas with high levels of air pollution (private motor traffic was the main target). At the same time, waste policy took on a new lease of life when strategies similar to those used in chemicals policy, e.g. economic incentive schemes and the inclusion of private actors, were considered. Nature and landscape protection policy also found new impetus in the mid 1990s (special inventories). With the revised Environmental Protection Act of late 1995, the Swiss state now has what can be described in the European context as very modern environmental legislation which provides, in particular, economic instruments for the implementation of environmental policy.

2.2 Basic Instruments

Up to the mid 1980s, Swiss environmental policy worked almost exclusively with classical command and control instruments. This can be clearly seen in the way that licensing, development and control decisions based on federal and cantonal legislation were not enacted as independent environmental policy procedures but were integrated as (new) features into existing procedures (building permit, planning permission and licensing procedures). With the coming-into-force of the new Environmental Protection Act of 1983 (1985) and the corresponding decrees, new instruments were also introduced. These include the environmental impact assessment (*Umweltverträglichkeitsprüfung – UVP*) which, despite being a component of existing licensing procedures, in its concrete form increasingly assumed a coordinating and integrating function both within and outside environmental policy which goes beyond individual decisions. Up to now (spring 1996), approximately 800 of these assessments have been carried out on large plants throughout Switzerland which are of particular relevance to the environment. The right to take legal action by environmental protection organisations has also resulted in the restructuring of environmental legislation procedural practice in that today (in most cases) such organisations play a significant corrective function in the initial licensing procedures for these plants. The organisations are involved as partners at an early stage in the procedure and are taken seriously as partners by both the national authorities and the project groups. An interesting new instrument which has been developed under the chemicals legislation is that of self-monitoring by manufacturers who have developed an autonomous and increasingly complex system for the monitoring of substances at company level: in keeping with the spirit of the principle of cooperation, manufacturing companies are today far more willing to provide access to test results for new chemicals as part of the registration procedure. Finally, it is possible to observe a trend towards the use of economic instruments at cantonal level (air pollution certificates in the cantons of Basle-Town and Basle-Country). However, the use of these instruments remains insignificant in practice.

During the first half of the 1990s, environmental policy widened its experience with cooperative instruments (particularly in the area of waste policy), negotiated solutions (in cases of conflict over large plants), materials-based eco-reports, company-related eco audits and with campaigns (air pollution, waste). These instruments have proven as successful as the environmental impact assessment in initiating ecological learning processes within the national authorities, environmental protection organisations and industrial and small business concerns. These learning processes could be seen to have led to irreversible changes in environment-related behaviour. Moreover, interpolicy cooperation instruments have also been adopted in individual areas (particularly in the area of agriculture and tourism). This was made possible in the area of agriculture by the fact that, as a result of the GATT Uruguay Round (1992) and internal pressure in the area of water protection policy, official agricultural policy switched to area-related subsidies and combined the direct payments to farmers with increasing incentives[4] for ecological services (Article 31b of the Agriculture Act). In this way, nature and landscape policy finally succeeded in implementing its concerns with respect to agricultural policy.

Environmental instruments in the area of planning also gradually assumed greater influence during the first half of the 1990s. What these instruments have

4 1993: 7.5%; 1994: 13.0%; 1995: 19.7%; 1996: 24.9% (Source: *Bundesamt für Landwirtschaft*, (Federal Ministry for Agriculture) 24. 01. 1996).

in common is that they focus on general regional pollution problems and aim at the implementation of a phased steering of the reduction in pollution levels through (usually quantitative) strategic aims. Typical examples here include the clean-air measurement plans which will involve the implementation of almost 1,000 coordinated individual measures by the cantons. Most of these measures are aimed at private motor traffic. The general local authority water management plan prescribed under the new water protection is a similar management instrument. This plan aims to put measures for sewage treatment in the larger context of a general water management strategy which involves drinking and rain water as well as still and flowing waters. The new Landscape Concept for Switzerland (*Landschaftkonzept Schweiz,* 1996) and the biotope protection inventories, which are to be included in local authority land use plans in accordance with the revised 1987 legislation for the protection of nature and habitats, are also worth mentioning in this context. Noise protection and contaminated sites policy, which is to be established in accordance with the new Environmental Protection Act of 1995, now also include a strong planning component. The use of these new planning instruments in environmental policy means that environmental policy is moving closer to regional development. It is possible, therefore, to refer to the "return of environmental policy to the territory".

Under the provisions of the new Environmental Protection Act, a few economic instruments are also to be introduced over the next few years (levies on heating oil containing sulphur, VOC and possibly other substances), as well as instruments for the promotion of technology and regional development and protection of compulsory catchment areas for waste disposal plants. No new regulation has been introduced in the area of environmental liability law.

2.3 Policy Style

A significant transformation in policy style has also been under way, particularly in the cantons but also at federal level, which can be seen as reciprocal to the development of these new policy instruments. The former attitudes and behaviour of the old police administration (particularly the water protection authorities) have been replaced by the cooperative style of younger, ecologically-minded officials who have considerably greater responsibility and competencies than was the case with their counterparts in the strictly hierarchical police-run environmental authorities of the previous era. Negotiated solutions began to be introduced as early as the late 1980s when emission limits had to be worked out in collaboration with representatives from industrial and small business sectors. The increasing willingness of these sectors to accept these limit values in the interest of equal treatment led to the disappearance of a previously widespread fear of contact with the authorities. In the early 1990s, the attitudes and behaviour of the 26 cantonal environmental authorities underwent fundamental change yet again as did those of the Federal Ministry for the Environment, Forests and Landscape (*Bundesamt für Umwelt, Wald und Landschaft – BUWAL*), which is organised in divisions with responsibility for individual environmental media. In many places, problem-specific cooperation networks emerged as part of what were known as project organisations, in which the representatives of the polluters, environmental organisations and the individual divisions of the environmental protection authorities cooperated with each other on a largely equal footing with the aim of finding a solution to the relevant problem. This new policy style is most prevalent in the area of waste policy (particularly in the application of the new beverages packaging decree of 1993), in nature protection policy (completion of land-use contracts between agricultural or nature protection authorities and farmers) and in industry-related environmental policy (environmental impact assessment procedure in highly complex, learning-oriented networks).

Long-standing observers of Swiss environmental policy should be united in the opinion that this undoubtedly positive change in policy style, which was brought about by the adoption of a range of new instruments, the advent of a new generation in the environmental authorities and new training programmes, should be seen as the most significant change in this policy during the first half of the 1990s. It could be part of a major trend towards new forms of public management which, however, has also benefited from a series of strong impulses from environmental policy.

2.4 Policy Evaluation

Systematic empirical analyses of all areas of Swiss environmental policy which also include the implementation activities of the cantons have yet to be carried out in Switzerland. There is also no systematic OECD evaluation on this policy. The first attempts at general evaluation can be found in the two reports on the state of the environment which were published by the Federal Ministry of the Environment, Forests and Landscape (BUWAL, 1990; 1993) and in various cantonal reports on the state of the environment, the second or third editions of which have previously been compiled (in particular by the cantons of Bern, Aargau, Zurich and Lucerne). It is not yet possible to predict the extent to which the initiative for the introduction of an integrated ecosystem-related environmental observation at national level, which was initiated by the Swiss Academy of Natural Sciences (*Schweizerische Akademie der Naturwissenschaften – SANW*), will succeed. It is worth noting, however, that a "Steering and Coordination Body for Environmental Observation" (*Lenkungs- und Koordinationsorgan Umweltbeobachtung – LEKUB*) has been established which aims to establish a comprehensive system of environmental observation in cooperation with the cantons and all other departments within the federal authorities which have relevant data available to them (Swiss Ministry of the Interior, 1995).

In contrast, summary evaluations of the efficacy of environmental protection legislation are available (in particular BUWAL, 1993 and Knoepfel, 1994) as well as systematic evaluations have been carried out on individual topics within the general areas. A case in point is the study for the evaluation of BUWAL which was published in 1991 by the National Council Commission of Enquiry (*Geschäftsprüfungskommission des Nationalrats*). Air pollution control, environmental impact assessment and rural water protection were systematically studied as part of this evaluation (Geschäftsprüfungskommission des Nationalrats, 1991). Studies on the efficacy of clean-air measurement plans (Knoepfel, Imhof and Zimmermann, 1995), the implementation of clean-air policy in the area of domestic heating systems (Balthasar and Knöpfel, 1995) and on the topic of soil protection (Zimmermann, Achermann and Knoepfel, 1996) exist or will be published shortly. The findings in these studies are taken into account in section 1 above.

In the international context, Swiss environmental policy can still be seen to take a leading position in the areas of air pollution and water protection where critical load limit values have been defined which, as a result of the "eco-centring" of the two acts, are considerably more stringent than those in other countries and in the EU. The adoption of the popular initiative for the protection of moorlands should also place Switzerland in a leading position in Europe with regard to the protection of this type of landscape. As opposed to this, Swiss environmental policy lies somewhere around the average mark for industrialised countries when it comes to noise protection, soil protection and energy-saving policy. In comparison to the EU states, the absence of regulations for free access to environmental information and for eco-audit in Switzerland is a striking deficit. Membership of the European Economic Area would have brought about some progress here, if only the federal authorities promoting the move had suggested the relevant adjustment. As is a well known fact, the electorate rejected this move in December 1992. The fear that as a member of the European Union, Switzerland could be obliged to cut back on its relatively high level of environmental protection is known to have played a role for the Greens in their opposition to Switzerland's membership.

With its new Environmental Protection Act of 1995, Switzerland also made some – tentative – attempts towards the use of economic instruments. However, the revised legislation did not succeed in introducing a CO_2 levy or increased taxation of energy carriers as part of the reinforcement of the energy components of environmental policy. At international level, Switzerland has generally aligned itself with the most progressive countries (Biodiversity Convention, CO_2 reduction, Montreal Protocol etc.). As opposed to this, however, due to internal pressure from the Alpine cantons which were concerned about their economic development, Switzerland did not sign the Alpine Protection Convention in 1995. This is all the more astonishing in view of the fact that in 1994 the electorate surprisingly voted in favour of the acceptance of an Alpine protection initiative[5] which proposed significant reductions in transit traffic through the Alps and the transfer of the transport to the railways. With this decision, the Swiss population contradicted an earlier international contract concluded with the EC (Transit Agreement, ratified in 1992).

5 Referendum of 20 February 1994. 51.9% yes (13 cantons and 6 half cantons).

3 The Capacity of Proponents of Environmental Policy

3.1 Public Institutions

In accordance with the basic principle of executive federalism, which applies equally to all public policies, the federal government is responsible for the formulation of programmes (drawing up and passing of laws, decrees and implementation guidelines) and the cantons are responsible for the implementation of this legislation etc. On the basis of their own legislation, the cantons can delegate individual implementation tasks to (larger) local authorities. This principle also applies in environmental policy, however some implementation tasks remain the task of the federal authorities (regulation of chemicals, testing of appliances and motor vehicles, inter-cantonal and international implementation tasks).

Personnel, financial and instrumental implementation resources are correspondingly distributed among the three administrative levels. According to my own estimations, at the beginning of 1996, 450 permanent positions were allocated for environmental policy requirements at central federal administration level, approximately 1,500 at cantonal level and 4,500 at local authority level. This total capacity of approximately 6,500 permanent positions (more than one permanent position per 1,000 inhabitants) is considerable, especially in view of the fact that half of this capacity was established over the last decade despite increasing overall budget restrictions. Public expenditure for environmental protection in Switzerland is estimated at approximately 4.3% of the total national budget[6].

When the Environmental Protection Act was first passed in 1983, the Federal Ministry for Environmental Protection, which had been developed from the old Swiss Ministry for Water Protection, underwent a period of rapid expansion. In 1989, this ministry was amalgamated with the former Federal Ministry for Forestry and Landscape Protection and in the wake of the change in direction was successively transformed from 1992 in accordance with the recommendations of the National Council Commission of Enquiry. This transformation involved, for example, the establishment of a new department for waste management, a horizontal division for environmental law and a horizontal unit for environmental impact assessment, and the general reduction of hierarchical structures. Today, BUWAL has a total of over 314.5 permanent posts and an annual budget of 528.4 million Swiss Francs (1996). During the first half of the 1990s, environmental policy administrative capacities were also expanded in other federal ministries. This applies in particular to the Federal Ministries of Health, Energy, Transport, Agriculture, Statistics, Regional Development, Civil Aviation and various sections of the Military Department and the Department of Foreign Affairs. A total of approximately 100 permanent positions are allocated for environmental requirements here.

---

6 Source: Statistisches Jahrbuch der Schweiz (1996: 394).

Even more impressive is the way in which environmental administrative capacities have been developed at cantonal level. Under the environmental protection legislation of the second half of the 1980s, the cantons were obliged to establish eight specialised offices (water, air, waste, materials, major accidents, soil, noise and nature and landscape protection). For historical reasons, the offices for nature and landscape protection are usually located outside the newly established cantonal environmental protection authorities and the other seven offices are not found within a single office in all the cantons. As a result of this, the implementation of environmental policy in the cantons shows a varied horizontal degree of centralisation. The cantonal environmental authorities which are structured on a successive and "law-oriented" basis (i.e. in accordance with the phased enactment of various federal decrees) today face increasing criticism. This originates not least from a strong trend in favour of client-oriented policy services. The first canton to establish a second-generation environmental authority was St. Gallen (Baudepartement des Kantons St. Gallen, 1995). In this case, a media-specific basic structure was replaced by a polluter-specific one. This enabled the acceleration of licensing procedures and the introduction of integrated general licenses and also makes it possible to take greater account of aspects of environmental land-use planning.

Considerable administrative resources have also been established in the area of environmental protection at the level of the approximately 3,000 Swiss local authorities in recent years. Thus, in addition to sewage-treatment personnel, local authority administrations in towns of approximately 10,000 inhabitants now also have environmental protection officers who work mainly in the areas of air pollution control and traffic, waste management, nature and noise protection. A conservative estimate puts the total number of permanent positions at local authority level at approximately 4,500. It should also be noted that at local authority level, in particular, environmental protection is often implemented in the form of voluntary official work performed by citizens in their spare time. Hence, the actual capacity at local-authority level is probably somewhat greater.

With the increasing density of regulations under environmental legislation, the courts have gained in significance. Cantonal administrative courts and the Swiss Federal Court have recently proven to be important guardians of the strict application of Swiss environmental protection law in the cantons and local authority areas. This applies particularly to regional development (buildings outside designated construction areas), quantitative water protection, nature and landscape protection (particularly forest protection) and – most recently – the clean-air measurement plans and noise protection. The Swiss justice system exercised a significant influence on the way in which greater coordination of procedures and corresponding reorganisation measures are being considered in the cantons.

3.3 Environmental Organisations

Table 2 shows the membership of the 20 environmental protection organisations in Switzerland which have the right to take legal action under environmental law enacted at the end of 1995. Although overlap in membership is very common, the membership of these organisation is, however, impressive (almost 10% of the population). It is only slightly lower than that of the trade unions[7] and significantly higher than that of the political parties[8]. The main environmental protection organisations formed an environmental coordination group in the early 1990s which regulates the topic-specific and project-specific responsibilities of the individual organisations. The larger associations have substructures based on cantons and linguistic boundaries. The willingness of the population to contribute to environmental protection in the form of financial contributions or active membership has remained constant since the mid 1980s at approximately 20%-30%; it is slightly higher in German-speaking Switzerland than in French-speaking Switzerland. The former considerable differences in the general political orientation of the different environmental protection organisations, whereby for example the predominantly radical position of the WWF contrasted with the rather conservative policy of the Swiss Association for Nature Protection (*Schweizerischer Bund für Naturschutz*), is generally a thing of the past, at least at national level. Differences can still, however, be found at cantonal and local level.

The environmental protection organisations are represented as recognised and scientifically competent partners at all levels of the Swiss state in the context of licensing, planning and legislative procedures. Moreover, they now increasingly take part in direct discussions with polluters from industry, commerce, agriculture and households and with representatives of state infrastructure policies (in particular: traffic). In the spirit of the principle of cooperation, they enter into contracts with groups of polluters, distribute labels (in cooperation with consumer protection groups), play an important role in practical environmental training, information and consultancy and are increasingly involved in the provision of ecology-oriented services.

The relatively weak position of the Swiss Green Party, which emerged from various groups in the mid 1980s, is striking when compared with the success of the movement in other central European states. This party, which is still vulnerable today to the tensions between the (linguistic) regions and individual cantonal sections, has been represented in the Swiss parliament since 1983; in 1991 it obtained 6.1% of the votes, however, in the last National Council elections of 1995, it lost votes to the Swiss Social Democratic Party. After an initial reluctance, the latter has assumed a strong profile on ecological issues since the early 1990s and is today seen as an important protector of ecological principles in

7 SGB: 429,024 members (1994); CNG: 134,167 members (1995); VSA: 127,103 members (1995) (= Total 690,294).

8 Total 360,000 members (1990), Statistisches Handbuch, Forschungszentrum für Schweizerische Politik, Bern, 1991.

Table 2. Register of Environmental Organisations with the Right to Take Legal Action (1995).

Organisation	*Membership*
Rheinaubund (Schweiz. Arbeitsgemeinschaft für Natur und Heimat) *(Swiss Association for Nature and Habitats)*	800
Schweiz. Vereinigung für Landesplanung (VLP) *(Swiss Association for National Planning)*	3,609
WWF (Schweiz) *(World Wildlife Fund)*	170,000
Schweizer Vogelschutz (SVS), Verband für Vogel- und Naturschutz *(Swiss Bird Protection, Association for the Protection of Birds and Nature)*	6,000
Schweizer Heimatschutz (SHS) *(Swiss Habitat Protection)*	20,000
Schweiz. Bund für Naturschutz (SBN) *(Swiss Association for Nature Protection)*	100,000
Schweizer Alpen-Club (SAC) *(Swiss Alps Club)*	85,000
Schweiz. Gesellschaft für Umweltschutz (SGU) *(Swiss Society for Environmental Protection)*	8,357
Helvetica Nostra *(Our Switzerland)*	1,981
Schweiz. Vereinigung für Gesundheitstechnik (SVG) *(Swiss Association for Health Technology)*	–
Schweiz. Liga gegen den Lärm *(Swiss Anti-Noise League)*	230
Schweiz. Vereinigung für Gewässerschutz und Lufthygiene (VGL) *(Swiss Association for Protection of Waters and Clean Air)*	600
Schweiz. Stiftung für Landschaftschutz und Landschaftspflege (SL) *(Swiss Foundation for Landscape Protection and Landscape Management)*	801
Schweiz. Energiestiftung (SES) *(Swiss Energy Foundation)*	3,200
Naturfreunde Schweiz (NFS)[1] *(Swiss Friends of Nature)*	29,932
Verband Schweiz. Abwasserfachleute (VSA) *(Association of Swiss Sewage Specialists)*	12,400
Aqua Viva (Nationale Aktionsgem. zum Schutze der Flüsse und Seen) *(National Action Group for Protection of Rivers and Lakes)*	–
Schweiz. Fischerei-Verband (SFB) *(Swiss Fishing Association)*	10,000
Schweiz. Verkehrs-Stiftung (SVS) *(Swiss Traffic Foundation)*	–
Verkehrs-Club der Schweiz (VCS) *(Swiss Traffic Club)*	100,000
Total	*552,910*

1 Membership numbers for 1993.

Source: direct survey of the different organisations (3/96).

the Swiss four-party government (the party obtained 23% of the vote in 1995). The consultancy procedures for the revision of the Nature and Habitat Protection Act and the Environmental Protection Act during 1995 also showed that there is ample potential for the continuation of the relatively progressive environmental legislation from the 1980s in the middle-class centre-right parties; attacks on the core of this legislation from small business and right-wing middle-class circles were successfully warded off.

3.4 Economic Proponents of Environmental Protection

The contribution of the Swiss ecology industry to the Swiss net national product is currently slightly lower than that of the agricultural sector (approximately 3%, see Pillet, 1988; Conus and Rossi, 1992; Walther, 1990). The ecology industry is gaining in significance in the areas of waste, water protection, air pollution control and noise protection and also represents an important area for potential development. This sector also involves services which are required as part of large-scale environmental impact assessment procedures, the implementation of the clean-air measurement plans and in the landscape and nature inventories. Hence, ecological products and services doubtless represent an area of growth in the Swiss national economy.

Major ecological learning processes have taken place in the other sectors of industry and commerce and – increasingly in the services sector over the past five years. These are reflected in new company-policy models, in company-internal ecological control mechanisms and in the conversion of product ranges (for example in the chemicals industry). It is possible to observe a virtual run on eco-labels and eco-certification of companies (in accordance with the ISO 9002 standard). This promises not only real improvement in company or product-specific eco-audits but also in the development of important markets for research and consultancy. Environment-oriented training and education programmes are correspondingly being tailored to the future needs of industry and service concerns. Through certification, we have succeeded in doing what eluded official environmental policy for some time, i.e. enabling ecological philosophies to penetrate the core of entrepreneurial strategies. Ecological messages in marketing are now very favourably received by consumers.

This trend can be seen to have reached the agricultural sector by the mid 1990s at latest. Supported by the above-mentioned direct ecological payments, which are intended to motivate farmers to switch to integrated production or even organic production, the agricultural sector and the large distribution chain has strongly promoted the ecological argument in product acquisition. A continued increase can be observed in sales of environment-friendly agricultural products now in early 1996. The two major wholesalers, Migros and Coop, are even reporting supply shortages in the area of organic products. The proportion of agricultural products originating from integrated production today lies at approximately 20%, that of organic products, which cost up to 150% more, is approximately 1.5%[9].

This willingness on the part of the consumer to purchase more expensive environment-friendly products is also reflected in the general willingness shown by the Swiss population to pay for environmental policy measures which has remained relatively high and unchanged in the European context since 1986. A slight decrease was observed during the recession of 1994 followed by a slight recovery 1995 (Knoepfel and Wälti, 1996). *Figure 1* demonstrates the high de-

9 Individual estimate 1996 from employee at the Federal Ministry for Agriculture (Frau Zajac).

Figure 1. Willingness of the Swiss Population to Adopt Environment-Relevant Behaviour.

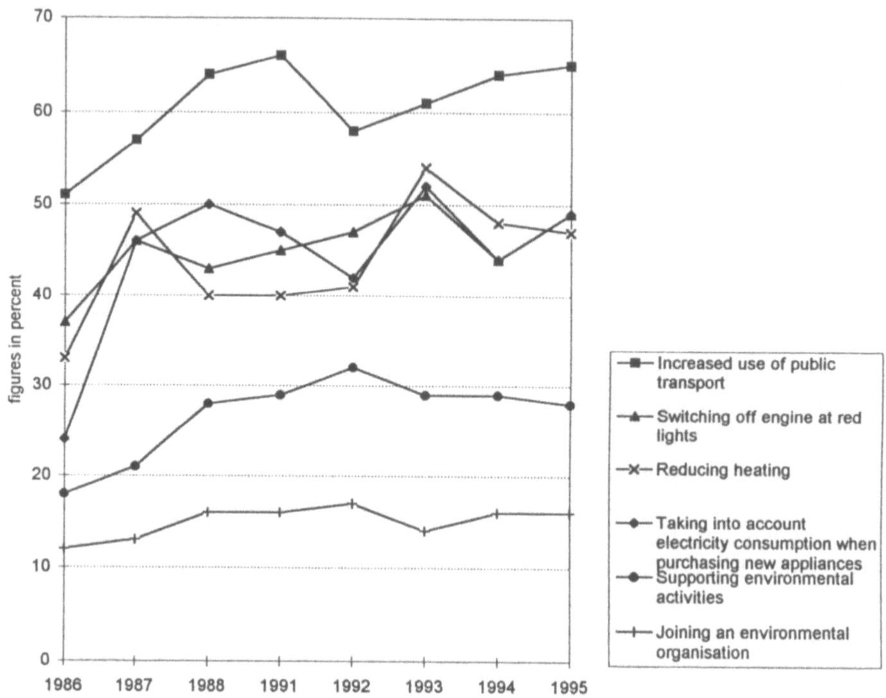

Source: Knoepfel and Wälti, 1996, UNIVOX.

gree of willingness among the population to alter their behaviour on a non-economical basis.

Hence, it is possible to assume that a very high economic potential now exists for ecological policy in 1996 as compared with that which existed in 1985 (coming into force of the first Environmental Protection Act). With respect to its contribution to the creation of wealth and jobs, ecology has today become an important economic factor and its elimination would involve significant economic costs. Hence, ecological policy has succeeded in economically guaranteeing its political potential.

4 The Changing Constellation of Proponents

It is possible to distinguish four development phases for environmental policy in Switzerland corresponding to the classification used by *Jänicke* and *Weidner* in this book. As is demonstrated below, the developments in Switzerland run for the most part parallel to those in federal German environmental policy.

- ***Dilution (up to 1975)***

  This embryonic phase, which can be identified in the first hours of life of every environmental policy, is characterised by its exclusive concentration on environmental quality. State intervention is limited to areas with excessive levels of pollution. This intervention can either consist of "improved" spatial-temporal distribution of the emissions (high chimneys, discharge of water into large drainage canals etc.[10]), or in the restriction of measures aimed at reducing emissions to particularly vulnerable areas (low-sulphur heating oil for towns and cities etc.). The protected object is always limited to particularly vulnerable locations with groups of persons or ecosystems at risk; these policies have no preventative component. Such strategies were also implemented in Switzerland up to the mid 1970s at urban and cantonal level; typical examples include the domestic heating controls which were limited to problem areas and the concentration of sewage treatment activities in local authorities with particular water shortages. The end of this period is marked in legislative terms by the Sewage Disposal Decree of 1975. This decree introduced emission limits for the first time which were implemented on a general level and irrespective of the previous level of pollution in the discharge canals of the sewage treatment plants. Around the same time, a similar insight was successfully imposed in the cantonal clean-air policy and in the (case-related) national environmental quality policy. The actor constellation in these dilution policies, which were always implemented by the police administration, is bilateral: you had the police authority on the one hand and the polluting "perpetrator" on the other.

- ***Dilution and End-of-the-Pipe Policy (1975-1986)***

  Although it only became an integral part of Swiss environmental policy in 1983, the "two phase concept" of environmental policy, which is still applied today, was introduced at cantonal and national level after the enactment of the above-mentioned sewage discharge decree. According to this school of thought, emissions limitations should be implemented throughout the country, irrespective of previous levels of pollution. In cases where despite application of these limit values, excessive pollution continues with the (increasingly quantified) critical load limit values being exceeded, the emissions restrictions must be reinforced. This policy was initially applied in water protection followed by air pollution control and finally in local and cantonal waste policy, which was increasingly converted to the principles of recycling and energy reharnessing. The origin of this policy can ultimately be traced back to an extended concept of ecosystems, which was able to demonstrate with increasing scientific precision that pollutants accumulate in ecosystem, can travel extensive distances (acid rain) and can combine to form actual material flows which can give rise to effects which go far beyond the original place where

10 Temporal measures (e.g. reduction of emissions during smog periods) were rare in Switzerland (unlike in France).

discharged or deposited. Hence, general emissions reduction strategies are required which are reflected in an increasingly broad range of emission limit values and in product and processing standards. These standards were developed, applied and gradually intensified in the approximate ten-year phase between 1975 and 1986. Finally at the end of this phase, this development found expression in legislative measures containing the definition of emission limit values for the individual media and economic sectors which were quickly enacted without political difficulties. The developments in this period are largely comparable with those in foreign legislation (in particular the federal German legislation – TA Air and TA Noise). However, in contrast to developments in the Federal Republic of Germany, in Switzerland cantonal environmental policy continued to be predominantly based on a bilateral actor constellation: The critical load limit values were adopted from (scientific) WHO recommendations and environmental protection organisations were not involved in their development; the emission limit values represented the outcome of negotiations between industrial and small business sectors and state authorities with minimum input from the other relevant organisations.

- ***"Intense End-of-the-Pipe Treatment" (1986-1992)***

With the coming into force of the most important decrees and corresponding immission limit values, an important mechanism was initiated in industry which ultimately led to the emergence of the eco-industrial sector described above. Emission limit values became an defining factor in competition policy. The polluters accordingly demanded equal application of the limit values and increasingly identified the economic advantages to be gained in achieving values which are below the legal maximum levels (energy and material savings, new technologies). In 1992, closed cycles and increasingly technologically sophisticated recuperation technology resulted in the, in part, significant tightening of emission limit values in air pollution control. Mandatory End-of-the-Pipe technologies became big business and eventually constituted the key impetus behind industrial and even small business renewal. The position of the environmental organisations was also strengthened during this phase not least as a result of their recognition as partners with a right to take legal action in mandatory environmental impact assessment licensing procedures. The increasing integration of personnel from these organisation with parts of the state administration and with university research teams resulted in their becoming familiar with the importance of emission limit values for effective environmental policy. This resulted in the unlocking of the previous bilateral actor structure, which towards the end of this period took on a definitive trilateral structure in the form of the "iron triangle" (Knoepfel, 1995: 14).

- ***Ecological Modernisation (1992-1996)***

During this last phase (for the present), the above-mentioned economic dynamism led to the political mutation of the topic of the environment: having originated in the form of a demand for economic resources management it became an actual component of economic policy. Supported, no doubt, by the conference in Rio in 1992 and the associated vocabulary of sustainable development, in Switzerland the realisation has spread that environmental policy involves more than the management of emissions. Hence, ecology is penetrating into the core area of economic behavioural maxims (Minsch et al., 1996). This demands of companies, and also of the state and its environment-damaging policies, that they take economic account of the consumption of renewable and non-renewable natural resources. Today, we are merely at the outset of this development. It is unlikely – in Switzerland too – that this message, which is as clear as it is simple, will succeed in spreading over this decade.

5 Capacity-building

5.1 Actor Capacity

The numbers, political weight and competence of official and social, ecologically proactive actors in Switzerland have reached considerable levels. On the side of the political-administrative system, a largely irreversible process was initiated involving the establishment of environmental policy capacities in areas which go far beyond environmental policy in the strict sense. Parliaments and authorities at federal, cantonal and local authority level doubled their capacities in this area in the decade from 1985 to 1995 – a process which is unique to this policy area. However, ecological policy is still relatively weakly represented in the area of state infrastructure projects and in the area of energy policy where it has led to corrective action but not to the kind of fundamental processes of renewal which can be observed in industry and commerce. This explains the previously mentioned low level of success in nature and landscape protection policy.

Industry, commerce and services also developed ecological capacities over the same period, a development which even the most astute observer would not have predicted in 1985. The previously described dynamism of the industrial system is as responsible for this development as a strongly proactive attitude of consumers. Surveys on household waste behaviour and attitudes show a high degree of acceptance of ecologically sound behaviour in this area. Old-style productivity-based agricultural policy was replaced overnight by a new policy which, unlike its predecessor, has a stabilising effect on the ecology situation. The central weak point is and remains the (strongly state-promoted) private and public transport systems. The most important area of conflict today is that between mobility and ecology.

5.2 System Capacity

The institutional and information structures available to official and social actors have also been considerably reinforced over the past decade. With the overall introduction of environmental impact assessment procedures at project level and different types of mechanisms for establishing consensus at planning level (particularly in energy policy), the possibilities for the real participation of environmental protection organisations, other citizens' action groups and individual persons have increased significantly. These procedures do not only represent decision-making and learning processes. Their positive effect on the ecological moulding of policy output can today be seen as proven.

Interestingly, this observation correlates with a clear reduction in the pressure exerted by the public on environmental policy authorities to "do more" on environmental policy issues. This observation is based on a series of corresponding findings from 1986 to 1995 (cf. *Figure 2*)

Figure 2. Pressure from the Public on the Authorities ("do more") Based on UNIVOX Surveys from 1986 to 1995.

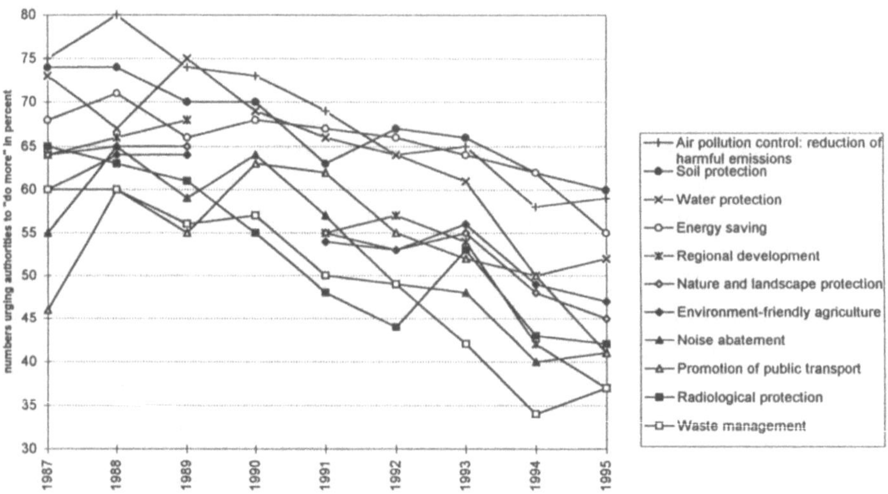

Source: Knoepfel and Wälti, 1996, UNIVOX.

Opportunities for involvement were also improved due to considerable efforts on the part of local authority, cantonal and national authorities to make data on the environment situation more widely available. The – media-based – system of environmental observation in Switzerland is in a leading position, in an international context also (Weidner, Zieschank and Knoepfel, 1992). Attempts at establishing integrated environmental observation have failed for the present. Systematic observation of the performance of environmental policy and non-environmental policy with respect to an ecological evaluation of state activity is only at the very early stages.

Scientific and technological environmental research at universities has expanded considerably since 1985. Several national research programmes and a focal programme of the Swiss National Science Foundation and the Swiss institutes for technology and eight cantonal universities played a key role in this development. Despite some important efforts, insights provided by social-scientific environmental research can still only be described as modest. The graduate and post-graduate studies programmes, which were developed in the second half of the 1980s and which are on offer today in almost all of the institutes for technology and universities, have made an important contribution to the development of environmental capacity. Ecological perspectives are also receiving far more emphasis in the traditional disciplines than was previously the case. Curricula for primary schools, second level and vocational schools have also undergone corresponding changes.

Despite the recession, now in the mid-nineties, the environment dominates the media unlike any other topic. To the delight of the increasing circle of readers and users, numerous new magazines, books, CD-ROMs and a variety of environment-relevant databases have been produced and published over the past five years.

6 Capacity for What?

Despite the considerable expansion of capacity in terms of actors and at the system level, it is impossible today to give a conclusive answer to the question as to whether this capacity will be sufficient to bring about the above-mentioned modernisation of environmental policy towards consistent resource management by the end of the century. Environmental economists demonstrate convincingly that important economic resources are becoming cheaper and not more expensive for production (energy and infrastructure). There is no sign of the formation of a solid coalition of actors in favour of consistent price increases for natural resources. The previously mentioned economic dynamism may initially have led to the use of economic instruments, however they are still very much part of state emissions management. At present, there is no room in Swiss policy for solid, widely-based coalitions of representatives from the industrial system, environmental organisations and state authorities which favour a system of artificial scarcity and increased pricing of natural resources in the interest of maintaining their regeneration capacities and in the interest of future generations. Only lone positive signals can be heard from individual political and social groups which are transmitted under the keywords of cost internalisation, true costs and energy tax.

Hence, now at the beginning of the 1996 year of recession, the signs tend to be more negative than positive: the globalisation of markets is forcing the implementation of cost-saving reductions in the area of mobility and the adoption of short-term survival strategies. The situation on the labour markets is used as an argument against more far-reaching environment strategies and for the stabilisation of the ecology industry which is based on emissions management. The ecology industry could also be seeing its position under threat as a result perhaps of a fundamental modernisation of industrial society. Hence the highly promising transition to ecological modernisation in Switzerland which was clearly in evidence in the first half of the decade is now beginning to falter.

References

Balthasar, A. and C. Knöpfel, 1994: Umweltpolitik und technische Entwicklung, eine politikwissenschaftliche Evaluation am Beispiel der Heizungen. Basel: Helbing & Lichtenhahn (Reihe Ökologie und Gesellschaft Nr. 8).

Baudepartement des Kantons St. Gallen, 1995: Reorganisation des Amtes für Umweltschutz.

BUWAL (Bundesamt für Umwelt, Wald und Landschaft), 1991: Zum Stand der Umwelt 1990. Bern: EDMZ.

BUWAL (Bundesamt für Umwelt, Wald und Landschaft), 1994: Zum Stand der Umwelt 1993. Bern: EDMZ.

Conus, J. and A. Rossi, 1992: Le business écologique: définition, structure, évolution et perspectives de l'industrie de la protection de l'environnement de la Suisse. Bern: Bundesamt für Konjunkturfragen.

Eidgenössisches Departement des Innern, 1995: Bericht des Lenkungs- und Koordinationsorgans Umweltbeobachtung (LEKUB) über seine Tätigkeiten der Phase I (Juni 1993-Juni 1995). Bern.

Forschungszentrum für Schweizerische Politik, 1991: Statistisches Handbuch. Bern.

Grolimund, P. and K. Peter, 1994: Integrierte ökosystembezogene Umweltbeobachtung, Konzept für die Einführung eines Beobachtungssystems. Zürich: Hochschulverlag AG an der ETH Zürich.

Knoepfel, P., 1994: Zur Wirksamkeit des heutigen Umweltschutzrechts, Umweltrecht in der Praxis/Le droit de l'environnement dans la pratique 8, 4, 201-236.

Koepfel, P., 1995: Approaches to an Effective Framework for Environmental Management, in: A. Carius, L. Hoettler and H. Merckur (eds.), Environmental Management in Kenya, Tanzania, Uganda and Zimbabwe. Berlin: German Foundation For International Devlopment, Public Administration Promotion Center.

Knoepfel P., C. Baitsch and A. Eberle, 1995: Überprüfung der Aufbauorganisation des Amtes für Umweltschutz des Kantons St. Gallen, Schlussbericht (i. A. des Baudepartements des Kantons St. Gallen). St. Gallen: AfU St. Gallen/Chavannes-près-Renens/IDHEAP.

Knoepfel P. and S. Wälti, 1996: UNIVOX-Umfrage Umwelt, Auswertung 1995. Chavannes-près-Renens: IDHEAP/Adliswil (GfS).

Knoepfel P. and W. Zimmermann (in coll. with G. Sailer and E. Matafora), 1992: Evaluation des BUWAL. Expertenbericht zur Evaluation der Luftreinhaltung, des ländlichen Gewässerschutzes und der UVP des Bundes, Schlussbericht, 18. November 1991. Bern EDMZ.

Minsch, J., A. Eberle, B. Meier, B. and U. Schneidewind, 1996: Mut zum Umbau. Nachhaltige Entwicklung durch ökologische Innovationen. Basel: Birkhäuser (forthcoming).

Pillet, G., 1988: Bilan des coûts économiques de la pollution (manuscript). Université de Genève, Centre universitaire d'écologie humaine et des sciences de l'environnement.

Schweizerische Kommission für Umweltbeobachtung (SKUB), 1993: Integrierte ökosystembezogene Umweltbeobachtung, Konzept für die Einführung eines Beobachtungssystems.

Walther, A., 1990: Die Folgerechnung von Umweltschäden des volkswirtschaftlichen Rechnungswesens am Beispiel der Schweiz (dissertation). St. Gallen: Hochschule St. Gallen.

Weidner, H., R. Zieschank and P. Knoepfel (eds.), 1992: Umweltinformation: Berichterstattung und Informationssysteme in zwölf Ländern. Berlin: edition sigma.

Korea

Young Suck Nam

1. Korea's Main Environmental Problems

1.1 Overview

Korea is one of the most densely populated countries of the world. Since 1963 the average economic growth rate has been nearly 9% (You, 1995). Environmental problems have been recognised since the 1960s. 1962, when the government began to implement its first Five-year Economic Development Plan, pollution in the industrial complexes - e.g. Ulsan, Onsan and the major urban regions - became an issue. Environmental problems abruptly came to the forefront in the 1970s. Environment-related illness and problems caused by the rapid growth of heavy industry could be observed. At the 1972 United Nations Stockholm "Conference on the Human Environment", a Korean representative stated that Korea's rapid industrialisation had resulted in "pathologic phenomena" and "considerable pollution" (Lee et al., 1993: 491).

Industrial pollution and the destruction of ecosystems became even more serious in the 1980s. The implementation of ambitious economic development programs throughout the country did not consider their impact on the environment. Compared with the cities of industrialised countries air pollution (SO_2 and TSP), particularly in the metropolitan areas (e.g. Seoul, Pusan), was extremely high in the early 1980s.

1.2 The Change in Environmental Quality since 1970

Air pollution. The SO_2 pollution levels in major urban cities are shown in *Table 1*. In 1980 the annual mean for Seoul was 0.094 ppm, nearly twice as high as the environmental standard (0.05 ppm). Due to the supply of low-sulphur fuel since 1982, and an additional obligation to use liquefied natural gas (LNG) as clean fuel for the large buildings in cities since 1988, SO_2 emissions have decreased remarkably by about 80%: In 1991 they complied with the environmental stand-

Table 1. Changes in SO_2 Levels and Supply of Low-Sulphur Oil to Major Cities (Unit: ppm).

City	*1984*	*1986*	*1988*	*1990*	*1991*	*1992*	*1993*	*1994*
Seoul	0.066	0.054	0.062	0.051	0.043	0.035	0.023	0.019
Pusan	0.050	0.042	0.052	0.039	0.038	0.033	0.028	0.023
Taegu	0.040	0.043	0.044	0.040	0.041	0.040	0.035	0.038
Kwangju	0.026	0.020	0.019	0.017	0.017	0.017	0.014	0.013
Ulsan	0.024	0.032	0.028	0.031	0.038	0.031	0.032	0.030

Source : Ministry of the Environment 1995

Table 2. TSP Pollution Levels for Major Cities (Unit: $\mu g/m^3$).

City	*1984*	*1986*	*1988*	*1990*	*1991*	*1992*	*1993*	*1994*
Seoul	210	183	179	150	121	97	88	78
Pusan	228	194	214	140	134	113	96	97
Taegu	224	140	155	134	109	119	105	93
Kwangju	132	133	100	109	100	104	75	64
Ulsan	177	172	238	122	104	102	98	99

Source : Ministry of the Environment 1995

Table 3. NO_x Pollution Levels for Major Cities (Unit: ppm).

City	*1985*	*1986*	*1987*	*1988*	*1989*	*1990*	*1991*
Seoul	0.034	0.033	0.033	0.033	0.027	0.029	0.033
Pusan	0.024	0.024	0.024	0.019	0.024	0.019	0.023
Taegu	0.024	0.024	0.024	0.023	0.017	0.018	0.021
Kwangju	0.015	0.018	0.017	0.014	0.011	0.014	0.013
Ulsan	0.027	0.030	0.031	0.026	0.027	0.022	0.024

Source : Ministry of the Environment 1992.

ard, and in 1994 they reached a level of 0.019 ppm. The ambient air quality standard for SO_2 was set at a level of 0.03 ppm/year in 1993. It is being discussed whether this strict level should be raised to that of developed countries (USA and EU). The very unhealthy air quality as regards TSP was improved similarly within a short period (see *Table 2*). *Table 3* shows the NO_x pollution in major cities. Here the rapid increase of cars has neutralised the effect of catalytic converters.

Acid rain pollution (which has been measured in the larger cities since 1987) has increased since the 1980s. But this is mainly due to transboundary air pollution from neighbouring countries such as China.

Water pollution. Most rivers in Korea are highly polluted. The situation for the four major rivers - the Han, the Youngsan, the Keum and the Nakdong - has been greatly aggravated since the 1970s. Only the Han river shows some improvement in recent years (see *Table 4*). The rivers near large cities such as Seoul and Pusan and the larger industrial complexes are so badly polluted that the water is unsuitable even for industrial use (*Table 4*). Water pollution in Korea has various causes: waste water from rapidly growing industries, domestic sewage caused by a growing population with a rising standard of living, livestock waste water due to increased demand, etc. The urbanisation of rural areas has brought additional sources of water pollution.

Table 4. BOD of the Four Important Rivers (unit: mg/l).

river	*1985*	*1986*	*1987*	*1988*	*1989*	*1990*	*1991*	*1992*	*1993*	*1994*
Han (Nor-yangjin)	4.7	3.6	4.3	4.3	3.4	3.4	3.9	3.6	3.1	3.3
Nakdong (Mulgum)	3.7	3.6	3.6	3.9	3.6	3.0	4.0	3.3	3.4	4.6
Keum (Puyo)	2.5	3.0	2.9	3.2	3.5	3.1	3.0	3.2	3.1	3.7
Youngsan (Naju)	5.2	5.2	4.2	7.0	6.6	6.7	5.6	5.6	4.5	7.3

Note: The standard of water quality for Biochemical Oxygen Demand (BOD) in Class 2 is 3 mg/l.
Source : Ministry of the Environment 1995.

Marine pollution. The pollution of coastal and marine regions accelerated during export-oriented economic development, resulting from an increase in coastal industrial complexes or overseas activities such as ship-servicing. The pollution of the Masan Bay by industrial waste water generated in "the Masan Free Export Zone" in the 1970s was so severe that the area was named the Dead Sea. But preventive marine pollution control measures carried out by the government since the 1980s have led to a slight reduction in the contamination of coastal areas. In the western and southern seas however, red tide phenomena occur. Red tide phenomena were observed some 38 times in this area in 1993 (Ministry of the Environment, 1994).

Solid Waste. As shown in *Table 5*, the volume of waste grew steadily until 1991, but declined in 1992. The reason for this was a change in measurement methods for domestic waste and a reduction in the volume of used briquettes and garbage. However, the volume of specified waste generated increased due to a change of classification methods in 1991. The volume of waste was previously measured on the basis of the capacity of trucks but today, in the regions where sanitary landfills are located, the volume of waste is measured by actual weight (Ministry of the Environment, 1994). The generation of waste per person and year was 657 kg in 1992 (341 kg domestic waste). Thus Korea belongs to the group of countries with an extremely large amount of waste per capita. The government plans to reduce the level of generated waste to 507 kg per capita in the year 2001.

64.5% of all solid waste was dumped in landfills and another 3.3% was incinerated in 1992. The recycling rate for domestic waste tends to increase and is expected to be further improved by a new volume-based collection fee system from 1995 onwards.

Table 5. Annual Generation of Waste by Type and Cause (Unit: tons/day).

		1988	*1989*	*1990*	*1991*	*1992*
	Total	124,127	135,666	145,374	158,676	144,535
Type	General	122,114	133,356	142,721	139,955	123,154
	Specified	2,013	2,310	2,653	18,721	21,381
	total	124,127	135,666	145,374	158,676	144,535
Source	Domestic	72,897	78,021	83,962	92,246	75,096
	Industrial	51,230	57,645	61,412	66,430	69,439
Vol. per Capita and Year (kg)		766.5	803	839.5	839.5	657

Source: Ministry of the Environment 1994

2. Development and Main Characteristics of Environmental Policy

2.1 Institutionalisation

In 1973, Korean environmental administration began with an understaffed Environmental Division in the Ministry of Health and Social Affairs. In 1975 it was separated into two divisions, the Air Quality Management Division and Water Quality Management Division. In 1977, an environmental management system was established under the control of the deputy minister of Health and Social Affairs. The Environment Administration (EA) was set up in 1980. It installed Regional Monitoring and Management Offices for pollution control in six areas. 14 Central Guidance and Monitoring Teams were established in 1984 to monitor industries that discharge pollutants. In 1986 the Environment Administration was expanded. The Regional Environmental Offices were now included, to take on exclusive responsibility for environmental administration in self-governing bodies.

In 1990, the Environment Administration was upgraded to a quasi-ministry. The EA obtained the status of a ministry in integrating diverse environmental functions. However, due to the lack of proper legal and institutional instruments, it could not exercise effective control and integrate policy. There were also many difficulties due to the office's lacking the competence to implement its own orders. Above all, because of insufficient intra-administrative co-operation, which reflects one of the present government's weaknesses, it could not properly execute its functions.

To overcome those weaknesses, the quasi-Ministry of the Environment was transformed into a real Ministry in 1995. Its functions include policy formulation, the implementation of orders and the establishment of comprehensive plans for environmental protection.

The Public Nuisance Prevention Act was enacted as early as 1963, as the first law relating to the environment. But, for the want of an adequate organisational and monitoring infrastructure, it was not actually implemented. The law had a more hygienic character, dealing mainly with air pollution. River pollution prevention areas were designated by the Minister of Health and Social Affairs and pollution safety standards were set up (Ku, 1981). When the environmental situation worsened, the new notion of "polluting factories" was introduced in 1971. The existing system of pollution prevention areas was then abolished.

Instruments such as permits, cancellation or removal orders for industrial facilities were introduced. However, in spite of efforts to amend the law in that direction, the act was not implemented properly because of inadequate organisational, budgetary and human resources. For this reason the Environmental Preservation Act came into force in 1977. The act introduced important measures such as the promulgation of environmental standards, environmental monitoring, emission standard setting and emission control, as well as various administrative sanctions for violations. In 1990 the act was replaced by a differentiated system of laws: the Basic Environmental Policy Act, The Atmospheric Environment Preservation Act, the Water Environment Preservation Act, The Noise and Vibration Regulation Act, The Noxious Chemical Substance Control Act, and the Environmental Pollution Damage Dispute Adjustment Act. The Basic Environmental Policy Act provides the general basis for individual laws and recognises the environmental norms laid down in the constitution (1980, extended 1987).

2.2 Basic Instruments

Regulation: As in most OECD countries administrative regulations, permits and standards are most relevant to Korea's environmental policy. National standards are set for air pollutants, specific water quality indicators (pH, BOD, COD, total nitrogen etc.) and chemical pollutants (Cd, Hg, Pb, PCB etc.). In principle, local governments have also the authority to enact their own ordinances for environmental protection, but until now no local government has done so.

Environmental Impact Assessment (EIA): In 1977 the Environmental Preservation Act introduced the Environment Impact Assessment System in the form of "prior consultation". The act was revised several times, and finally the Environmental Impact Assessment Act was enacted in 1993. Under this act, 59 project types, for example, housing, energy, and tourist resort development projects, are subject to a prior environmental impact assessment.

From 1982 to 1993, 1,145 projects were subjected to assessment consultations. The main organisations which undertake large projects are government controlled enterprises, such as the Korean Electric Power Corporation, central development institutions, such as the Korean Land Development Corporation, and local government. According to data collected by the Ministry of the Environment (for 1992) public enterprises and development agencies in particular did not take environmental matters seriously into account.

Planning: Korea has a tradition of effective economic (five-year) planning. Compared with most OECD countries, environmental planning too is relatively well-developed. The medium-term Masterplan for Environmental Improvement (1992-1996) was laid out in 1991 (see *Table 6*). Its budget amounted to more than 1% of the Korean GNP in the first two years. The contribution of the private sector to the total costs of the plan was calculated to be 45% (Ministry of the Environment, 1994a: 6). Environmental long-term planning in Korea is based on the Basic Environmental Policy Act (1990), which defines criteria for diagnosis, prognosis, goal-setting, operation, monitoring, and budgeting (Lee, 1995: 152).

Table 6. Targets of the Master Plan for Environmental Improvement.

Programme	*1991*	*1996*
Improving the quality of drinking water resources		
- Proportion of 1st-rate drawing posts for drinking water (to total drawing posts)	34%	70%
Recovery of water quality in rivers		
- treatment rate of sewage	33%	65%
Sanitary treatment of solid waste		
- proportion of sanitary treatment of solid waste from households (to total treatment)	27%	90%
Maintenance of clean air		
- concentration of SO_2 (Seoul City)	0.043 ppm	0.033 ppm
Enlargement of green areas for rest and recreation		
- proportion of natural park area (to total land area)	7.5%	10%
- area of parks in the cities	680 km^2	800 km^2

Source: Ministry of the Environment 1994a

Other environmental protection measures have also been formulated in plans. The Comprehensive National Plan for Waste Treatment (1993) has given itself the target of a recycling rate of 30% by 2001 (1992: 7,9%). In the same year, a Five Year Plan for Energy Saving was formulated, intended to reform the existing transportation and construction industries, including the goal of reducing the proportion of energy-intensive branches in the manufacturing sector. The Target Fuel System lays down fuel efficiency targets for 2000.

Charges: Charges and fees are frequently applied instruments in Korea. The Emission Charge System was introduced in 1983 for businesses which violate environmental standards. The Minister of the Environment can order that emitters should pay emission charges based on the type of pollutant, the amount and time-span of excess emissions. Emission charges include basic charges which are fixed according to the size of the firm, and treatment charges which vary directly with the amount of excess discharge of pollutants. Emission charges are imposed on firms discharging specific types of air pollutants and water pollutants. Improvements will be made in 1996. Firms, then, will have to pay charges for air and water pollution even if they are within the established tolerance limits (Director General, 1995).

The Environmental Improvement Charge System: This is an economic incentive system which is intended to promote measures for environmental improvement and to raise the funds required. Its goal is to secure a financial basis for investments and to induce low emissions by applying the polluter-pays principle. Targets for its imposition are the owners of buildings or vehicles which cause relatively large amounts of pollution. However, problems have arisen in implementing this system (Kim, 1993).

Economic instruments for domestic waste: In the beginning the waste disposal fee levied on discharges was very low and fixed according to the property tax for buildings and the size of land required for a building, independently of the amount of waste produced. The rapidly increasing volume of waste created financial and other problems. And so the Volume-Based Collection Fee System for domestic waste was introduced in 1993. It was imposed on households and small-scale firms according to the volume of their waste. First executed in 33 selected trial areas, it has been implemented nation-wide since January 1995. According to the Ministry of the Environment, "a considerable reduction in waste has been observed throughout the country" (Director General, 1995). However, the effectiveness of the system is controversial.

In 1992, when the Act Relating to Promotion of Economy and Reutilisation of Resources was introduced, the Deposit-Return System for products containing toxic materials or causing mass waste was established in order to reduce waste by encouraging the retrieval of reusable items. At present it applies to 11 items, including batteries, beverage cans, air conditioners, or TV sets. The Waste Treatment Charge System was introduced in the same year to curb the consumption of products and containers which cause problems in handling waste (Director General, 1995).

When reserve forest lands are converted for other uses, a reforestation fee and a conversion fee are imposed (Ministry of the Environment, 1992: 81).

Eco-Mark system: This system was introduced in 1992. It is designed, according to differentiated criteria, to certify consumer products which are manufactured by environmentally sound processes, and which can be used in environmentally sound ways. In June, 1994 this system included 162 products and 68 companies. The Eco-Mark Committee, a decision-making body for the management of the program, consists of representatives from business as well as consumer organisations, environmental experts and journalists. A private association, the Eco-Mark Association, was founded in 1994 to facilitate the implementation of the programme (Director General, 1995). In 1992, a Labelling Scheme for Energy Efficiency (e.g. for cars and electric appliances) was also introduced.

2.3 Political Structure

The Korean model of development was based on an authoritarian state which was able to maximise economic growth by strict government control (You, 1995). The environmental policy structure, too, was characterised by a strong centralism. After the political changes since 1987, a reorientation towards a certain decentralisation took place. Presently, central government acts in carrying out development projects, and administrative affairs relating to the environment are turned over to local government. The direct election of local governments in June 1995 was an important step here.

3. The Main Proponents of Environment Protection

3.1 Environmental Policy and other Political Institutions

Environmental Institutions of the Central Government

The Ministry of the Environment is mainly responsible for the conservation of the natural and human environment and for pollution control. As the central office, it has to plan, integrate and evaluate environmental measures, considering both the environment and natural resources.

The recently reorganised institution consists of the more general Planning and Management Office and the Environment Policy Office and - organised according to environmental media - the Nature Conservation Bureau, the Water Quality Management Bureau, the Air Quality Management Bureau, the Waste Management and Recycling Bureau and the Water Supply and Sewage Treatment Bureau. But each part has problems with integration and the proper management of its functions. Thus its structure needs to be reorganised once again.

The central office conducts the main affairs in the Ministry of the Environment. It has at the same time a difficult integrative task with respect to the many environmentally relevant activities located in other ministries. For example, the Ministry of Construction and Transportation takes charge of the rivers, water resource management and the implementation of the National Land Use Control Act. The Ministry of Home Affairs is responsible for marine pollution. The Ministry of Commerce and Industry takes charge of matters relating to the use of energy. This involves safety control in nuclear power plants. But the Ministry for Science and Technology then takes charge of the treatment and disposal of nuclear waste. The Ministry of Agriculture, Forestry and Fisheries deals with measuring environmental pollution in the areas of agriculture and fisheries. It is also responsible for forestry conservation and management. The Ministry of Culture and Athletics undertakes measures connected with designating, and managing natural monuments such as exotic animal and plant species (Ministry of the Environment, 1994).

The Regional and Local Environmental Administration

Environmental Management Office. The Regional Environmental Office was reorganised and Environmental Management Offices were set up around the four large rivers (Han, Nakdong, Keum, and Youngsan). Regional Environmental Management Offices were also established in Taegu, Chonju, and Wonju. Environmental Branch Offices have been instituted in densely industrialised or highly polluted areas (Inch'on, Ch'angwon, Ch'unch'on, Ch'ungju, Sunch'on, and Cheju), where Environmental Management Offices do not exist.

The main duties performed by an Environmental Management Office are: planning/evaluation and management. In the former case the functions involve inter alia public environmental information and education, planning and implementation of environmental conservation programs, Environment Impact Assessments, conservation and management of natural ecosystems and the environment, participation in the conference on national land use programs, licensing, registration and control of businesses regarding anti-pollution measures, registration of firms using toxic substances, imposition of waste charges. In the latter case, the functions of an Environmental Management Office involve the quality and quantity of drinking water resources, monitoring and control of river pollution, operation of sewage treatment plants, gathering environmental statistics, implementation/operation of an environmental pollution monitoring network for air, soil, acid rain, etc.

Environmental administration of local governments. Local Governments under the Ministry of Home Affairs also have special administrative branches for environmental affairs. Seoul and each special city - under the direct control of central government - have an Environmental Landscape Bureau, and each district has a Health, Social Welfare & Environment Bureau. In the case of a small city, country or ward, environmental protection divisions have been set up.

The local governments take charge of their own environmental policy tasks laid down in the environmental law, but also carry out duties assigned to them by the Minister of the Environment. A number of environmental affairs have been turned over from central government to local government after July 1992.

Environment-related tasks entrusted to local government are: the setting-up and implementation of environmental protection programs within their jurisdiction; pollution control regulation, control of the establishment and management of facilities which discharge air and water pollutants; the imposition and collection of emission charges and the various types of non-compliance fines; construction of sewage treatment plants; matters relating to the management of water quality in lakes and farmland.

At present, the environmental administrations of local governments still lack the professional skills required to conduct environmental policy and research. They also lack the funding necessary to carry out the more extensive tasks and are lacking in the integrative functions of environmental protection and resources management that an environmental administration should possess. For these reasons they are not able to execute their tasks properly.

The local assemblies are the highest decision-making organisations at local level which are able to reflect public opinion and are important in determining the environmental policy of the self-governing bodies. In the process of democratisation and decentralisation of the country, the first local assembly was established in 1992. As legislatures they are able to enact regulations within their own jurisdiction. As supervisory bodies they can observe and criticise the local administration. But the development of local self-government is possible only if the local assembly actively uses the functions described above.

Parties. In Korea, the parties played only a minor role and were not empowered to set independently environmental policy goals until the political changes of the late 1980s. At present the party system still has to find a stable structure. The ruling New Korean Party (the former Democratic Liberal Party) focuses rather on the promotion of a pleasant and stable natural environment. The National Assembly Party and the Democratic Party emphasise a more preventive environmental policy. A green party does not exist.

Environmental NGOs and Citizens' Action Groups

Since the political changes in 1987, the environmental movement has been growing rapidly. 1,219 resolutions and petitions against environmental pollution, as counted in the official environmental statistics, were registered in 1988 (1,153 in 1992; Ministry of the Environment, 1994: 512). Some of these protests have been successful. A group of citizens, for instance, brought about the closure of three factories which were polluting the Kuem river. Locally-based environmental associations have been organised in major industrialised cities such as Seoul or Ulsan.

There also exist more then ten nation-wide environmental NGOs. The oldest such organisation is the Korea Nature Conservation Association whose membership includes professors, teachers and journalists. It has been active in such fields as wildlife conservation. The Korea Environmental Preservation Organisation consists of business people and environmental engineers. It has been publishing a number of booklets and a monthly magazine. The Anti-Pollution Movement Association and the Friends of Nature are the two most well-known environmental activist groups. The Anti-Pollution Movement Association for instance has been engaged in the investigation of pollution damage (Ministry of the Environment, 1992: 90). Other important NGOs are the Korea Consumers Federation which has a nation-wide information network and is also engaged in environmental protection issues. The women's rights movement or the traditional water conservation movement should also be mentioned. However, the participation of NGOs in political decision-making is still unusual.

The Basic Environmental Policy Act, enacted in 1990, has expressly introduced the opportunity for local inhabitants to participate in an Environment Impact Assessment. But the people's demand for stronger environmental policies is not reflected in clear methods. An effective institutional mechanism for participation still has to be developed, to enable NGOs to play an active role in environmental protection.

3.2 The Role of the Press in Environmental Protection

Compared to the period before 1987, the number of reports in the media on the environment has increased rapidly in the 1990s. According to the Ministry of the Environment, the rate of environmental reports appearing in daily newspapers was 414 in 1980, 1,313 in 1988, 5,331 in 1990, and 8,884 in 1992.

This rise probably indicates that it has become increasingly possible for the media to report on environmental issues. The press and television, and their campaigns, are playing the leading role in creating public awareness of the environment. In particular, television is presenting reports on environmental problems in Korea and the rest of the world more than ever before.

3.3 Target Groups (Polluters)

The number of firms discharging pollutants has been steadily increasing. In 1987 58,310 such firms existed. In 1993 the number was 86,183. This includes 25,808 firms discharging air pollutants, 28,834 firms discharging water pollutants, 28,365 producers of noise, and 3,176 producers of vibration.

Many enterprises still do not observe emission standards because awareness of the necessity of environmental protection is still low. The proportion of capital investment for environmental protection is also generally low (1988: 0,8%; You, 1995). However, some large enterprises such as the Dusan and Pohang Iron and Steel Company are becoming more concerned about environmental protection and are investing heavily in the development of environmentally friendlier technologies.

4. Towards a New Strategy?

At the beginning of the 1980s, the environmental administration. concentrated on affairs such as end-of-pipe pollution control, with the dominant policy style of reacting to problems that had already occurred. As domestic and foreign interest in environmental protection grew, the government designated the year 1990 as "the first year for environmental protection" and deliberated on comprehensive plans for policy development. Now, in connection with the Five Year Economic Plan, plans relating to the environment have also been developed:

The first goal of the environmental concept is to prevent environmental pollution. Therefore the government intends inter alia to increase in stages the number of those regions and technologies which are under an obligation to use LNG and gasoline in order to reduce air pollution. It has also strengthened the Environmental Impact Assessment System and introduced the concept of "Environmental Sustainability Appraisal". On this basis, the government intends to introduce a system that evaluates the security and the sustainability of each project.

Second, the government is to promote the development of environmental technologies. This was done through a special law (1994). It will thus designate 20 projects by 1997, such as low-pollution incineration techniques, or green gas control techniques. It also intends to strengthen the financial support of R&D for such technologies. Tax incentives and a special environmental technical management system will also promote the introduction of such technologies.

Government is also to expand financial resources for environmental investment in the public sector. By extending the scope of various forms of environmental charges, it intends to implement its policy more efficiently and to raise the funds required. It also intends to introduce a "special account for environmental improvement".

Environmental standards will be improved in stages. It is also intended to set environmental standards appropriate to the situation in each region. There will be inspection of water quality in each river and water system. The government is also considering the introduction a "water quality information centre" providing information through an on-line information system.

Generally speaking, this is not yet a change of paradigm from end-of-pipe treatment towards efficient resource management. But the conditions for policy learning have improved and may finally lead in this direction.

5. General Evaluation

According to Jong-Il You, "Korea has developed an industrial structure which is intensive in materials and energy use and employs technologies that are highly polluting as a result of the virtual non-existence of environmental regulations. Consequently, Korea's rate of increase in pollutant emissions has been higher than its world record-setting growth rate. In addition, Korean agriculture is one of the most intensive in fertiliser and pesticide use. The problems of municipal, toxic and nuclear waste have become critical, with increasing difficulty in finding new landfill sites or storage facilities. But Korea is not without its success stories, even in its environmental record. Its record on curbing population growth is one of the best in the world, its reforestation programme is one of its more spectacular successes, and it has done a decent job of providing clean water and sanitation" (You, 1995).

It may be added that air pollution measured in Korean large cities in the early 1980s was extremely high in international comparison. But within a few years pollution from sulphur dioxide and TSP has been significantly reduced. Korea introduced relatively early on (1987/8) regulations for low emission cars and catalytic converters. River pollution is high and has not improved in the 1990s (see *Table 4*). Only one third of municipal waste water was cleaned in the early 1990s. But it is planned to increase sewage treatment to 75% by the year 2000. The recent measures taken against the urgent solid waste problem might be worth mentioning in comparison with the majority of industrialised countries. The proportion of urbanised land grew by 18% between 1982 to 1990 and is predicted to increase further. But Korea is one of the few countries in the world with a soil protection law (1995) and a soil monitoring system (Ministry of the Environment, 1992; Lee, 1995).

As regards government institutions, non-governmental organisations, press reporting, or legal resources, environmental policy capacity has increased rapidly since 1987. The main restriction may be the high status of economic growth within the hierarchy of objectives both of government and the people (You, 1995). There is also a great need for knowledge and professional training in the expanding field of environmental policy and management. National environmental monitoring and reporting (Ministry of the Environment, 1994, 1995) is of a high professional standard, though it is mainly dealing with pollution, and not so much with structural conditions. Intrapolicy integration between different environmental administrations is still insufficient. Interpolicy integration between environmental policy and industrial, energy, transport, construction, or agricultural policy is still at a low level. Without significant changes in such sectoral policies the rapid growth of pollution problems will not be halted.

On the other hand, there may also be some potential advantages in the Korean way of development: the tradition of long-term planning, the tradition of strict government intervention, the excellent economic situation, combined with democratic reforms and a relatively high acceptance of change. Parallel economic and political modernisation especially could represent an opportunity also for ecological modernisation.

References

Director General for International Co-operation, Ministry of the Environment, Korea, 1995: Sustainable Consumption as a National Agenda. Seoul.

Jeong, H. and Y. Nam, 1993: A Study on the Development of Sustainability Assessment Models and Their Application, KETRI/1993/RE-08. Seoul: KETRI.

Kim, K., 1993: A Study on the Instruments of Financing Public Environmental Investments. Seoul.

Ku, Y., 1981: The Theory of Environment Law. Seoul.

Lee, D. et al., 1993: Human Environmental Theory. Seoul.

Lee, Y. H., 1995: Umweltschutz, Bodenschutz und Planung. Vergleichende Aspekte von Umwelt-, Bodenschutz und Planung in der Republik Korea und der Bundesrepublik Deutschland. Technische Universität Berlin: Fagus-Schriften Bd. 6.

Ministry of the Environment, 1992: National Report of the Republic of Korea to UNCED 1992. Kwacheon.

Ministry of the Environment, 1992a: Environmental Yearbook. Kwacheon.

Ministry of the Environment, 1994: Environmental Yearbook. Kwacheon.

Ministry of the Environment, 1994a: Environmental Protection in Korea, Kwacheon.

Ministry of the Environment, 1995: Environmental Yearbook. Kwacheon.

Nam, Y., 1993: A Study on the Analysis of Environmental Policy and the Assessment of Sustainability of Provincial Governments, KETRI/1993/RE-09. Seoul: KETRI.

Nam, Y., 1994: Promotion of Environmental Sustainability Assessment, KETRI/1994/RE-19. Seoul: KETRI.

Nam, Y., 1995: Promotion of Korean Environmental Impact Assessment, Environmental Forum 18.

You, J.-I., 1995: The Korean Model of Development and its Environmental Implications, in: V. Bashkar and A. Glyn (eds.), The North, the South and the Environment. Ecological Constraints and the Global Economy. London.

CHILE

Eduardo Silva[1]

Introduction

Chile is a relative latecomer to environmental capacity-building, having been under military rule during the period when most advanced Latin-American nations and many European countries initially developed their capacity (1972-1980) (Brañes, 1991). Unfortunately, unrelenting economic crisis during the 1980s and 1990s has significantly eroded the real capacity of environmental institutions in Latin America (as well as in Africa, as evidenced by the case-study of Nigeria in this volume). Thus it is ironic to note that in contrast to their economic and institutional decay, Chile, after a history of deep political conflict and sharp economic swings, has emerged as a model of economic and political stability for the developing world. It now boasts stable economic growth, a fiscally responsible state, rising per capita GDP, and consolidated political democracy. These elements suggest that Chile has reached a sound footing for the development of capacity for environmental policy-making. This paper explores the extent to which Chile has delivered on that promise.

In Chile, the twin processes of healthy economic growth and democratisation have opened political space for increased attention to environmental issues. Since 1990, policy makers and organised civil society have systematically built up the nation's capacity for environmental policy-making. Effort has concentrated on the development of basic legal, institutional, participatory, and perhaps also integrative capacities. In the process, a joining of forces among state, business and agricultural, and environmental groups has resulted in a focus on highly visible pollution and emissions. Policy has concentrated on dilution, high chimney strategies, and end-of-pipe technology. Given the recentness of the process – these are all fledgling initiatives – one can only assess their objectives and their potential strengths and weaknesses in achieving them.

1 Thanks to Patricio Silva, Frank Fischer, Patricio Rodrigo, the Corporación para la Preservación y Conservación del Medio Ambiente (Alejandra Aburto and Carola Urrutia), and the Center for International Studies of the University of Missouri-St. Louis.

By 1994 it was clear that initiatives to diagnose fundamental problems, craft basic legislation, and to strengthen public institutions were at heart weak. Moreover, the resulting structures make consideration of strategies to implement a more holistic conception of sustainable development problematic. For example, extraction of natural renewable resources is not considered. Environmental problems are treated as if they were unrelated to other policy areas, such as economic and social policy. There are no incentives for the substitution of highly toxic industrial processes. Participation in the policy-making process is tightly restricted, to groups which support the status quo.

This outcome might seem natural enough given Chile's level of economic and political development. But the shortcomings of Chile's environmental capacity-building efforts were not the inevitable result of any particular level of economic and political development. Rather, they were the consequence of a political struggle during the first two years of democratic government. On the losing side were state and societal actors who favoured the creation of a more broadly encompassing legal framework, stronger institutions, more diverse sources of participation, and the establishment of norms and principles that would facilitate the elaboration of policy geared toward operationalising the concept of sustainable development.

This perspective on Chilean environmental policy guides the organisation of this chapter. The first section covers the period of military government, and the major environmental problems that were aggravated during that period. The second section examines the democratic governments' responses to these issues. A third section concludes with an evaluation of the direction of policy and the political accommodations (and the ensuing policy style) that gives the policies their orientation.

1 The Legacy to the Environment of Military Rule, 1973-1989

In Latin America, Chile emerged from military rule as a successful example of market-oriented economic restructuring among developing nations. The economic model emphasised exports based on extraction activities in agriculture, fisheries, and timber, as well as minerals, principally copper. It privatised the financial sector, reduced the manufacturing sector, and encouraged commerce, especially import-export transactions. To spur on economic restructuring, the military government drastically reduced the scale of government and public ownership of the economy. Within the confines of this neo-liberal model, Chile's economy has done quite well. As of the mid-1980s, it has enjoyed sustained GDP growth rates; its fiscal accounts are in order; it has low inflation, rising investment rates, a sound financial system that exports capital to neighbouring countries; and vigorous exports keep its external accounts in order.[2] Yet, sustained

2 For the general political economy of Chile see, Drake and Jaksic, 1995; Oppenheim, 1993; Hojman, 1993.

economic growth – born in the heady, freewheeling days of neo-liberal economic restructuring under a military dictatorship – came with a significant environmental price tag.[3] This section describes these environmental problems, the military government's response to them, and the early development of the environmentalist movement in Chile. They are the necessary background to the efforts in environmental capacity-building of the current democratic period.

1.1 Chile's Main Environmental Problems

The Fishing Industry. Chile possesses a 6,435 km-long coastline which is very rich in marine resources. During the military government, Chile's fishing industry expanded dramatically from a catch of about 1.5 million metric tons in 1971 to 2.5 million in 1979 and just above 6 million in 1989. The average catch per year between 1989-91 was 5.9 million metric tons, about twice that of the period between 1979 and 1981. While this is good for industrialists and national accounts, it does not bode well for the long-term prospects of Chilean fisheries. According to the World Bank, 3.5 million tons constitutes the maximum sustainable average for Chilean coastal waters. The current size and age of the catch suggest over-fishing and the consequent decline of fisheries (Arsenberg et al., 1989; Cárdenas, 1985; Gómez-Lobo, 1991).

The Forestry Sector. Forestry was another sector that grew at spectacular rates during the period of military government. Between 1984 and 1990, it expanded from 4.8 to 7.5% of GDP and accounted for 750 million dollars worth of exports in 1991 (World Resources Institute, 1994). The development of the forestry sector has had both benign and deleterious environmental effects. Most of the timber industry is based on plantations of exotic species, predominately pine and some eucalyptus. On the positive side, the early plantations grew up on abandoned eroded soils in south-central Chile. As a result, they helped to recover degraded agricultural and ranching land (CONAMA, 1992). On the negative side, the emphasis on plantations within the timber industry devalued native species. Thus, as plantations have expanded they now replace natural forests. Moreover, the growing international demand for short-fibre wood chips abroad has led to wholesale clear-cutting of natural forests (Silva, 1996b).

The Mining Industry. Mining, dominated by copper, remains Chile's principal export-earning activity, accounting for nearly half of the nation's hard currency income (US$ 4.4 billion in 1991). Chile is the second largest producer of copper in the world, and the publicly owned company (CODELCO) owns one quarter of the mines and provides three quarters of the industry's output. Between 1980 and 1990, CODELCO doubled production from 600,000 to 1,200,000 metric tons (World Resources Institute, 1994). The mining industry, concentrated in the

3 Of course, the dictatorship did not create all of those environmental problems. Many of them already existed. Its laissez faire policies, however, aggravated them and created a few new ones, such as over fishing, natural forest destruction due to substitution for plantations, and intensive agrochemical use to name a few.

deserts of Northern Chile, contributes substantially to rural and urban pollution in that region. Smelting, refining, and the extraction of commercial byeproducts account for the bulk of it in the form of sulphur dioxide, fine particulate matter, and arsenic. The resulting air pollution poses significant health risks for local populations. Toxic liquid and solid wastes pose serious problems downstream, in coastal waters and urban areas (Aburto et al., 1995; CONAMA, 1992: 70).

The Agricultural Sector. Chile's free-market model emphasises agro-industrial exports, dominated by fruits from Chile's fertile central valley, which have expanded from 36% of agricultural exports in 1970 to more than 70% since 1988 (World Resources Institute, 1994). Intensive use of agri-chemical inputs have sharply increased contamination of soil and water, and pose serious health problems for agricultural workers and their villages, as well as other downstream effects. There is a lively and unresolved debate on their effects on consumers (CIPMA, 1991; CONAMA, n.y.). Soil loss, erosion, desertification, sedimentation, and salinisation are not considered, by some accounts, as serious problems for the modern, technologically intensive agricultural sector in the short term. However, lack of attention to these issues could pose serious problems 20 years hence. In contrast, these conditions actively threaten marginal lands worked by impoverished peasants (CIPMA, 1991). Desertification and salinisation are serious problems in northern Chile (CONAMA, 1992).

The Built Environment. The urban-industrial areas of Chile suffer from a myriad of problems, of which air pollution and sewage are among the most pressing. While the issues described below are common to many cities in Chile, they are best documented for Santiago, the capital city. It is estimated that about 4.5 million people inhabit greater Santiago, out of a total Chilean population of just over 13 million. Santiago lies in a valley and is surrounded on three sides by the Andes mountains. A great deal of the nation's manufacturing plant is located in the area, and air pollution has reached alarming levels as a result of point and non-point source emissions. Since 1987 the vehicle stock has increased at an average rate of 10% per year (CONAMA, 1992). By one estimate, in 1994 industry and public transport spewed approximately 16 metric tons of particulates per day into the metropolitan region's atmosphere; while industry added about another 110 tons of SO_x per day (InduIndustria, n. y.).[4] Moreover, industry-based point sources routinely discharge insufficiently treated toxic wastes or organic materials into streams and canals, and coal-fired utilities contribute to sulphur dioxide emissions. The health effects of this pollution include a drastic rise in the incidence of respiratory disease, especially in winter, where a temperature inversion noticeably worsens the situation. For example, in 1988 there were 300,000 additional cases of bronchial-pneumonial disease (CONAMA, 1992).

Urban areas in the rest of the country have similar problems. Northern cities are particularly affected by the mining industry, due to uncontrolled smelting and refining processes which produce intolerably high emissions of arsenic, sulphur

4 The same study notes that the average concentration of particulates in Santiago oscillates between 105 and 110 micrograms per cubic meter per day. This is double the acceptable standard of the USA, which is 50. In critical winter months the index can surpass 500 micrograms per cubic meter.

dioxide, nitrous oxide, copper particles. In addition to these problems, inadequate sewage treatment and industrial discharges from mining, manufacturing, food processing, fish plants, and the pulp and paper industry contaminate the water supplies of urban areas. Open canals that receive these toxic and unsanitary discharges are often used to irrigate truck farms. This contributes to outbreaks of typhus, hepatitis, and, until recently, cholera (Gutiérrez, n.y.; CONAMA, 1992).

Hydraulic Resources. Chile generates 33,680 cubic meters of internal renewable water resources per capita. These are all generated within its borders, and remain within them. Surrounded by the Andes mountains and the Atacama desert, Chile does not benefit from river flows from other countries as do most of the other cases in this study. Chile's internally generated renewable water supply is either comparable or superior to that of many South American, European, and African countries. However, Chile's rate of withdrawal is very high compared to most of Latin America and Africa. In 1975, the only year for which data exist, that rate was 1,623 cubic meters per capita, or 4% of total water resources. Notably, European nations tend to have far higher rates of withdrawal. Chile's figure, however, can be misleading. In 1975 Chile was in the trough of a deep recession. With economic recovery and expansion, one might expect that water use has risen significantly. Indeed, it is estimated that the uncontrolled economic development (in mining, manufacturing, fish processing, the pulp and paper industry, intensive agriculture, and timber) places great stress on Chile's fresh water resources and coastal areas. Increased industrial and agricultural production and the expansion of cities have created a water shortage. By one estimate, agriculture uses up 89% of fresh water resources, industry 5%, and domestic sources consume 6%. Uncontrolled economic development and urban growth have also exacerbated the pollution of Chile's rivers and coast (Eco-XXI, no. 4, 1993; CONAMA, 1992; World Resources Institute, 1994).

Biodiversity. Soil erosion and salinisation in the North and coastal central Chile, deforestation, plantations, extensive bovine and ovine grazing further south, pollution: all of these processes have affected the habitats of flora and fauna. By 1985, 69 species of Chilean flora found themselves in danger of extinction. Of these, 11 were classified as endangered, 26 were placed in the vulnerable category, and 31 were considered rare. At least six species of mammals and birds also found themselves in danger of extinction. An additional two species of mammals and six of birds were in the vulnerable category (CONAMA, 1992). Moreover, approximately 13.7 million hectares are covered by Chile's system of protected areas. This is equivalent to about 19% of Chile's landmass. This puts Chile far ahead of most Latin-American and many European countries (with the notable exceptions of Ecuador, the Dominican Republic, Venezuela, Austria, and Germany), as well as many African nations including Nigeria. Countries with similar levels of protection include, the United Kingdom, Switzerland, and Botswana. The system is managed by the Corporación Nacional Forestal, a semi-autonomous government agency.

1.2 Capacity-Building During the Military Government

Chile's military government, bent on recasting Chile in a neo-liberal mold, essentially repressed the environmental issue area at a time in which European and North American nations were rapidly building capacity in that area. Yet, as pollution became more acute, it could not suppress the emergence of an environmental movement. Moreover, as international pressure mounted, it could not avoid creating rudimentary mechanisms for building participatory, legal, and institutional capacity. Of course, the military government ensured that they remained weak and ineffectual. Nevertheless, the non-governmental organisations, laws, and government agencies that appeared constituted the matrix of actors and institutional structures that affected environmental capacity-building under the subsequent democracy.

Participatory capacity for the environment in Chile began with the birth of environmentally-oriented non-governmental organisations. For nearly three decades these were the main proponents of capacity-building for the environment. The Chilean environmental movement began in 1963 with the founding of the Comité de Defensa de la Fauna y Flora (CODEFF), which remains one of Chile's most important environmental non-governmental organisations (NGOs). Although this group of ecologists, dedicated to the preservation of biodiversity, was largely ignored at the time, the Stockholm conference on the environment in 1972 and the creation of the U.N. Environmental Programme gave further impetus to the movement (Rojas, 1994). In 1974, at the beginning of the military government, a group of ecologists founded the Instituto de Ecología. During the military government this organisation actively lobbied the authorities for the creation of a legal and administrative framework within which to address Chile's environmental problems.[5]

Many more environmental NGOs sprang up in the late 1970s and 1980s as a result of political repression that purged centrist and left-leaning academics from the universities. The Centro de Investigación y Planeamiento para el Medio Ambiente (CIPMA) – another of Chile's leading environmental NGOs – was founded in the early 1980s. So were the Grupo de Investigaciones Agrarias (GIA), the Centro de Estudios y Tecnologia (CET), El Canelo de Nos, and the Instituto de Ecología Política (IEP). Rather than focus on strict conservationist concerns, these organisations linked development, social justice, and environmental issues (Rojas, 1994). Of these NGOs, CIPMA remained in the political centre and, through a series of congresses over the years, played an important role in consensus-building among the different groups, and in political agenda-setting. GIA and CET tended to lean more to the left, while IEP, Canelo de Nos, and CODEFF took more radical ecological stances along new-left lines.

While civil society organised cautiously, the military government either ignored the environmental impact of its development model or, when pressure

5 Interview with Dr. Juan Grau, noted Chilean environmentalist who won the UN Global 500 award in 1987, 3 March, 1995, in Aburto et al., 1995.

mounted, gave it very low priority. The government's main goal was rapid economic growth at any and all costs, and with minimum government intervention. As a result, between 1973 and 1984 environmental problems worsened significantly and received little or no attention, as evidenced by the lack of legal development and institutional capacity-building (Grau, 1989). Nevertheless, growing world attention to environmental issues, limited lobbying, and the consolidation of the Chilean environmental movement when the regime was politically vulnerable induced the military government to respond in ways that built legal and institutional capacity. It must be understood, however, that the military government's purpose was to co-opt parts of the environmental movement in order to check the growing force of domestic and international political opposition to the government.

The first significant measure of legal capacity building occurred in 1980. In answer to the strenuous efforts of key individuals in the Instituto de Ecología, the military government gave the issue legal status in the Constitution of 1980, which is still the law of the land today. It stipulated that citizens have the right to live in an environment free of contamination, and that it is the duty of government to guarantee that right and to oversee nature conservation. Although these were mere pretty words during the dictatorship, they gained new meaning under democracy.

The military government's second measure involved institutional capacity-building, and followed on the heels of first Scientific Congress for the Environment organised by CIPMA in 1983. The event brought together 377 environmentally concerned scientists and academics (a good turn out given the lack of political freedom in Chile). It proved to be a milestone in the development of participatory capacity because it was a first step in uniting hitherto isolated environmental organisations. They met at this congress to diagnose the country's main environmental problems and offer policy recommendations. Despite government control, media reporting revealed an agitated scientific community that offered solid scientific analyses about the critical condition of Chile's environment. In short, the environmental movement began cautiously to flex its muscles in an effort to gauge its strength.

The military government – now in a period of political vulnerability due to the nation's economic collapse and high levels of mobilisation against it – responded by creating the National Commission for Ecology in 1984. Its main task: the formulation of a national environmental policy. That goal, however, remained unaccomplished despite the best efforts of its well-intentioned staff (Grau, 1989; Rojas, 1994). This was mainly due to the closed nature of the policy-making process. This allowed the military government to bury the issue in "legislative commissions".

Despite this rather pro forma approach to environmental capacity building, the military government aggressively pursued one environmental problem. While Chile is a party to most global environmental conventions (World Resources Institute, 1994), there was one in particular that the military government took seriously: the ozone layer convention of 1985 and the CFC control convention of

1987. This problem affected Chile directly. A large "ozone hole" over Antarctica, which Chile had virtually no role in creating, presented very real hazards. In addition to adverse health effects on humans, Chileans worried over the consequences of higher doses of ultraviolet radiation on fish, timber, and fruits, the mainstays of Chile's economy. As a result, Chilean diplomats worked hard for the success of the Montreal Protocol and follow-up meetings (White, 1993; Brañes, 1991; Silva, 1996a).

2 Capacity Building for the Environment under Democratic Rule

For the proponents of environmental action, the return of democratic rule in March 1990 signified a notable change for the better in the political opportunity structure. This was due as much to the composition and strategy of the victorious opposition forces as to the greater openness inherent in democratic political systems. Growing international pressure for the creation of environmental policies in developing countries added to their strength. The result was a concerted effort to build legal, institutional, cognitive, and participatory capacity for environmental policy-making.

Democracy was inaugurated with the administration of Patricio Aylwin, a Christian Democrat, and candidate of a broad centre-left coalition of political parties that had formed the moderate, reformist opposition to General Augusto Pinochet since 1983 (Drake and Jaksic, 1995; Oppenheim, 1993). The Concertación por la Democracia's (CPD) programmatic platform, dominated by the centrist Christian Democratic party (PDC) and including non-revolutionary leftist parties, principally the Partido por la Democracia (PPD) and reformed socialists (PS), championed virtually all of the socio-economic and political issues that the dictatorship had either exacerbated, ignored, or repressed. Thus, environmental problems, received attention along with the issues of political democracy and social justice (Concertación de Partidos por la Democracia, 1989). A second CPD administration that began in March 1994 under Christian Democratic president Eduardo Frei, Jr., continued to address these concerns.[6]

Domestic factors were not the only ones giving urgency to environmental issues. Demands from external actors also put pressure on the Aylwin administration. Given the advanced state of Chile's market-oriented economic restructuring, it was among a select group of candidates for inclusion in an expanded North American Free Trade Agreement (NAFTA). The NAFTA debate with Mexico showed that entering into such an agreement with the United States demanded attention being given to environmental issues.

The military government's lack of attention to the environment meant that the new government had to begin at the beginning. It had to build elemental legal, institutional, cognitive, and participatory capacity to deal with serious pollution

6 For Eduardo Frei, jr. administration's environmental program see Concertación de Partidos por la Democracia, 1993, 1994.

hazards in addition to natural renewable resource depletion and threats to the nation's biodiversity. This section analyses the Aylwin administration's environmental capacity-building efforts and the limits of existing capacity.

2.1 Legal, Institutional, Cognitive, and Participatory Capacity-Building Efforts: The Environmental Framework Law and Priority Setting

The Environmental Framework Law. The establishment of a legal and institutional basis for environmental policy-making was one of the Aylwin administration's most significant achievements. This was achieved with the passage of the Environmental Framework Law in March 1994 (Silva, 1994).[7] Additional objectives of the law included the establishment of a few basic instruments with which to carry out that policy, and to ensure that subsequent laws would provide a framework of action applicable to individual economic sectors (Aburto et al., 1995; World Resources Institute, 1994). Four principles inspired the framework law: prevention, the polluter pays, gradualism, and participation.

The preventive principle addressed the problem of how to minimise the environmental degradation resulting from future economic activity. To operationalise this principle, the framework law established mandatory environmental impact reporting (EIR). It aims to force all new national and foreign investment projects to incorporate the costs of compliance with environmental standards. Since this is the most detailed section of the law, one can conclude that the EIR emerges as the principal instrument of Chilean environmental policy. In addition to improving the quality of investment projects, it is expected that the reports will contribute to cognitive capacity-building. They should generate valuable environmental data and applied knowledge. This is a key factor, since Chilean policy makers lack fundamental information with which to make policy. It is also hoped that investment in environmental safeguards will maintain Chile's competitive advantages in world markets (CONAMA, 1994; Aburto et al., 1995).

The polluter pays principle places the costs of non-compliance with environmental law and regulation on the perpetrator rather than the state; a clear effort to force companies to internalise the costs of pollution. In theory, the norm is quite explicit: any legal person, private or public, that pollutes above prescribed levels must pay for that pollution's clean up. Only in exceptional cases will the government provide funds for such efforts. In practice, however, the wording of the relevant clauses place the burden of proof on the prosecution rather than the polluter, making it easier for the latter to shirk responsibility. The polluter pays section of the law suggests two additional instruments of environmental policy: traditional command and control practices and market based incentives, such as tradable pollution permits (CONAMA, 1994).

The principle of gradualism addresses two issues. On the one hand, it recognises the need to prioritise environmental problems. On the other hand, gradual-

7 For the institutional arrangements between 1990 and 1994 see, Brañes, 1991.

ism urges political caution. Conflictual measures should be avoided at all costs. This principle appeals to the pragmatic realism on which many of the policy makers pride themselves. In their view, strong measures generate too much conflict and produce legislative gridlock rather than solutions (CIPMA, 1992).

The framework law established the institutional foundations of environmental policy-making with the creation of the *Comisión Nacional de Medio Ambiente (CONAMA)*, a decentralised national environment commission. CONAMA is fundamentally an interministerial coordinating committee of 10 ministries of state with environmental functions. It is chaired by the minister of the General Secretariat of the Presidency (SEGPRES), the strategic and tactical nerve centre of the presidency. The other ministries are Economy, Public Works, Agriculture, National Properties, Health, Mines, Housing and Urban Affairs, Transportation, and Planning (the later is also the link to international cooperation). CONAMA has relatively few powers of its own. Each of the sectoral ministries retain their full range of rights and responsibilities with respect to environmental issues relating to their sector.

CONAMA has two main functions: oversight of the implementation of environmental impact reporting, and arbitration of disputes that may arise between ministries or between CONAMA's and a ministry's view of a problem. For most purposes, environmental policy is left to individual ministries and private sector consultants. This is especially true for the regulatory body of the law to which the crucial issue of standards were deferred (CONAMA, 1994; Partido Socialista-Partido por la Democracia, 1992). Nevertheless, CONAMA does have a modest technical secretariat charged with developing some national-level guidelines for the process. Its focus to date has clearly been in the area of pollution control. CONAMA will also promote environmental education and manage funds for specific projects.

CONAMA is a small institution. Its staff is limited by law to 62 persons, an executive director (who has the right to speak at meetings of the ministerial committee), 20 directors, 25 professional staff, 2 technicians, 9 administrative secretaries, and 5 more or less menial positions. It is a decentralised institution with 13 semi-autonomous regional commissions for the environment (CONAMA, 1994).

CONAMA also receives funds from international financial and bilateral development institutions to help build institutional capacity. In 1992, the World Bank opened a tranche of US$11.5 million for the institutional development of CONAMA's predecessor. It was recently renewed for the present incarnation of CONAMA. The loan strengthens the technical secretariat, funds data-gathering research, provides training in environmental assessment, and supports projects for decontamination in mining and industry. US AID has also provided funds for building a data base to establish a foundation from which to monitor pollution trends. These capacity-building projects should help Chile to overcome some of its major institutional obstacles to environmental management. These include a shortage of environmental professionals, insufficient information, and the lack of a clear institution for monitoring and controlling pollution.[8]

A consultative council in CONAMA builds channels for the development of participatory capacity. It offers space for societal participation in the formulation of environmental policy. The president of CONAMA (the Minister Secretary of the Presidency) heads the consultative council. The president of the republic nominates two representatives of the following organisations on the basis of lists of five candidates proposed by them: two scientists from universities, two business representatives and two from labour. The NGO community may also have two members on the council, as may the sector of non-university academic organisations. There is also a representative of the president of the republic. Each of the Regional Commissions for the Environment will also have a consultative board made up of two scientists, two representatives of NGOs, two each from business and labour, and one representative of the regions' political head – the Intendente. Societal participation is also contemplated in overseeing environmental impact reports. The latter must be made public (safeguarding sensitive commercial and proprietal information) and citizens and organisations may challenge it during a period of 60 days (CONAMA, 1994).

In addition to these measures, the framework law establishes the legal basis for the protection of Chile's biodiversity. It accomplishes this by reaffirming the state's commitment to the maintenance and expansion of its National System of Protected Areas. In 1991 it covered about 18.8 million ha, or 18.5% of national territory. The bulk of the actual territory covered by these protected areas lies in the sparsely populated extreme South of Chile (XI and XII regions).[9] The law also gives the state responsibility for classifying species according to the degree of their endangerment.

Setting Priorities among Pressing Problems. In addition to pushing through the environmental framework law, the Aylwin administration addressed specific, critical environmental problems inherited from the dictatorship. These dealt primarily with air pollution associated with state mining companies. In a further attempt at institutional capacity-building, the presidency encouraged the creation of an Environmental Unit in the Ministry of Mines in order to implement environmental management. The Unit draws heavily on private sector consultancies for its handling of the task.

The Unit has concentrated on incorporating environmental design in new projects and decontamination programs for existing problems (InduAmbiente, 1994). With respect to the former, and in conjunction with the Regional Environmental Commissions, the Ministry of Mine's Environmental Unit has started a pilot program for the design of EIRs.[10] With respect to decontamination programs, Supreme Decree no. 132 of 1993 outlines a decontamination plan for

8 Author interview with Patricio Rodrigo, former director of planning and budget of the Ministry of National Properties, 29 May 1994; and World Resources Institute, 1994.

9 The system consists of National parks (8.358 million ha); national reserves (5.6 million ha); natural monuments (14.5 thousand ha).

10 "Informe de gestión ambiental del Ministerio de Minería (1990-1993)", Marzo, 1994.

the world's largest opencast copper mine, Chuquicamata, owned by the government. Between 1993 and 1998 sulphur emissions should be cut from 252,000 metric tons per year to 162,000; particulates from 9,720 metric tons per year to 3,240; and arsenic discharged to air should fall from 225 metric tons per month to zero by 1996. Supreme decrees 180 of 1994 and 252 of 1992 call for drastic reductions of those airborne toxic wastes in other state-owned mines. Supreme Decree no. 185 of 1991 sets guidelines for emissions of those substances for the entire nation, but principally the mining sector as a whole, public and private, because smelters and refineries are the principal polluters.[11] In some instances, the government reportedly released funds to implement the decontamination plans. For example, in 1992, Codelco and Enami, the two state-owned companies, allocated a total of about US$ 300 million each to comply with decree 185 by 1999.[12]

The government has also addressed the issue of air pollution and sewage in Santiago. With respect to the former, its main actions have centred on pulling nearly 3,000 ageing buses off the streets, and by closing especially problematic industries on days of emergency – in winter, when pollution indicators are at their worst. According to one source, air quality in non-winter months was "acceptable" in 1994, although the measures taken so far are not up to the task of further, and much needed, improvement.[13] The Inter-American Development Bank has provided loans for addressing sewage problems. Overall, the Aylwin administration quintupled spending on the environment as compared to the military government. By one estimate, it spent approximately US$ 33 million on CONAMA and specific decontamination projects (World Resources Institute, 1994).

While the focus has clearly been on pollution abatement, democratic governments have made an effort to control depletion of renewable natural resources as well, albeit to a lesser extent. The most widely publicised venture addresses the problem of desertification. In the early 1990s, CONAF, with financing from UNEP, started a National Desertification Plan. Desertification is a problem that mostly affects marginal lands with high incidences of rural poverty. As a result, the plan emphasises the linkages between economic development and the environment. The plan, in its first phase, involves data-gathering for the first through fourth regions of Chile in order adequately to diagnose the problem before making policy recommendations. CODEFF has been doing the research for CONAF and has so far completed a profile of the fourth region (Lagos, 1993; Torrico 1994). CONAF has also started reforestation projects. In a similar vein,

11 República de Chile, "Decreto Supremo 132", Ministerio de Minería, 7 June 1993; "Decreto Supremo 180", Ministerio Secretaría General de la Presidencia, 18 October 1994; "Decreto Supremo 185", Ministerio de Minería, 13 January 1992; and "Decreto Supremo 252", Ministerio de Minería, 30 December, 1992.

12 "Codelco: US$ 300 millones para descontaminación", La Nación, n.y.; and "US$ 285 millones invierte Enami en refinerías", El Mercurio, 19 October, 1992.

13 Interview with Ana María Ibacache, Head of the Environmental Unit of the Ministry of Mines, 2 March, 1995. Interview conducted for this research project by the Corporación para la Preservación y Conservación del Medio Ambiente.

CONAMA, in conjunction with the Ministry of Agriculture, has proposed a National Plan for Soil Conservation. Again, the first step is information-gathering to build a data base on the degree of degradation and its causes. The next step would be to make policy recommendations (CONAMA and Ministerio de Agricultura, 1994).

2.2 Policy-making Style and the Limits of Capacity-Building

If one focused primarily on the *form* of Chile's capacity building efforts one might draw some rather optimistic conclusions; for in the abstract, the measures taken suggest a number of positive aspects. For example, the framework law appears to promote flexibility in the choice of policy instruments. By leaving actors free to choose their own instruments, Chile might avoid some of the more negative aspects of command and control techniques. Second, CONAMA is a decentralised agency with regional branches, which seemingly gives Chile a chance to overcome the vices of centralisation. Third, CONAMA's interministerial structure should facilitate the development of integrative capacity, which would pave the way for higher stages of capacity building for the environment. Fourth, the gradual principle, the structure of participation, and the consensual policy-making style of the CPD stress a non-conflictual approach to policy-making. It is often hypothesised that such styles have a positive effect on policy innovation, and that implementation will be easier and more successful. Fifth, environmental policy makers have also embarked on the generation of cognitive capacity. A crucial step, for basic environmental data at national level are sadly lacking (as evidenced, for example, by their absence from many of the tables in this volume). Sixth, if one focused only on the formal relationships between the state, industry, green organisations and the media, one might conclude that Chile is poised to enter the stage of intensive end-of-pipe treatment. All of the necessary features seem to be present. There is a reciprocal relationship between the state and industry; and a one-way relationship between the state and green organisations and the media on the one hand, and between the latter two and industry on the other. The rhetoric in Chile even suggests that the stage of "ecological modernisation" may not be far off.

Such overly optimistic interpretations of Chile's environmental capacity-building efforts must be tempered. Looking beyond the formal relationships, actual capacity for the environment appears to be weaker than it might seem at first glance. It also becomes clear that existing capacity is being under-used. This less optimistic evaluation results from an examination of environmental policy-making within the deeper context of the general political situation in Chile, the CPD's overarching goal of rapid economic growth, and the policy-making style of democratic Chile. For the most part, these factors have induced CPD administrations to create weak environmental institutions dependent on the private sector. Those same institutions sharply curtail participation by environmental groups. This was done to place fetters on their eventual utilisation by the proponents of more thoroughgoing environmental policy.

In this context, the CPD's present strategy emphasises dilution, high chimney, and end-of-pipe technology for only a few pressing emissions and pollution problems. The government is mainly applying these measures to problems over which it has easy control: the state owned mining industry. It is also addressing a few problems it can no longer ignore, such as waste and air quality control. However, when business interests are directly involved, the crucial issues of regulations and standards are displaced to ministries where business has disproportionate influence, and where environmental groups have little or no input.

The weakness and under-utilisation of capacity for the environment in democratic Chile rest on three interrelated factors. These are the internal structure of CONAMA, the strategy of the green organisations involved in the policy-making process, and the strength of polluters in the context of democratic Chile's general policy-making style. Let us start with the last point. Chile's policy-making style is characterised by close cooperation between policy makers and the private sector, carefully limited attention to the interests of other social sectors, and tightly controlled channels of participation for them. This pattern emerged from the process of the transition to democracy which involved explicit bargains between the opposition and the dictatorship (the CPD), the business community, and the armed forces. The latter two wanted assurances that if the CPD won elections it would retain the neo-liberal socio-economic model. The CPD, partially to ensure the transition to democracy and partially out of conviction that it was the only way to sustain rapid economic growth (its highest priority) acquiesced (Silva, 1995).

To give credibility to its promise, CPD governments conspicuously included business sector representatives in the policy-making process. Since the private sector provides the investment that underpins economic growth, it gets an ample and preferential hearing. Policy makers recognise the interests of other social groups, such as labour, peasants, and middle classes – as well as non-business interpretations of problems – but always firmly subordinate to private sector preferences. Therefore, the degree of their participation in the policy-making process is more circumscribed than that of business. In short, the government sets the policy agenda and then interacts closely with business groups in policy formulation and implementation. Other social groups – or points of view – may get a hearing, but to a significantly lesser extent and with less substantive impact (Silva, 1996c, 1992).

The structure of CONAMA faithfully reflects this policy-making style. The fact that SEGPRES chairs CONAMA is central to understanding both why existing capacity is less than it appears to be in a formal sense, and why it is under-utilised. These outcomes are related to the central functions of SEGPRES. It is the strategic and tactical right arm of the president and its principle trouble shooter. Or, more bluntly, SEGPRES is the presidency's chief council and watchdog. Of course, much of its orientation depends on the priorities of the presidency.

In the current alignment of forces in the CPD, existing capacity for the environment is under-utilised because of two key functions that SEGPRES fulfills within CONAMA. First, it ensures the protection of business interests and preferences before those of other groups, and second, it narrows the points of access for opposing ideas, or excludes them outright. This is especially apparent in the structure of the channels of participation which will be discussed shortly.

In effect, at the insistence of SEGPRES the technical secretariat of CONAMA relies almost entirely on consultants from the private sector mining industry and from CIPMA, which interacts closely with the private sector in general, and the mining industry in particular (Silva, 1994). In fact, ever since 1985 CIPMA had embraced the philosophy of close collaboration with the business sector. For example, it turned the 4th scientific congress of 1993 into an agenda-setting forum for the debate on the framework law. The professionals who attended the congress were overwhelmingly from the business sector, right-of-centre sectors of the academic community, and NGOs with an outlook similar to CIPMA's. These forces, with the blessing of SEGPRES, elaborated the gradual principle of environmental policy and ensured the creation of a weak coordinating institution (CONAMA). In other words, SEGPRES also made sure that business and business-oriented environmental NGOs were the principal forms of outside participation in the formulation of the environmental framework law, to the exclusion of proponents of different ideas both within government and in civil society.

In addition to the problem of under-utilisation of capacity, CONAMA is a weak institution to begin with. These weaknesses stem from several sources. First, its central staff, that of the technical secretariat, is very small and limited by law. Expanding it will be difficult because, all else being equal, it would entail a significant political battle for any incumbent administration that wished to do so. Given the relatively low priority of the environmental issue area it seems unlikely that the conflict would be worth the effort. Second, in keeping with its small staff, the technical secretariat's functions are also limited by law. It is mainly responsible for the implementation of a single policy instrument – the environmental impact report. Its other functions centre on building cognitive capacity. Third, the bulk of project funding comes from international sources. This undermines national policy-making by ceding the choice and prioritisation of problems, as well as project duration, to the foreign sector.

CONAMA's structure and the limited functions of its technical secretariat also undermine the presumed integrative capacity of the institution. In practice, they delegate control over policy – standards and regulation – to individual line ministries. This weakens proponents of the environment because environmental problems are not the highest priority within these ministries. Moreover, the ministries are highly responsive to business interests. They are dominated by "iron triangles" of bureaucrats, the interests they regulate, and legislators. In effect, given the technical weakness of the ministries when it comes to the environment, one might well expect that new environmental standards will not be very demanding, for it is the private sector's environmental consultants who dominate the arena.

The fine print of the polluter pays principle in the environmental framework law bears out these expectations. In the bill, business lobbies were able to shift the burden of proof from the polluter to the prosecution, thus significantly weakening the provision. This picture is reinforced by the fact that the actual task of developing model EIRs was delegated to the Ministry of Mines. The fact that so much of the mining industry is in government hands suggests that policy makers might have a freer hand. However, this ministry too is heavily dependent on the private sector-CIPMA nexus for its interpretation of environmental issues.

Key business organisations, such as the Society for Industrial Development and the National Mine Owners Society, have organised environmental committees chaired by high-profile businessmen. Their task: to develop a private sector definition of environmentalism, one that fits in best with their existing production processes – hence the development of a strategy that stresses dilution, high chimney, and end-of-pipe technology. Their technical staff work with CIPMA and similar organisations to provide the government with data and solutions. According to their publicity, large-scale companies that are part of Chile's most important conglomerates are gearing up to invest in the latest pollution abatement technology.

Many policy makers, business-oriented environmental NGOs and academics view this development with a great deal of optimism. They have faith in the capacity of the price system and of markets to push modern corporations to adopt innovative technology. But if this has become an incipient trend in some developed nations, it is probably related to the fact that the high cost of polluting is influencing business' voluntary adoption of materials management and substitution. Those costs include relatively tough standards, regulations, and penalties, as well as stricter labour laws and higher wage rates, steeper taxation, costlier litigation, and so on. But this is not the case in the developing world. Since these conditions are all laxer, companies can extend the product cycle of older industrial processes with high chimney and end-of-pipe technologies. Moreover, developed countries that have invested heavily in the development of such technologies (such as Japan) offer attractive incentives to buy their pollution control devices. Conversely, they may simply mandate the purchase of such technological packages in bilateral economic cooperation agreements.

Moreover, with respect to the "greening" of business, one must distinguish rhetoric from action. In practice, at least in Chile, and in the United States as well, corporations work behind the scenes to weaken standards. They also lobby hard to ensure that standards only apply to the presumed effects of a single substance. This means that the vital issue of the cumulative effects of many hazardous substances gets little attention, with the concomitant limitations on building cognitive capacity. In short, these actions lower the costs of compliance for business, while raising its ability to publicise their "greening".

The structure of CONAMA's consultative council (the official space for citizen participation) is further evidence of the weakness of institutional capacity and of its under-utilisation. The environmental framework law created legal structures for participation that reinforce the strong state-business nexus, as well as the collaboration of business-oriented environmental NGOs. It leaves little political space for alternative views, much less their forceful and binding advocacy. In short, the structure of participation stunts participatory capacity building.

What is that structure? Most importantly, SEGPRES chairs the Consultative Council and the president of the republic directly appoints a second government representative. Moreover, of the non-business representatives, the scientists are nominated by the rectors of universities, notoriously sensitive political posts, subject to approval by the president of Chile. The representatives of labour are also subject to approval by the president. Although not subject to presidential approval, NGOs, despite their great diversity, only get two representatives, just like all other "interest groups" in the environmental issue area (CONAMA, 1994). To all intents and purposes this guarantees that one of those will be business-oriented, and preferably CIPMA.

Over-reliance on the private sector also has potential drawbacks for the development of cognitive capacity. The government expects private sector EIRs to generate a lot of the nation's environmental data. This may have at least two consequences. First, the question of which data gets collected will depend heavily on the nature of future investment projects, as opposed to any more systematic effort at data collection. Second, the data generated may not be very reliable. Companies may manipulate data in order to shift blame onto other industries, or to make their own business look more "environmentally friendly" to boost sales. Signs of this are already evident.

Strict adherence to the gradual principle further reinforces the weakness of Chile's institutional environmental capacity building efforts and its under-utilisation. In fact, to a great extent, the gradual principle deliberately steered the construction of weak environmental institutions. An overly timid interpretation of the gradual principle also affects under-utilisation of existing capacity because it binds policy makers to a strategy of small incremental gains in restricted problem areas. The gradual principle guarantees that patently non-conflictual problems and solutions receive attention first. More difficult, or intensely politicised issues are postponed, and it is expected that resulting regulations, drafted in line ministries dominated by "iron triangles", and out of the public eye, will be none too stiff (CIPMA, 1992). Furthermore, the gradual principle helps explain why the environmental impact report receives so much attention – it involves future investments related to standards that have not yet been set.

In summary, as currently structured, Chile's collaborative policy-making style has led to weak institutional capacity building, which in turn shows all of the signs of being under-utilised. Those institutions involved in policy-making style will most likely lead to the design of a combination of command-and-control and market incentive approaches to curb urban-industrial pollution – air and water – through the application of end-of-pipe technology. This was clearly the trend in the measures taken during Patricio Aylwin's administration discussed above. Moreover, the government follows a strategy that emphasises the path of least resistance, content with small, incremental gains. For example, it tackled state enterprises and services first, leaving the thornier issue of dealing with the private sector (in mining as well as other industries) for later. Controls over renewable natural resource extraction – including intensive agriculture – will mostly lag well behind. For example, while a fishing law was enacted in 1990, it leaves many loopholes for fishing industrialists. A natural forest law has remained blocked in congress since 1992.[14]

3 Conclusion: Prospects for Change

Redemocratisation in Chile changed political opportunity structures in favour of environmental proponents more than at any point before 1989. For one, the environment figured in the policy agenda of the victorious opposition coalition. Moreover, democratic administrations have been more sensitive to external conditionality than was the military government. As a result, Chile has built significantly more capacity for the environment over the past five years than was the case during 18 years of military rule. It has an environmental framework law and is drafting the corresponding regulations, it has established basic environmental institutions, it has generated a cooperative policy-making style, environmental groups have greater access to policy-makers than at any other time, they are free from repression, data collection on key environmental problems has begun, and in some cases the government has taken measures to reduce emissions and pollution. The specific content of these achievements, the CPD's strategy, and the web of relationships among key actors, suggests that Chile is headed towards a capacity building stage that emphasises dilution and end-of-pipe technology.

This paper has also argued that given the CPD's overarching strategy and policy-making style, Chilean policy makers have constructed weak institutions and then underutilised them. The general timidity of CPD governments has led them to miss the opportunity to build stronger institutions and mechanisms for societal participation in environmental policy-making, the point being that it will be at best difficult to move to intensive end-of-pipe treatment within the current situation, to say nothing of materials management, substitution, and the far more complex task of planning for sustainable development.

Optimists might argue that a number of factors influencing fuller utilisation of existing capacity could propel Chile into the intense end-of-pipe treatment stage in the medium term. For example, as CONAMA's technical secretariat and the member ministries build cognitive capacity they will lay the foundations for more thorough action. If the technical secretariat takes a stronger role in coordinating the member ministries it could begin to build integrative capacity. Close collaboration between the CONAMA's technical secretariat and the environmental unit

14 For natural forest policy see, Silva, 1996b.

of the Ministry of Mines might produce a quality EIR instrument, with attendant effects on stiffer regulations and standards and systematic data collection. Such effects could be reinforced if private sector environmental consultancy firms work well with the line ministries. Environmental NGOs that work closely with business, such as CIPMA, might ease the private sector's worst fears about the environmental movement. As a result, they might find greater opportunities to shape polluters' perceptions of possible solutions in more progressive directions.

Such changes, however, would require a significant shift in Chile's political opportunity structure: the electoral victory of a CPD government (or some other party or political coalition) with a different overarching strategy, one that took a firmer hand toward business. Such a government might speed up the transition to the intensive end-of-pipe treatment stage by using indicative planning to force a more rapid response from business, by increasing funds for environmental agencies, by prosecuting select flagrant violators, by building integrative capacity, and so on. Unfortunately, such political forces, to the extent that they exist, seem to be nowhere near being able to form a government.

The shift to more intensive end-of-pipe treatment would also require a change within the private sector as well. This refers back to the problem of disentangling public relations exercises from real intentions. Business in Chile has worked hard to "green" its public image. This is especially apparent in their pronouncements about their willingness to work positively in favour of the environment rather then to obstruct government efforts. Yet in practice, it has done everything it can to dilute legislation, to structure institutions in such ways that increase the power of the private sector in decision-making institutions, to delay the appearance of regulation, ensure lax standards, to build in exemptions and extensions, to make prosecution of violators difficult, to exclude challengers to their position from the policy-making process, and the list goes on. As was previously shown, the private sector has enjoyed a significant measure of success in all of these endeavours. In short, business now touts its green rhetoric in relative freedom from governmental pressure. It remains to be seen what the private sector's response will be when, within the same collaborative framework, it is eventually confronted by a government that is not quite so accommodating. It might opt for an openly obstructionist stance, which, to date, has been its response in other issue areas. In any case, until now, the private sector's actions place severe obstacles in the path of generating integrative capacity, and of moving to other stages of environmental capacity-building.

In short, these factors suggest that between 1990 and 1994 Chile missed an opportunity to build a higher level of legal and institutional capacity for the environment. The CPD's strategy closed the door on the creation of stronger, more extensive institutional capacity building. While it is true that such capacity would also have been under-utilised at first, it would have left institutional structures in place for reaching higher stages of environmental action as cognitive capacity expanded. As things stand now, costly political battles over the creation of such institutions would have to be fought anew. Given the relatively low priority of the environment on the national policy agenda, that is unlikely.

In fact, one could argue that the Aylwin administration deliberately created weak institutions. It may have done this to foreclose the political opportunities of an alternative environmental policy-making group, and its NGO allies in civil society, should an administration more amenable to its view ever come into office. This interpretation is based on the fact that within the CPD sharp political conflicts characterised the policy formulation stage of the environmental framework law. The way in which those conflicts worked themselves out strongly suggests that the limits placed on environmental capacity-building were a response to challenges from a group of environmental policy makers within the administration who were at odds with the strategy of the higher echelons of the administration.

This group advocated greater degrees of change from the neo-liberal socio-economic model. Institutionally, it favoured the creation of a Ministry of Environment to increase the independence of environmental policy-making from other line ministries and their penetration by business groups. However, in the interest of developing integrative capacity, such a ministry would also operate in close coordination with other line ministries. The policy-making group also wanted to encourage research and development in technologies that went beyond dilution, high chimney, and end-of-pipe strategies. They envisioned an institutional structure capable of generating policy instruments and incentives to advance the shift to materials management and substitution strategies. In addition, they also displayed a keen interest in the sustainable use of natural resources that included grassroots development projects as a means to address both ecological concerns and poverty. Moreover, to further counteract the power of business, they argued for strong citizen participation in both the policy-making process and decision-making about technological packages. They did not, however, propose to exclude business from the decision-making process, just a greater independence from its influence.[15] Moreover, given Chile's political realities, they fully realised that initial utilisation of the capacity they sought to build would fall short of its potential. But they consciously wanted to leave the door open for the future.

This power struggle within the CPD and its outcome has important implications for the future of capacity building for the environment in Chile. First, building environmental capacity for ecological modernisation seems a long way off. The struggles over environmental policy reveal that this is not because Chile lacks the proponents for such a path, either from within or outside government. Rather, they lack the necessary political allies within higher levels of the political leadership. In other words, one crucial element of the political opportunity structure necessary for them to flourish is not in place. The fact that they were deliberately thwarted, and that the ensuing weak capacity was created to shackle more progressive environmentalists should any future government support them, suggests just how difficult the road to ecological modernisation will be in Chile.

The power struggles over environmental policy in Chile hold a second implication for the study of capacity-building for the environment. When one exam-

15 For further details see, Silva, 1994.

ines the proponents of environmental action it may be useful to disaggregate them according to their strategies and to link them to coalitions with other social and governmental actors. This may sharpen the analysis in understanding first the current structure of legal, institutional, and participatory capacity; and second, the possibilities for changing those structures. Environmental NGOs are not homogenous in their outlook. Governments can exercise their power to include in the policy-making process those that agree with their general strategy and to exclude those that do not. For there is no reason why some important environmental NGOs may not be supportive of a limited, high chimney and end-of-pipe approach, and make common cause with industry.

If, as is the case in Chile, the proponents of environmental measures that lead to greater change than strategies which emphasise dilution and restricted application of end-of-pipe technologies have so few avenues of direct influence, what strategies are they likely to use to be heard? What are the implications of those strategies for the style of environmental policy-making in Chile? In all probability, they will follow a defensive, conflictive strategy by exercising watchdog functions. Blocking new economic projects by challenging their environmental impact reports seems the most likely path of action. As is already the case, causes of confrontation may range from biodiversity preservation, to pollution, to claims of exclusion from participation in decision-making, to the impacts of economic development projects on the cultural identity and life chances of native peoples. Demonstrations, news media exposés, civil disobedience, legal action will most likely be their weapons. Thus, Chile will probably be characterised by a two-tier policy style: an official one that gives privileged treatment to collaboration between business and government over the preferences of other groups; and a conflictive one where mobilisation and legal action challenge their decisions. With luck, this two-tiered policy-making style might slowly put Chile on the path to the intensive end-of-pipe treatment stage. It is not likely, however, to put it on the path of ecological modernisation.

References

Aburto, A, C. Urrutia and P. Rodrigo, 1995: Informe: Desarrollo de la política ambiental en Chile, March 1995, mimeo.

Arsenberg, W., M. L. Higgins and R. Asenjo, 1989: Environment and Natural Resource Strategy in Chile. U.S. Agency for International Development/Chile.

Brañes, R., 1991: Institutional and Legal Aspects of the Environment in Latin America. Washington, D.C.: Inter-American Development Bank.

Cárdenas, 1985: Pesquería chilena, dimensión oceánica y desarrollo, Ambiente y Desarrollo, 1, 3, 101-118.

CIPMA (Centro de Investigaciones y Planificación del Medio Ambiente), 1991: Propuesta para iniciar un proceso participativo hacia una política nacional de medio ambiente, Ambiente y Desarrollo, 7, 1 .

CIPMA (Centro de Investigaciones y Planificación del Medio Ambiente), 1992: Gestión ambiental en Chile: Aportes del cuarto encuentro científico sobre el medio ambiente. Santiago de Chile: CIPMA.

CONAMA, 1992: Chile: Informe nacional a la conferencia de las Naciones Unidas sobre medio ambiente y desarrollo. República de Chile: Ministerio de Bienes Nacionales.

CONAMA, 1994: Ley de Bases del Medio Ambiente. República de Chile.

CONAMA, n.y.: Propuesta. Plan Nacional de Conservación de Suelos.

CONAMA and Ministerio de Agricultura, 1994: Plan Nacional de Conservación de Suelos.

Concertación de Partidos por la Democracia, 1989: Programa de gobierno, mimeo.

Concertación de Partidos por la Democracia, 1993: Bases Programáticas para el Segundo Gobierno de la Concertación, mimeo.

Concertación de Partidos por la Democracia, 1994: Los compromisos ambientales del gobierno de Frei, Ambiente y Desarrollo 10, 1.

Drake, P. W. and I. Jaksic (eds.), 1995: The Struggle for Democracy in Chile, 2nd edition. Lincoln: University of Nebraska Press.

Gómez-Lobo, A., 1991: ¿Es sustentable el desarrollo pesquero en Chile?, Ambiente y Desarrollo, 7, 1, 17-20.

Grau, J., 1989: Ecología y Ecologismo, 2nd edition. Santiago: Ediciones Oikos.

Gutiérrez, R., n.y.: Contaminación: Las aguas bajan turbias, Ercilla, 2.936.

Hojman, D. E., 1993: Chile: The Political Economy of Development and Democracy in the 1990s. Pittsburgh: University of Pittsburgh Press.

InduAmbiente, 1994: El Aire en la Balanza, InduAmbiente 2, 10.

InduIndustria, n.y.,: Gas Natural: La revolución que viene.

Lagos, E., 1993: Plan Nacional contra la Desertificación: Tiembla el desierto, Chile Forestal, no. 210.

Oppenheim, L. H., 1993: Politics in Chile: Democracy, Authoritarianism, and the Search for Development. Boulder: Westview Press.

Partido Socialista-Partido por la Democracia, 1992: Posición sobre la ley del medio ambiente, Santiago de Chile, mimeo.

Rojas, A., 1994: The Environmental Movement and the Environmentally Concerned Scientific Community in Chile, European Review of Latin-American and Caribbean Studies, 56.

Silva, E., 1992: Capitalist Regime Loyalties and Redemocratisation in Chile, Journal of Interamerican Studies and World Affairs 34, 4.

Silva, E., 1994: Contemporary Environmental Politics in Chile: The Struggle over the Comprehensive Law, Industrial and Environmental Crisis Quarterly 8, 4.

Silva, E., 1995: The Political Economy of Chile's Regime Transition: From Radical to Pragmatic Neoliberal Policies, in: P. Drake and I. Jaksic (eds.), The Struggle for Democracy in Chile, 2nd edition. Lincoln: University of Nebraska Press.

Silva, E., 1996a: "Environmental Policy in Chile", in: F. Fischer and M. Black (eds.), Greening Environmental Policies. London: Paul Chapman Publishing (forthcoming).

Silva, E., 1996b: Conservation, Sustainable Development, and the Politics of Native Forest Policy in Chile, in: G. McDonald, D. Nielson and M. Stern (eds.), Latin-American Policymaking in International Perspective. Boulder: Westview Press (forthcoming).

Silva, E., 1996c: From Dictatorship to Democracy: The Business-State Nexus in Chile's Economic Transformation, 1975-94, Comparative Politics (forthcoming).

Torrico, J. M., 1994: Un problema que avanza: Desertificación en la IV región, Ambiente y Desarrollo 10, 1.

White, R. R., 1993: North, South, and the Environmental Crisis. Toronto: University of Toronto Press.

World Resources Institute, 1994: World Resources, 1994-95: A Guide to the Global Environment. New York: Oxford University Press.

CHINA

Yu-shi Mao

1 The Overall Situation

With 22% of the world total, China has the largest population in the world, but its geographical size ranks only third, covering 6.4% of the world's landmass. It is a mountainous country with just 12% plains, and one third lying over 3,000 m above sea level. Deserts, gobies, high mountains and deep valleys make up much of the landscape. The hostile environment of a large part of China drives its population to the Southeast, where 50% of the population live in 13% of the country, and where the population density of $480/km^2$ exceeds that of all major countries except Bangladesh. This part of the country has become one of the most congested regions in the world. The share of urban population increased from 19.4% in 1980 to 29% in 1994. High population density is one cause of the heavy burden on the environment, especially for a country which relies so heavily on agriculture. During the period 1949-78, China strove hard to industrialise its economy, more or less in isolation from the rest of the world, relying on cost-cutting and centralised economic planning, without much care for the efficiency of resources utilisation. The result of such a pattern of development exacerbated environmental and ecological problems, as shown by heavy deforestation, degradation of large areas of grassland due to over-grazing, and severe soil erosion, causing a cycle of floods and droughts.

Since then, sixteen years of economic reform have followed, in which China's industrial output has grown 7 times over (an average annual increase of 13%), resulting in a dramatic increase in industrial pollution. Thus China's environmental problems possess characteristics in common with those both of a developing and of a developed country, i.e. over-exploitation of natural resources and heavy industrial pollution. On the other hand, as the Chinese economy becomes stronger, it is gaining more economic power to address environmental damage. The government and the people have made impressive progress in this respect. All in all, although the situation is serious, there is hope for the future.

The following provides a brief picture of China's environment situation.

1.1 Population

Among developing countries, China is a good example of population planning, where the birth rate and death rate have both been decreasing, resulting in a natural annual population increase of 1.12% and a total fertility rate of 2.0. [1] In Shanghai, population growth is tending towards the negative, while in some economically under-developed regions it is still fluctuating at around 1.8%. The population had reached 1.2 billion on February 15, 1995 (China Daily, February 1995). The major concerns now are as follows: Firstly, how long can the current population policy be maintained and remain effective? Would population return to a high growth rate without this policy, especially in the less advanced rural areas? Secondly, the gender ratio is out of proportion: in new born babies, the ratio of male to female has gone up to 113:100 (Li, 1995). And thirdly, the sharp decrease in fertility since pursuing the policy of "One Child per Family" is resulting in a distortion of the age structure, so that twenty years hence the percentage of those of working age among the total population will decrease to a dangerous level.

1.2 Depletion of Natural Resources

Deforestation. Forest coverage decreased from 13% in 1950 to 12% in 1979, and then rose again to 13.9% in 1994, but the quality of forest has deteriorated seriously. During the Second World War (1937-45) and the ensuing Civil War (1946-49), few trees were planted, with the result that in 1950 most of the forests in China consisted of mature trees. After half a century's deforestation, however, mature forest has become so scarce that it will be completely exhausted within just 5-6 years (Qu and Li, 1992). The government has launched an afforestation campaign, but resources are limited and inefficiently utilised. The central problem in restocking forests in China is the question of ownership of the forests. With more reliance on market forces for afforestation, security of ownership will attract the private sector to invest in tree planting, which could provide a high return of capital, due to the high price for timber products as a consequence of the scarcity of forests.

Loss of Farmland. Loss of farmland has been a pressing problem for China, which uses 7% of the world's farmland to feed 22% of the world's population. Desert occupies 15.9% of Chinese territory, and is spreading at an annual rate of 1,500-2,000 km^2. Soil erosion eats away 5 billion tonnes of fertile soil every year (Wang, 1994), which results in an economic loss of 30 billion yuan (US$ 3.5 billion). Urbanisation is reducing available farmland by 0.53 million hectares a year (Qu, 1992). Intensive farming uses more than twice as much chemical fertiliser per hectare as is the case in the United States and this unbalances the

1 The fertility rate has been fluctuating around 4.53 (1973), 3.58 (1975), 2.31 (1980), 2.61 (1981), 2.86 (1982), 2.20 (1985), 2.57 (1986), 2.70 (1987), 2.09 (1990), 1.54 (1992) (Lin Fu-de, 1994).

chemical composition of soil (WRI, 1992). Water and solid waste pollution also reduce the area of available farmland.

Loss of Grassland. China is 23.4% grassland, but due to over-grazing and poor management 39% of it has suffered a decline in grass quality and land productivity (SPC, 1991). Overall average output of meat per unit area is only one twentieth of that in the Unites States (Qu and Li, 1992). There have been 26 million hectares of grassland turned into farmland, but one third of it has become desert.

Fresh Water Crisis. China's total precipitation amounts to 6 trillion m^3, of which river overflow and underground water takes 2.7 t m^3 and 0.87 t m^3 respectively; the per capita fresh water resource is 2,400 m^3/year, only one quarter of the world average. As a result of deforestation and climate change, the already dry northern China is becoming still more arid. Annual precipitation in Inner Mongolia has declined from 400-450 mm in the 1960s to 250-300 mm in the 1980s (SPC, 1991). There are 300 cities currently suffering shortages of fresh water, and 80 million people do not have enough drinking water.[2] Over-extraction of underground water in northern China has lowered the ground-water level year by year, and caused subsidence in large areas. The yield of farmland is badly affected by persistent drought.

1.3 Industrial Pollution

Air pollution is a serious problem; China burned 1.1 billion tonnes of coal in 1994, which is 75% of the primary energy supply. Annual SO_2 emissions have gradually increased to 18 million tonnes in 1993, and only two power plants are equipped with desulphurisation equipment. Acid rain damages forests, farmland, fisheries and buildings in southern and Southwest China, as the environmental background there is acidic, and coal mined in this area has a high sulphur content, while in the North the environment is alkaline, and thus less of an acid rain problem has presented itself. Discharge of particulate matter has decreased from 11 million tonnes in the mid-1980s to 6.2 million tonnes in 1993 (NEPA, 1990-1994). The concentration of NO_x in the atmosphere of northern cities averages 59 $\mu g/m^3$; in southern cities it averages 40 $\mu g/m^3$. With the increasing number of public and private motor vehicles, NO_x emissions and concentrations will rise strongly.

Regarding water pollution, annual waste water disposal is 35.6 billion tonnes, of which 21.9 billion tonnes are industrial waste, and 72% of which is treated. The annual variation in pollutants, such as COD, heavy metals, arsenic, cyanide and volatile phenol in discharged water is around plus or minus ten percent, or basically stable. But all rivers, lakes and ponds around large cities are heavily polluted. In coastal waters, water quality is declining; the incidence of red tide has increased from once or twice a year in the 1980s to thirty or even fifty times a year in the 1990s, hitting the ocean fishery industry very hard (NEPA, 1990-

2 News Broadcast by China Central TV Station on 23 March 1995 celebrating World Water Day.

1994). Fish resources have declined extremely fast as a result of both pollution and over-exploitation, some highly prized species in particular are almost extinct.

The discharge of industrial solid waste has been slowly increasing from 0.57 billion tonnes in 1988 to 0.6 billion tonnes in 1993, while household waste is increasing rapidly as a result of the improved standard of living. The recycling of solid waste has developed well, not because of people's voluntary contributions, but because cheap labour lowers the cost of waste collection, rendering recycling a profitable business. Urban waste incineration is booming, since it saves land resources, which are very scarce in China. Another environmental problem is urban noise, a result of concentrated habitation and frequent personal communication, especially in situations of high population density. It is estimated that two-thirds of urban areas endure a background noise level of above 55 decibels, and most of the environmentally related complaints of urban residents concern excessive levels of noise.

Although in Chinese history there has been a lot of original thought dealing with the symbiosis between human and nature, its modern concept of environmental protection was born only after China sent its delegation to the United Nation's Conference on Human Environment in 1972. This Conference alarmed the Chinese with its environmental concerns, but during the period of the Cultural Revolution (1966-76), no serious measures could be undertaken to address environmental damage. In 1973-74 some water treatment projects were constructed along the upper reaches of the Sang-Gan River, a major water source for Beijing, but only in the name "of protecting the health of Chairman Mao" could the project be implemented. After the end of the Cultural Revolution, and especially following the reforms begun in 1979, environmental protection has gained public and government attention, and political, financial, educational and legal action have been put into full swing.

2 Main Characteristics of Environmental Policy

2.1 A Command Economic System Produces More Pollution Than a Market System for the Same Output of GNP

China has been gradually casting off the old command economic system, which is a system with division of labour but no correct price signalling, leading to a misallocation of resources between sectors. Very high energy intensity per unit of GDP is an example of the low economic efficiency of the Chinese economy.[3] Poor utilisation of resources leads to heavy pollution, which is a general phenomena in all command economies, such as the former East European states. Since economic reform started in China in 1978, some important steps have been taken,

3 According to the Bulletin on Energy Policy Research, 1994, No.11, energy intensity in GNP for 1980 was 7.6 tce per 10,000 yuan, and for 1993 it was 4.54 (both based on 1990 purchasing power of yuan). The corresponding figure for US was 16.3 BTU per dollar for 1993 which amounts to roughly a quarter of the figure for China (CERS, 1994).

such as the removal of price controls in some sectors, the development of the private sector, tax reforms, and the abolition of the dual exchange rate in 1994. Price distortions have been gradually removed, and economic efficiency has been improved; the average elasticity of energy consumption relative to GNP declined from more than one in the 1970s to 0.6 in the last five years. The percentage shares of primary, secondary, and tertiary industries in GNP have changed from, respectively, 28.4%, 48.6%, and 23.0% in 1978 to 21.1%, 51.8%, and 27% in 1993 (China Statistical Yearbook, 1994). Thanks to structural change and efficiency improvements in resources utilisation, the increase in pollution is lagging some way behind economic growth. The distortion in the economy now lies mainly in some monopoly sectors, where the market does not operate – railway, telecommunications, electric power, and petro-products etc. – but to a far lesser extent. The Chinese reform process has an important lesson to teach: that pollution can be reduced by changes in the economic system. The market system is efficient in its use of natural resources as compared with other systems. Its incentive structure of fair competition and self-interest ensures production of the same quantity of saleable material using less resources. However, this is true only on the production side; on the consumption side the market promotes wanton waste of resources, and China is also "advancing" rapidly in this direction.

2.2 Public Ownership Aggravates Environmental Damage

In the old system, almost all firms in all sectors were owned by the people or the government. Such a system had been believed to be able to avoid major difficulties within society. In the public ownership system nobody really assumes the responsibilities of ownership. The lack of clearly defined ownership and property rights is the major cause of deforestation and degradation of grassland. This is the old problem of the "Tragedy of the Common". Public ownership also renders economic instruments less effective. If a charge or tax is imposed, the manager of the publicly owned firm hands over a sum of money to the tax office; but it amounts to a transfer of money from one pocket of the government to another. It happened more than infrequently that the firm delayed paying the charge, even defaulting on their liability.[4] The PPP (Polluter Pays Principle) in particular becomes enigmatic because the polluter can scarcely be identified. The consumer is the people, the producer is the people, and the government is also the people. So who pays whom? In principle, the discharge fee is a privatised social cost and should be offered to those who suffer damage. But in China not a penny thus collected has been paid to those who suffer from pollution; all the funds have been re-channelled back to the factories which discharge the pollutant (about 70%) and to environmental administration offices (about 30%),[5] as the factories

4 For example, many of the firms in Xining city default on the discharge fee. Only 13% of the fee was collected in the first and second quarters of 1994 (China Environmental News, 17 Sept. 1994).

5 The Environmental Protection Law of People's Republic of China requires that the discharge of pollutants beyond a certain level of concentration will incur a discharge fee. This clause has evolved into a detailed set of rules, and has been repeatedly emphasised. The use of the fund thus collected has

who create pollution are compensated for their investment in pollution control. It is recognised that environmental problems arise precisely because no property rights to the environment can be identified, and the internalisation of social cost into private cost is a fundamental idea adopted in environmental protection. In the case of public ownership, even enterprises themselves have no clearly identified property rights, and thus the environmental problem becomes even more complicated. However, since the late 1980s plural ownership has begun to emerge; the non-public sector has grown at an accelerated pace, and now generates more than half of the total GNP in China. For these non-public enterprises, economic tools are rather effective. A recent development is the relinquishment of common access to a part of forest-land and grassland, and recognition of the individual peasants' rights to use them. In all these places where the right of use is clearly defined, the conservation of natural resources has been greatly improved.

2.3 The Weak Role of Legal Instruments

In China's long history, the nation was for most of the time governed by a system of administrative officialdom, at the top of which was the emperor. The legal system was weak compared to instructions from the emperor, and often yielded to it. With such a traditional and cultural inheritance, legal instruments play a much less important role compared to those in western countries. In order to protect the environment, law is obviously required, and so dozens of laws and regulations have been formulated in China. However, because of China's bureaucratic tradition, law enforcement is without teeth (He and Liu, 1995). The inexperience and incompetence of judges make some of their decisions questionable.[6] When an issue of environmental conflict is brought before the court, it sometimes takes years to settle.[7] As legal instruments are weak, people often resort to other means to protect themselves from environmental damage, such as protest, complaints, deputations, even fighting leading to bloodshed. Every year there are several such major cases, all of them induced by severe and immediate damage, while only those due to chronic damage may be debated in court. But the environment is a public good, and since most people will not take on the costs of legal action, but will wait for others to take the lead and then take a free ride, local government should take a leading role in legal action.

---

been limited to environmental projects (80%) and administration (20%), but the actual proportion has been 70:30.

6 There were two similar cases of deforestation reported in typical law suits on environmental issues in China (Xie, 1994), one offender was sentenced to death, while the other received two years imprisonment. Another case is a mistaken verdict on a group of people who protested about pollution damage and were given custodial sentences under an indictment of sabotage, but were later released.

7 It has been reported that a winery discharged pollutants and polluted underground water, and was then sued by the neighbouring residents. The case took four years to settle (He and Liu, 1995).

2.4 Regulation is the Major Means of Environmental Protection

Environmental policy can be divided into four categories, namely law, regulation, economic instruments, and public participation. In China, because of public ownership and its cultural inheritance, law and economic instruments are not as effective as in other countries, while regulation and public participation are more popular. The political system in China is such that the Party controls personnel arrangements down to grass roots level, so the instructions from higher ranking officials determine much of the behaviour of ordinary governmental officials and managers of state-owned firms. If these instructions were ignored the status of low ranking officials might be seriously affected and this, in addition to their salary, is one of the major sources of their material welfare. The distribution of consumer goods in China, such as chauffeur-driven cars and decent houses, even banquet expenses relies not only on the market but depends also upon the rank and status of officials, who have the right to consume at the public's expense; and this right is in proportion to their rank. This is why instructions or regulations from high ranking officials are important. In general, policy on environmental protection has a high priority among upper-level authorities. This is a result of economic development and global participation, i.e. international exchange of environmental policy and imported knowledge in this area. Indeed, most of the success in environmental protection in China is promoted by the administrative power of Government. However, China is undergoing rapid change, especially in its economic system; more and more market functionality is being introduced and more and more non-public enterprises have emerged, thus economic instruments for environmental protection will play a more active role.

At the other end of the chain it is local people who have the desire to protect their own environment, and they are becoming another independent force promoting environmental protection. Public education in this field is becoming popular and the subject appears frequently in newspapers, journals, films, and curriculum text books in primary and secondary schools. Public participation is a growing factor in this respect.

2.5 Economic Theory Employed in Policy Design is Inadequate

If there are several different ways to achieve the same objective, a cost-benefit analysis should come into play; economic theory should be employed to optimise resource utilisation in the cause of environmental protection. For example, the rate of discharge fees should be designed according to social cost, otherwise costs from misallocation of resources will be incurred. The current rate of the SO_2 discharge fee in China is RMB 200 (US$ 24) per tonne, but the damage and loss due to SO_2 discharge are an estimated RMB 30 billion annually for a total discharge of 18 million tonnes (Qu, 1992), representing a difference factor of eight between the discharge fee and the average per tonne cost of the emissions. It is also well established in economic theory that unifying marginal costs to reduce the discharge of a specific pollutant in all cases can minimise the total cost of pollution mitigation. But this principle has not been fully recognised. In the case of funding for environmental projects, the issue of who should shoulder the costs has been much debated, and no conclusion has been reached.[8] In practice, more than 90% of capital input for environmental protection has been paid by local and central government or the ordinary tax payer, and not by polluters. Although the Chinese economy is changing rapidly, the funding structure for environmental projects has not changed correspondingly, except in some sectors, where there is high foreign investment. If economic instruments are to be utilised, the PPP must be maintained.

2.6 Economic Development Advances Environmental Protection

One of the important experiences China has to tell of is that environmental protection needs various kinds of resources which only economic development can provide. Developing countries have often been advised that they should keep their environment healthy, and not let it be spoiled only to have to clean it up at a later stage, because that would cost much more. But almost all developing countries fail to follow these "words of wisdom". The reason is that poor countries simply do not have enough resources. In terms of economics, the utility provided by one dollar when one is poor has a much higher value than the utility provided by the same dollar when one becomes rich. In dollar terms, preventing pollution at an earlier stage costs less, but in terms of utility, preventing pollution at a later stage costs less. Capital investment on environmental projects in China was only 0.51% of GNP in 1981 and increased to 0.60% of GNP in 1990, while GNP itself increased by 136% during the same period (Zhang, 1992; CSY, 1991). In 1994, investment increased further to 0.7% (China Daily, April 1995). If we look at the environmental spends for individual provinces in China, or take a cross-section for analysis, we can find the same rule in operation: that rich provinces spend more than the poor provinces, and public awareness is greater, environmental standards are higher. It seems therefore reasonable to expect that environmental problems can be solved only when countries become affluent; poor countries can hardly improve their environment without external assistance. Economic development is the most important precondition for environmental protection.

2.7 Environmental Management is Generally Poor

China is following developed countries, and has established multi-dimensional measures to address its environmental problems. Written documents – laws, regulations, economic tools and so on – are pretty exhaustive, but they are not effective in practice. For example, collection of the pollutant discharge fee, which

8 In the Introduction by Qu Ge-ping to the book Report on China's Investment for Environmental Protection it is stated that the question of who should pay for investment, the State or the discharger of pollutant, is a controversial issue (Zhang and Wang, 1992).

is in any case far lower than it should be, has been difficult;[9] regulation requires "three simultaneous points", i.e. that the equipment to treat pollution must be designed, constructed and operated simultaneously with the design, construction and operation of the main project, but many projects, especially small ones, have not complied with this regulation, causing incidences of serious pollution; the operation of already constructed equipment for pollution treatment is generally poor. For example, the nationwide average operating time of waste water treatment equipment is only 44.9% – in other words, for more than half the time this equipment has been lying idle;[10] although the law for the protection of wild animals has been in force for some years, people still eat a lot of wild animals.[11]

2.8 A Policy of Marginal Improvement Helps to Reduce the Resistance to Implementation

Marginal improvement is a practice of distinguishing between old and new enterprises and adopting different environmental standards, so as to introduce more stringent requirements and gradually raise environmental quality without much resistance. For example the "three simultaneous points" are adopted only for new projects. The practice of marginal improvement has been adopted in many countries, developed and developing. But in China it has been generally employed not only on environmental issues but also in overall economic reform, for example, in the dual pricing system, in which the old price is retained for old consumers, while the new price – which is usually higher – is offered to new consumers for the same product. Such a system can introduce a new, less distorted price than the old one, without harming vested interest groups, thus smoothing the reform process. However, this dual standard practice can only be adopted as a transitional measure; if it is institutionalised or formulated as a permanent rule, the principle of optimal resources allocation will be violated, because the cost of reducing the same quantity of pollutant for old factories is less than for new ones, and so the principle of unified marginal unit cost for pollution reduction is violated.

---

9 In 1993, eight investigating groups of NEPA were sent to provinces to check the collection of discharge fees. The result was that cities collected more than 70% of the fees, while small counties collected only about 50%. Up to June, 1993 uncollected fees totalled 1.41 billion yuan (US$ 0.17 billion) or 35% of the total fees that should have been charged (NEPA, 1994).

10 The nation-wide average performance efficiency of waste water treatment equipment is only 31%. (Zhang, 1992: 7) NEPA laid down in August 1992 a penalty for those who do not operate this equipment properly.

11 As reported by "China Environmental News" in a picture showing four pangolins intercepted by policemen while being transported to a restaurant. The pangolin is under 2nd grade State protection. A lot of similar reports have surfaced in newspapers (24 November 1994, China Environmental News).

3. The Main Actors

In China, the structure of the State resembles the structure of a family, i.e. every member in the unit is taken care of by the central authority, to which it owes loyalty: the State. Enterprises are founded and invested in by the State; households earn their wages from the State and serve the State. So there can hardly be real independent organisation outside of Party and Government in China. In the wake of China's reforms, and in addition to private enterprises, some truly non-governmental organisations have begun to emerge. And in the governmental context we can already discern different units acting independently on environmental issues with different interest considerations, although they are not genuinely independent organisations.

3.1 Government

The National Environmental Protection Agency (NEPA) was established at central government level in 1984 and reports directly to the State Council, China's executive arm. NEPA is an organisation with a lower rank than the ministries, but it oversees the national environmental situation, formulates policy, supervises its implementation, coordinates environmental work in different ministries, and participates in international cooperation. At provincial, county, even township level there are over 2,300 similar organisations under the administration of NEPA. The number of staff in this system has been increasing, from 60,400 in 1989 to 81,300 in 1993 (NEPA, 1990-1994). There are also environmental protection offices within ministries, provincial sector bureaux and large enterprises. The number of environmental staff in enterprises alone exceded 170,000 in 1993. In order to support authorities in policy formulation and implementation, the State Environmental Protection Commission (SEPC) of the State Council was established in 1984, bringing together the leaders of more than 30 ministries and bureaux, and meeting quarterly. At the end of 1991, the State Council established a 17-member senior scientific advisory group to help in the formulation of national environmental policy. In respect of legislation, the National People's Congress established the Environmental Protection Commission in 1994 (its name changed to the Environment and Resources Protection Commission, or ERPC in late 1994) which is also composed of 17 members. In theory the organisation of the National People's Congress should be independent from the government, but as stated above, in the Chinese context all institutions bear the stamp of government to some extent. Some members in SEPC work at ERPC simultaneously. On the international front, with assistance from Canadian Paramount, The China Council for International Cooperation on Environment and Development was established in 1993, which provides a unique channel of communication between top international experts on the environment and development on one side, and Chinese experts and high ranking officials on the other.

If one takes a close look at these governmental organisations, one finds that these organisations pay a great deal of attention to pollution control but much less to conservation of resources. For example, the NEPA's office of natural resources employs only about one tenth of its total staff, simply because pollution problems are usually urgent issues, while resources conservation is not an immediate problem; soil erosion, desertification and deforestation have existed for decades, even centuries. The real threat to China is depletion and destruction of its natural resources, which are the very sources of life for a large agricultural country.

3.2 The Judicial System

The judicial system should be independent from government, but in the Chinese case all these institutions are under Party control, so they are mutually dependent. Such a unified system can reduce transaction costs; the number of lawyers and judges required is much smaller. But if the mainstream made a wrong decision, rectifying the situation would be difficult. Another characteristic of the dependent judicial system is weak enforcement in relation to policy design.[12] This is why China has a more or less comprehensive legal system on environmental protection but a weak record of enforcement.

3.3 The Mass Media

The mass media are also a division of government. For example, the "China Environmental News", one of the largest national newspapers in the world to specialise in environmental protection issues, with a circulation of 400,000, was founded and funded by NEPA; the director is nominated by NEPA's Party Committee. In the past decade there has been a blossoming of environmental publications in China. There were nine local environmental newspapers in 1989 (NEPA, 1990-1994); and nearly one hundred environmental journals, from academic to popular, from scientific to literary, are being published. In the most important newspaper in China, the "People's Daily", there is a news story on the environment every two to three days on average, illustrating the high public interest in environmental issues. The Environmental Publishing House alone published between 56 and 111 books annually in the period 1990- 1993, and more than 2,600 books were published on environmental topics between 1970 and 1990 by Chinese publishers. In addition, TV programmes and films on environmental issues have been on the increase, and text books containing environmental information have been widely available from primary to high schools.

12 In order to enhance the capacity for enforcement of environmental law, the SEPC and ERPC jointly organised ten investigating groups to monitor enforcement levels in ten provinces from May to October, 1994 (Xie, 1994). The major findings were as follows: some leaders at the local government are striving hard to develop the economy and ignoring the environment; serious pollution is being created by township industries and there is depletion of natural resources; law enforcement is weak; there is too small an input of resources to protect the environment.

International exchange of literature and scholars has greatly advanced Chinese awareness and knowledge of the environment.

3.4 The Industrial Sector

As usual, the industrial sector often tends to find ways around environmental protection regulations, so as to reduce their production costs. Therefore, pollution prevention is really a battle between government departments and the industrial sector. As is required in China, large state-owned enterprises have a specific environmental office dealing with pollution prevention activities, and their environmental awareness is greater than that of small businesses, especially the township factories, which are the major polluters in China (Xie, 1995). Township factories are non-state-owned enterprises which take production costs more seriously than publicly owned factories. They absorb large quantities of surplus labour in rural areas, but with meagre financial capacity, primitive technology, and poor management. Local government leaders sometimes choose development rather than environmental protection, thus leading to even more difficulties in curing pollution. However there is an encouraging factor on the industrial scene, the newly developing "green" businesses, including manufacturers of environmental clean-up equipment (water treatment devices, etc.), environmental consulting firms, organic food producers. Particularly the manufacturing of environmental clean-up equipment is developing very quickly.[13] In Yi-xing city, Jiangsu Province, a special zone for environmental equipment manufacture was established in 1994 and international businesses were called on to participate. This will provide a channel for the introduction of advanced environmental technology from abroad into China.

3.5 Public Participation

Environmental campaigns of various forms have been promoted, including tree planting, a special TV programme called "Century Environmental Travel" (nationwide video reporting on environmental issues), environmental general knowledge quizzes, essay contests, radio and TV lectures. These forms of public participation are organised by the Party and Government.[14] According to Chinese law, popular organisations must be attached to a particular governmental department at ministerial or sub-ministerial level, so these organisations can hardly be independent: their budget and personnel management are under government control. Therefore, non-profit NGOs under no external control are

---

13 The State Council put out a Circular: "Notes to Encourage Actively the Development of Environmental Industry" on November 5, 1990.

14 In chapter 20 of China's Agenda 21 there is a full description of this element, including the roles played by women, children, youth, minority nationals, workers and labour unions – and also academics. But little room was left for independent voluntary organisations. Greenpeace International has tried to establish a branch in China but has not succeeded.

very rare in China, but under the influence of international trends, truely non-governmental NGOs are emerging and will evolve further. A recent development is the annual assessment of 32 major provincial capitals according to an index of environmental indicators and the publication of the results. This method of increasing public awareness has provoked widespread interest. Individuals can write letters of complaint or suggestions, or visit an environmental office to voice their concerns about the environment. The number of letters received every year varied from 53,000 to 60,000 in the period 1989-1993, and the number of visits was between 78,000 and 85,000 (Environmental Statistics).

4. Capacity-Building

4.1 Legal Framework

Since 1978 China has passed a more or less comprehensive system of environmental laws. In 1978 and again in 1982, articles on environmental protection were added to the Constitution; in 1979 China formulated a provisional Environmental Protection Law and finalised it in 1989. Other laws concerning forest, water and mineral resources, land, pollutant discharge, wildlife and various environmental standards have been passed. Laws to protect groundwater and fragile ecosystems are in preparation. Individual provinces and cities have drafted local regulations and standards which are more stringent than the national ones. In the last five years, there have been five national regulations issued: "Implementation of Regulations on Water and Soil Conservation", "Circular on Banning the Trade in Tiger Bones and Rhinoceros Horn", "Implementation of Regulations on Protection of Aquatic Wild Animals", "Implementation of Regulations on Prevention of Atmospheric Pollution", "Regulations on the Prevention of Oceanic Pollution due to Land Discharges". To date, four environmental laws, eight resource management plans, more than twenty administrative regulations on the environment and resource management, and more than two hundred and sixty environmental standards have been laid down as the legal basis of environmental protection (China's Agenda 21, 1994). China has signed a number of international treaties and protocols on environmental matters, including the "Montreal Protocol on Ozone-depleting Substances", the "Basel Convention on the Control of Transboundary Movements of Hazardous Wastes and their Disposal", "United Nations Convention on Biodiversity", "Nations Framework Convention on Climate Change".

4.2 Human Resources

University-level study of the environmental sciences has been established since 1970s, and there are 79 universities and colleges with departments in environmental studies with a total of more than 11,000 students. Some 5,500 students graduate from these programs annually. Doctorate programmes are offered in 38 faculties and masters courses in 107 faculties. The total number of postgraduate students has reached 500. Some 80 polytechnic schools, together with 160 vocational high schools in environmental technology have 12,000 students (Zhang, 1992). Environmental education as a part of other courses (Chinese, chemistry, biology, etc.) is required from primary school to high school; specific environmental courses are sometimes offered as an option in high schools. Since 1992 teachers have received training in an environmental curriculum for primary and high schools. In 1985, 1988 and 1989, environmental training courses were offered to provincial and municipal governmental leaders as well as to managers of large industries. Since 1993 an international LEAD program (Leadership on Environment and Development) has been in operation to train young environmental officials for promotion to higher ranks. However, a large imbalance still exists between the huge population, vast territory and the number of environmental professionals.

In-service training is essential, especially in the environmental sector, not only to enhance the competence of those who do not have academic degrees, but also to update the knowledge of those who already do have degrees, as the body of knowledge on the environment is quickly expanding. The annual number of persons receiving in-service training has increased from 103,000 in 1989 to 476,000 in 1993 (Environmental Statistics), which means that every environmental worker has had the opportunity of receiving on average more than one training course per year, since the total number of environmental workers in China is about 300,000.

Among some 600 universities and colleges in China, 50 are carrying out research in environmental studies including water treatment, safe disposal of hazardous waste, acid rain, climate change and techniques for monitoring the environment. These researches are carried out in cooperation with various academic and governmental research institutes, and sometimes with foreign institutes. The Chinese Academy of Sciences is undertaking an ambitious programme, the Chinese Ecological Research Network (CERN), designed to coordinate and strengthen its long-term environmental research programs. Its budget is US$ 1.75 million for projects and US$ 10.5 million for construction and capital expenditure in the current five-year plan (1991-95). However, the general problem restricting research activities is lack of funding.

4.3 Main Action to be Taken

The main activities which would enhance capacity for addressing China's environmental problems are the following:

Collection of discharge fees. Charging for pollution discharges has been underway since 1982. Under the system, charges are levied only if emission does not meet the standards required; the charge is not in proportion to the total amount of pollutant emitted. For these reasons, it is in effect a charge for noncompliance with a standard. When a factory meets the standard, no charge is imposed; and so there is no incentive to encourage the factory to further reduce its emissions. The charges cover pollution of water and air, solid waste, and noise. Total charges levied in 1984 were RMB 0.63 billion (US$ 75 million), and in 1993 they were RMB 2.68 billion (US$ 0.32 billion), representing an increase of 80% in real terms. The sum of charges levied constitutes about 16% of the total national cost of treating pollution in 1993. Usually the charge rate is much lower than the marginal cost of clearing the pollution, so factories often choose to pay the charge rather than to stop pollutant emissions. Obviously, some factories reduce their discharges and comply with the standard out of more than economic considerations. The charge system is being improved both in its design and implementation, especially in order to reduce the number of cases of default. The charges for SO_2 and for sewage are based on total discharges, and have been in effect since 1992 and 1993 respectively.[15]

Certificate of discharge permission. In addition to the rule of "three simultaneous points" starting from 1988, 18 cities in China introduced a regulation that factories discharging waste water must apply for a certificate of discharge permission from the government. In 1991, 150 further cities followed, and 12,500 factories have applied for the certificate (Wang, 1991). Since 1991, 16 cities have introduced a certificate for atmospheric pollutants, mainly TSP and SO_2, but it is being extended to other kinds of discharge. This is a first step towards total quantity control and tradeable discharge rights, which are being tested in some cities.

Large projects to reduce pollution. Provincial governments, sometimes with the help of central government, have set aside special funds to treat pollution in specific areas where population is dense and pollution has caused heavy economic losses. These areas include Bai-Yang-Dian, the Tai Lake, Dian-Chi and the Huai River due to water pollution, and Ben-Xi, Tai-Yuan, Bao-Tou and Lan-Zhou due to air pollution. These cases of pollution are the results of long-term negligence in environmental protection and cannot be ameliorated with local government resources alone.

Promote public awareness. Various strategies for promoting public awareness have been pursued. For example: the use of a label to denote environmentally friendlier products including food, refrigerators, energy efficient devices and less seriously polluting chemicals; publishing an annual list of the 3,000 factories which produce most of the pollutants in China; ranking the big cities according to their environmental index; publicising recommendations for the best environmental protection techniques; setting integrated environmental targets for the leaders of local government; mobilising people to plant trees on March 12, the National Day of Afforestation.

---

15 The normal charge for SO_2 is RBM 200/ton (US$ 25/ton); and RMB 1.0/ton (US$ 0.125/ton) for sewage.

5. Policy Evaluation

Although the annual number of environmental accidents, people's letters of complaint and visits to departments have stabilised, the agreed assessment is that China's environmental situation is still worsening overall, with the exception of some specific cities, factories, and areas of natural beauty, where the State pays particular attention to addressing any problems that may arise. But the efforts China has put into environmental protection thus far have been very impressive, otherwise China's environment would have been in a far worse state than it is today, given the fourfold economic growth in real terms over the past decade and a half.

5.1 Too Great a Dependence on the Environmental Awareness of High-Ranking Officials

The success of environmental policy depends very much on how much importance government leaders attach to it, and much less on the legal framework, the coherence of related policies and public participation. There have been many cases of pollution which should have been solved through the Environmental Protection Law and related regulations, but which continued for months, even years, until government leaders came to intervene.[16] To mobilise all positive actors in environmental protection, rights and responsibilities should be shared between independent players, and the decision-making process should be transparent. It would seem that some additional force must be added alongside governmental authority: the non-governmental organisation. The idea that non-governmental forces must of necessity oppose the government and act as troublemakers should be quashed. Grass-roots environmental concern is a real and reliable source of sustainable development.

5.2 International Participation is of Vital Importance for China

The modern idea of environmental protection in China is an imported idea. The concept of a legal system, economic tools, and much of the knowledge on environmental protection have come to China through international exchange. Various international conventions, regardless of whether China has signed up to them or not, are a significant force in promoting environmental protection in China. On the Chinese side, an open policy is essential for addressing environmental issues; on the international side, inviting China to participate is equally essential.

16 Many cases of victims' visits or letters of complaint were only solved after leaders of local or central government issued an instruction. (See a special section, Letters and Visits of People's Complaint, in the "China Environment Yearbook"). When a Deputy Director of NEPA was interviewed by a newspaper reporter and asked what the most important factor in protecting the environment was, the answer given was not the law or regulations, but the environmental consciousness of government leaders (Liu, 1995).

5.3 Continuous Economic Growth is a Key Success Factor in Environmental Protection

Although economic growth has brought about more industrial and agricultural pollution and fostered the depletion of natural resources, it has also helped China to strengthen its capacity to relieve poverty, control population, raise educational standards and better satisfy various aspects of human aspiration. Economic growth awakens a wish to protect the environment and provides the capacity to do so, thus it becomes a necessary condition for environmental protection. Any policy that ignores economic growth would be harmful to the environment. However, growth is not sufficient; environmental health can be kept at an acceptable level with the help of grass-roots participation in addition to official initiatives based upon knowledge and organisational experience.

5.4 Continuing Economic Reform, Especially Reform of Ownership and Anti-Corruption Measures, Will Advance Environmental Protection

The fundamental direction of China's economic reform is towards the use of more market mechanisms, which greatly increases the efficiency of resources utilisation and reduces waste and pollution. In turn, the market is based upon exchange of ownership. Clearly defined ownership is therefore a prerequisite for market functionality. The previous ownership system in China, public ownership, is in fact an ambiguous ownership. When a foreign investor wishes to purchase a State-owned factory, he cannot find out who has the final word and who is to be paid for the transaction. Various kinds of economic instrument for environmental protection become futile under ambiguous ownership – since costs or penalties are imposed on the whole nation and not on any one individual, nobody notices or cares. In the course of China's economic reforms, numerous private enterprises have burgeoned, but the transfer of state-owned enterprises into private hands has never been easy. In the transition period from a command or planned economy to a market economy, the rules governing the distribution of national income are undergoing drastic change, and lot of opportunities for speculation are appearing, of which a particular phenomenon is corruption. It not only creates unfair distribution, but also undermines market rules and institutes loopholes in the legal system and in economic instruments in all kinds of policy areas, thus impairing in end effect any regulation of the environment.

5.5 Economic Considerations Must be Incorporated into Environmental Policies

China has great power to mobilise its people towards a specific target, but their efficiency can be low, in that more resources are used up than if the goal were achieved through a rationally designed policy and well-organised implementation. China is still a poor, developing country, living within the constraints of a low supply of capital and skilled human resources, and so efficiency is of special importance. In this regard, better trained civil officials and a political system which facilitates communication between the elite and the highest authority are necessary.

References

China Daily: Article "Population Tops 1.2 Billion", February 15, 1995, Article "Population Increase Should Sound Alarm, February 23, 1995, and Article "Halting Pollution should be Priority of Industrial Plan", April 17, 1995 (all in English).

China Energy Research Society (CERS), 1994: Bulletin on Energy Policy Research, No 11.

China Environment News: Article "Local Government Default Discharge Fees, 17 September 1994, and 24 November 1994.

China Statistical Yearbook, 1991, 1992, 1993, 1994. Beijing.

Environmental Statistics, (various years): Environmental Statistics 1989-1993. Beijing: National Environmental Protection Agency (NEPA).

He, W. and Y. Liu, 1995: Why a Simple Suit Case Prolongs Four Years?, China Environment News, February 23, 1995.

Li, N., 1995: High Female to Male Ratio and Its Marriage Consequences, Population Science in China, No. I, 16-20.

Lin, F., 1994: Trend of Rate of Fertility in China After Adjusted for Marriage Age, Population Science in China IV, 10-20.

Liu, Z., 1995: An Interview with the Deputy Director of NEPA, Prof. Ye Ru-qiu, China Environmental News, 16 March, 1995.

NEPA (National Environmental Protection Agency), (various years): China Environment Yearbook 1991-1994. Beijing.

NEPA (National Environmental Protection Agency), 1994a: China Environment Bulletin 1993. Beijing.

Qu, G., 1992: China's Environment and Development.

Qu, G. and J. Li, 1992: Population and Environment in China. China Environmental Science Publishing House.

State Council, 1994: China's Agenda 21. White Paper on China's Population, Environment, and Development in the 21st Century (in English).

SPC (State Planning Commission), 1991: State Report of Environment and Development, presented at the United Nations Environment and Development Conference, 1992 (in Chinese and English).

Wang, X., 1994: The Grim Situation of China's Farmland and the Relevant Policy. A Report to the People's Congress.

Wang, Y., 1991: Speech at the National Conference on Certificate for Waste Water Discharge, in: China Environmental Yearbook 1992, p. 112.

WRI (World Resources Institute), 1992: World Resources 1992-93. Oxford, New York: Oxford University Press.

Xie, Z., 1994: Typical Law Suit Cases on Environmental Issues in China. China Environmental Sciences Publishing House.

Xie, Z., 1995: Report of Investigation on Environmental Law Enforcement, in: Newsletter on Environmental Work, March 1995.

Zhang, K., 1992: A Speech on the National Conference on Environmental Education.

Zhang, K. and Y. Wang, 1992: Report on China's Investment for Environmental Protection. Quinghua: Quinghua University Press.

NIGERIA

Fatai Kayode Salau

1 Main Environmental Problems and Environmental Quality Changes since 1970

Nigeria has an area of 923,800 km^2, and is, with a total population of 115.7 million (in 1992), the most populated country in Africa. 64% of the economically active population of 43.1 million were employed in the agricultural sector. The contribution of agriculture to economic growth has experienced a continuous decline in the last two decades. Its share of Gross Domestic Product (GDP) dropped from 55% in 1965 to 36% in 1990. The share of the industrial sector increased correspondingly from 12% to 38% in the same period.

The size and structure of industry in the country have been subjected to rapid change in the last two decades. Industrialisation began in the 1960s with labour-intensive industries for the production of light consumer goods such as beverages, food products, wood and wood products. The substantial financial resources which accrued to the government from the export of oil in the course of the 1970s enabled investment in capital-intensive processing industries, such as petro-chemicals, cement, and ferrous metals.

1.1 Industrial Pollution

The increasing number of capital-intensive industries brings with it a significant pressure on the environment. New types of environmental hazards unknown in traditional labour-intensive industrial processes are becoming rampant as the residues from toxic materials employed in the new industries are discharged into the environment. Effluents are discharged into the surface water and gullies without prior treatment. The chemicals and cement industries continue to pollute the atmosphere with dust emissions. An estimate puts the level of dust emissions from a cement factory in Calabar in 1990 at about 1.25 metric tonnes or 25 bags of cement per day. Petroleum production, which has accounted for the lion's share of industrial production since the 1970s has caused the devastation of the environment in the vicinity of the oil fields. A total of 2,380 oil spillages, releasing two million barrels into the Nigerian environment, were registered between 1976 and 1988. In the thirty years up to 1994, 18,000 billion cubic meters of toxic hydrocarbon gas are said to have been pumped into the Nigerian environment at a rate of 1.6 million cubic meters per day (Guardian, 13 July 1989; Wall Street Journal, 28 June 1994). Commercial energy production increased rapidly in the 1970s but decreased in the 1980s due mainly to the economic crisis (*Table 1*).

Table 1. Production and Consumption of Commercial Energy.

	1970		*1975*		*1980*		*1985*		*1992*	
	Production	Consumption	Production	Consumption	Production	Consumption	Production	Consumption	Production	Consumption
Total	54,461	1,888	89,177	3,549	105,827	7,624	77,075	10,288	96,473	16,728
Solids	41	45	171	154	121	124	97	98	66	47
Liquids	54,203	1,626	88,440	2,829	104,190	5,992	74,180	7,403	91,565	11,848
Gas	102	102	368	368	1,213	1,213	2,610	2,610	4,567	4,567
Electricity	115	115	198	198	302	295	188	177	275	267

Source: United Nations Energy Statistics Yearbook.

1.2 Forest Depletion

Like most small-scale industries, the Nigerian domestic sector still depends to a large extent on wood as the dominant domestic fuel. 90% of the rural population, with a very low income, uses wood as fuel. A serious environmental implication of the use of wood as fuel is the indoor pollution which arises therefrom. The extraction of fuel wood has contributed greatly to the process of deforestation. The annual percentage change in production of fuel wood between 1977 and 1987 was as high as 4.1%, that was about 0.8 percentage points above the average for all countries in sub-Saharan Africa. It should be stressed that the demand for fuel wood does not explain alone deforestation in Nigeria. The large demand for land for the siting of industries and for the building of infrastructure such as highways has, among others, hastened the process of deforestation and desertification. Over 300,000 hectares of forest area were cleared between 1975 and 1980 during construction of road infrastructure. As the FAO estimates show, the total forest and woodland area of 16 million hectares in 1975 had decreased by 5 million hectares in 1992 (*Table 2*). According to expert opinion, the whole tropical forest in Nigeria will have been destroyed at the end of this century if the present rate of deforestation should persist.

Table 2. Land[1] Use and Population in Nigeria

Year	*Arable land and Permanent Crops*		*Forest & Woodland*	
	1,000 ha	as percentage of land area	1,000 ha	as percentage of land area
1975	30,000[2]	32.9	16,400[2]	18.0
1980	30,385[2]	33.4	14,900[2]	16.4
1985	31,130[2]	34.2	13,400[2]	14.7
1990	32,300	35.5	11,900[2]	13.0
1991	32,335[2]	35.5	11,600[2]	12.7
1992	32,385[2]	35.6	11,300[2]	12.4

1 Total area: 92,377,000 ha; Land Area: 91,077,000 ha; (100 ha = 1 km^2).

2 FAO estimate.

Source: Production Yearbook, Rome, several editions.

Mechanised agriculture, dams and large-scale irrigation projects have in the last twenty years caused a steady rise in demand for land, including forest land. For the implementation of the "Green Revolution", launched in the Fourth National Development Plan (1981-1985), a total of 250,000 hectares of land was to be made arable within a period of four years. This ambitious "Green Revolution" programme continued the "Operation Feed the Nation Scheme" which was started in 1976. The administration of General Babangida continued the agricultural development programme under the name "Back-to-Land-Scheme" and created for this purpose the National Agricultural Land Development Authority (NALDA). Only two years after it began operation (1991-1993), NALDA had already developed a total of 14,850 hectares of farm land while 10,924 hectares of bush had been cleared, the greatest part of which is in the Nigerian Savannah zone. The percentage share of arable land and permanent crops in the total land area increased from 32.9% in 1975 to 35.6% in 1992 (*Table 2*).

1.3 Rural Water Pollution

The impetus for the ambitious agricultural development programmes in Nigeria is the rapid change in the consumption pattern of the population, as the high demand for wheat breads since the mid-1970s shows. The main objective of the "Green Revolution" policy was to boost agricultural productivity through the promotion of improved agricultural technology, most especially in the drought-ridden areas of Northern Nigeria. The introduction of intensive agricultural practices (mechanised irrigation) has caused an increase in the consumption of mineral fertilisers in the last two decades: 2.6 kilograms per hectare of arable land and permanent crops were consumed in 1976 as against 13.3 kg/ha in 1991. Investment in fertilisers and high-yielding seeds consumed 62% of capital investment in the Fourth Development Plan. Fertiliser prices were highly subsidised to establish a uniform price throughout the country.

1.4 Urbanisation

The pull from new industrial growth centres and the push from the deterioration of the rural situation has led to rural migration and rapid urban growth since 1970. The fastest rate of urbanisation is recorded in Lagos (at a rate of 5% per annum) where about 70% of the 3,000 manufacturing firms recorded in Nigeria are located. With a landmass of 3,577 square kilometres, Lagos State represents only 0.4% of Nigeria's total land area; however, about 10 million people, i.e. 12% of the total population are resident here. Because of the concentration of manufacturing activities there, over 50% of the country's electrical power is consumed in Lagos State. An estimated total of about 540,000 cubic metres per day is drawn from surface water and used as municipal water. An additional 140,000 m^3/day are drawn from underground water for municipal use in both the rural and metropolitan areas. The Lagos State Waste Management Authority collected a total of 3,194,148 tonnes of domestic refuse in 1989 and 3,718,684 tonnes in 1990. Other growing urban areas are Ibadan (Western Nigeria), Port Harcourt, Onitsha, Aba Enugu (Eastern Nigeria), Kaduna, Kano, Jos (Northern Nigeria). As in Lagos, the local authorities of these areas are not equipped with adequate regulations, facilities and professional staff, and are therefore unable to manage the waste generated by industries and households. As a result, urbanisation has translated into a deterioration of environmental quality since the late 1960s.

2 Development and Main Characteristics of Environmental Policy

2.1 The General Political Context

Nigeria attained political independence as a federal state on 1 October 1960. The military has ruled the country for most of the time up to now. There was only a partial military withdrawal during the Second Republic (1979-83), headed by Alhaji Shehu Shagari, when an American-type Presidential system was introduced.

The constitution of 1979 (the reviewed Constitution of 1989 has not yet come in to force as planned for 1992) is quite distinct from its American model in many respects. The choice of the authors of the Constitution for an Executive President with "an independent right to govern" reflects the political situation at the time of the constitutional debate, namely military rule and a unitary form of government. Environmental problems were not an issue in the debate. The constitution gives the federal government power to legislate on sixty-five enumerated items in an exclusive list which includes areas such as minerals, mineral oils, natural gas, nuclear energy etc. Since there is no mention of the word "environment" in the Nigerian constitution, major environmental concerns such as waste and water management belong among residual matters to be placed under the states' jurisdiction. But the federal government has been usurping the states' legislative power since the enactment of the "Supremacy and Enforcement of Powers Decree" in 1970. The Decree No. 1 of 1984 which stipulates that the federal government could "make laws (...) with respect to any matter whatsoever" further strengthened federal supremacy over the states.

A salient feature of the Nigerian economy since 1970 is its over-dependence on oil income. Because of Nigerian oil's low sulphur content, demand in the United States rose after the enactment of the Clean Air Act of 1970; most of the petroleum is still exported to the US market. The huge revenues from the export of oil have speeded economic transformation and transformed the political structure inherited from the colonial government. The Petroleum Production and Distribution (Anti-Sabotage) Decree 35 of 1975 which stipulates the death penalty or imprisonment without trial for "criminal" acts which disrupt the production or flow of petroleum reveals the importance of oil for the political system also. Oil has nevertheless contributed to the crisis in the Nigerian state and thus created favourable conditions for military intervention in politics.

The period after independence in 1960 to the present day has seen the publication of four National Development Plans (1962-68; 1970-74; 1975-80; 1981-85) and three National Rolling Plans (1990-1992; 1993-1995; 1995-1997). Economic planning changed at the start of the Second Republic in 1979 from being a voluntary exercise to being a constitutional obligation of the federal government, to enable more participation of different interest groups and interested citizens in the preparation of the plan. In spite of this, due to paucity of data, the economic, political and social objectives of the National Development Plans are still defined only vaguely.

2.2 The Starting Phase of Environmental Policy

Whilst the internal pressure must indisputably be stressed as the most important impetus for the development of environmental policy in Nigeria, the ambition of Nigerian politicians to play a leading role in the African continent on this major international issue has spurred on and shaped many government environmental policy actions since the United Nations Conference on the Human Environment at Stockholm in 1972.

In 1989 Nigeria launched the "National Policy on the Environment" (NPE), the first environmental programme in Africa. The country has been instrumental in fostering regional cooperation between the African countries on environmental issues. Since 1988 it has taken a radical stance on the problem of the export of hazardous wastes from industrialised countries to the African continent and was the first West African country to sign the Basel Convention on Transboundary Movement of Hazardous Wastes in 1989. On the whole, Nigeria has signed about thirty conventions on the improvement and protection of the global environment since 1970. Implementation of these treaties has been less spectacular. Less than one quarter of the twenty environmental treaties of the Economic Community of West African States (ECOWAS) signed by the Nigerian federal government has been ratified and less than a quarter of those ratified treaties has been implemented (Adewale, 1996). This result shows that the influence of the regional organisation ECOWAS on Nigerian national environmental policy is still minimal.

The political upheavals from the mid-1960s to 1970 diverted the attention of both federal and regional governments from the rapidly increasing environmental problems. After the Civil War the principal policy focus was placed on rapid economic growth and not on the increasing environmental menace, even though some environmental concerns could be discerned in the Programme of "National Reconstruction" published in 1970. The Third National Development Plan (1975-80), published after the Conference in Stockholm, re-emphasised the notion present in the Second Development Plan (1970-1974), that poverty and economic underdevelopment and not population growth are the major causes of the Nigerian environmental problems that need to be urgently addressed.

Even though the government planned to introduce Environmental Impact Assessments for all development projects in the Fourth National Development Plan Period (1981-85), there were no laws passed to this effect throughout the Second Republic (1979-1983). However, in response to a report from a Parliamentary Committee on the environmental consequences of oil production set up by the civilian government, a bill to establish a "Federal Environmental Protection Agency", modelled on the US Environmental Protection Agency, was proposed to the National Assembly in 1981. Its aim was to coordinate the fragmented environmental activities of the Federal and State institutions and to provide the first national framework law within which the states could initiate environmental legislation to prevent and combat various forms of pollution. But in spite of this relatively early attempt to institutionalise environmental policy at the national level a Federal Environmental Protection Agency was not set up until 1988.

In 1984 the Environmental Sanitation Programme of the so-called War Against Indiscipline (WAI) was introduced by the military government of the Buhari/Idiagbon. It was targeted at the improvement of hygiene and aesthetic in villages and cities. But the programme was also intended to provide both the federal and the state governments with favourable political conditions for rigorous environmental law reforms. This is shown at the federal level in the establishment of the National Committee on Ecological Problems (NCEP) in 1985 and in the review of the Factory Decree in 1987. At state level, the Lagos State Town and Country Planning Regulations of 1986 provide that Environmental Impact Analysis Reports be produced for some development projects. WAI, which compels all citizens to clean their homes and working environment on the last

Saturday of every month from 7 a.m. to 10 a.m. is still observed in the whole country, albeit under another designation. During the three-hour cleaning exercise, movement of vehicles is prohibited, except vehicles conveying persons who are on essential duties. A financial incentive is given by the federal government to the cleanest state in order to enhance environmental discipline.

It was not until the later part of 1980s that recognition of the adverse effects of the government's development strategies (secondary environmental problems) and the determination to curb transboundary movement of hazardous wastes prompted the government's actions in the area of industrial pollution. A new policy orientation towards Nigerian environmental problems started, following the report "Our Common Future" by the World Commission on Environment and Development (1987), and after the "Koko episode" (1988). The first national Environmental Framework Law, Decree 58, was passed in 1988 after the illegal dumping of about 4,000 tons of toxic industrial wastes from Italy was discovered at Koko, a village in Bendel state/Southeastern Nigeria. "Koko toxic waste episode", according to the Nigerian President, General Ibrahim Babangida, "became the crucial test of the will of the government and the people of this nation to protect the national environment in a more comprehensive and enduring manner than had been done hitherto" (Babangida, 1989: 16).

The government's new approach is documented in the National Policy on the Environment (NPE) 1989 and incorporated into the *National Rolling Plan (1990-1992)*, inaugurated at the beginning of the 1990s. The NPE sets out the goals and implementation strategies in eighteen priority areas, including population, air pollution, agricultural chemicals, toxic and hazardous substances etc. According to the Guidelines for the National Rolling Plan the government is determined to pay "more attention to the integration of environmental considerations into our economic and social planning such that side-effects that will give negative impact on our environment are fully anticipated and minimised". A detailed capacity-building national plan with an estimated cost of about 63 million US$ over the five years between 1993 and 1998 was prepared after the UNCED in Rio (details in UNDP, 1994).

2.3 Basic Institutions and Regulations

The Environmental Framework Law of 1988, Decree 58, established the Federal Environmental Protection Agency (FEPA) and placed it under the supervisory control of the Minister for Works and Housing. 23 of the 30 Nigerian States had created their own Environmental Protection Bodies by 1993. Comprehensive control of the atmospheric pollution emanating from factories and power generating stations was embodied for the first time in a national law (FEPA decree). Decree 42 of 1988 on Harmful Waste (Special Criminal Provisions etc.) makes the illegal import, transport and storage of hazardous wastes a criminal offence, punishable by life imprisonment. Stringent fines and long-term imprisonment are laid down in Decree 58 (section 35 or 36) as a measure to combat contravention of the government's environmental standards. Decree 58 also makes provision for non-penal measures such as grants, agreements with polluters etc., to combat industrial pollution (section 5). Awards of compensation, reparation and restitution for environmental damage, omitted in the FEPA-Bill of 1981, are now included in the Environmental Framework Law. Other important pieces of legislation are the National Environmental Protection Regulations (Effluent Limitations; Pollution Abatement in Industries and Facilities Generation Wastes) from 1991, and the Environmental Impact Assessment Decree No. 86 of 1992.

Legislation on environmental protection remained uncoordinated even after 1988. A considerable number of preventive, regulatory, and penal measures by the state and federal governments to regulate the use of natural resources, control pollution and enhance environmental quality is still scattered about in many legal norms which have no direct relationship to environmental control. The institutionalisation of FEPA was followed in 1990 by a report entitled "Towards the Development of an Environmental Action Plan for Nigeria" prepared by the Government in collaboration with the World Bank. The policy instruments proposed in this document were subsequently expanded in the so-called "anticipate-and-prevent strategies" for achieving sustainable development. These strategies cover, among others, the following areas: environmental education, environmental awareness and environmental legislation. A predominant characteristic of the new policy paradigm is the strong support for incentives and other fiscal tools to facilitate compliance by the target enterprises with the environmental laws and spur on investment in pollution abatement equipment and cleaner technologies. The introduction of discharge permits from 1992 was also proposed. In the area of natural resource conservation, the use of financial incentives to encourage the development as well as utilisation of alternative sources of energy such as smokeless coal briquettes, biogas, natural gas, etc. is to be strengthened. Some innovative institutional changes envisaged after 1990 to integrate environmental concerns into development planning were the establishment of the Sustainable Development Unit in the federal and state Ministries of Budget and Planning, and the adoption of a new system of national accounting by the Ministry of Finance and Federal Office of Statistics to reflect the impact of economic development on the environment and natural resources.

2.4 Policy Style

Policy style as it relates to industrial pollution shows some ambivalence. Successive Nigerian governments have shown on one hand a strong propensity to enact stringent environmental laws. There has been on the other hand a recognition that it would be inconceivable for the environmental institutions to enforce the laws effectively, due to their low capacity as regards human and material resources. Consequently, the so-called policy of "persuasion through dialogue" was introduced at the beginning of the 1990s. "Persuasion through dialogue" implies that industries are to report regularly on the voluntary industrial pollution abatement measures. To facilitate such dialogue, the Lagos State government set up Working Groups on environmental management in the following nine industrial branches: textiles, paints, ceramics, cement, food processing, wood and wood processing, pharmaceuticals, iron and steel and plastics. Criticism has been levied against the government's policy, especially as it relates to the oil industry, on the grounds that the Head of the Petroleum Inspectorate, who is absolutely independent of the FEPA, has supervisory authority over the operation in this sector, while FEPA is accorded only a "supportive role" in the end-of-pipe treatment measures in cases of oil spills. The "Polluter-Pays Principle", cited in many government documents as the basic principle underlying environmental policy is totally ignored in the laws that regulate petroleum drilling and production activities.

3 The Main Actors

3.1 Environmental Policy Institutions

For the most part, measures taken during the period of the Third Development Plan form the basis of environmental policy and institutions in the country today. In 1975 the area of responsibility of the Federal Ministry of Works and Housing was extended to cover issues of the environment. The Federal Ministry of Works and Housing, later renamed the "Federal Ministry of Housing, Urban Development and Environment", existed for only three years because of the jurisdictional conflicts between it and other governmental institutions (Mabogunje, 1995: 13). Between 1978 and 1988, the Environmental Planning and Protection Division (EPPD) in the Federal Ministry of Works and Housing was responsible for the environment-related activities of the federal government. No intention was shown by the government to use EPPD as an instrument for controlling emerging industrial pollution. Therefore, each tier of government and government department dealt with environmental issues as related to its own area or specific activities. The problem of oil pollution was addressed by the Federal Ministry of Petroleum Resources, while responsibility for protection of the environment from the use of pesticides and fertilisers was vested in the Federal Ministry of Agriculture. Lagos State was the first to establish its own separate Ministry of Environment in 1979. In subsequent years, many other states followed its lead in founding new institutions or ministries in charge of the environment. The Ecological Fund was established by the federal government in 1981 and allocated 1% of the Federation Account for the purpose of ameliorating environmental damage caused by ecological disasters. A Presidential Committee on the Administration of the Funds for Oil Mineral Producing Areas under the Ministry of Finance was created and received 1,5% of the Federal Account to provide infrastructure, education, job opportunities and environmental clean-up in all oil-producing areas.

Since 1989, many environmental policy institutions have emerged, for instance the Special State Committee on Environment, which is to facilitate the coordination and cooperation of states with the local authorities; the Federal-State Conference of Environment Directors, which is chaired by FEPA and was created to further federal/state cooperation; the National Council on Environment (NCE), which was created in 1989, comprises all Ministers of the federation responsible for environmental matters and is responsible for the harmonisation of policies and implementation at all government levels. The members of the NCE meet annually under the chairmanship of the federal Minister in charge of the environment. The creation of a Federal Ministry of Environment and a Joint Consultative Committee on Environment and Population involving the private sector and government at all levels was envisaged in the National Policy on Environment.

Presently, the Federal Environmental Protection Agency (FEPA), the EPPD, and the Natural Resources Conservation Council (NARESCON) are the major Nigerian environmental institutions. The responsibility for the protection and development of the environment in general and environmental technology is vested in FEPA.

NARESCON was established (Decree No. 50 of 1989) one year after FEPA and is, like FEPA, an integral part of the Presidency. NARESCON comprises eight members including, among others, the President, who functions as the Chairman, the Ministers in charge of agriculture and water resources, foreign affairs, works and housing, and health. It is to formulate a national policy for natural resources conservation, provide policy guidelines and necessary directives on the sound management and utilisation of renewable resources, coordinate the activities of all international agencies, government institutions including FEPA, universities and research institutes as well as non-governmental organisations involved in natural resources conservation. Other institutions with environment-related responsibilities are the Federal Ministry of Agriculture, Water Resources and Rural Development; the Federal Ministry of Works and Housing; the Ministry of Petroleum and Mineral Resources; the National Agency for Food and Drugs Administration of the Federal Ministry of Health and Human Services.

There was an abortive attempt to merge FEPA and NARESCON in 1992. The reorganisation brought about a power struggle within the administration between August 1993 and February 1994 because of the refusal of the incumbent director of FEPA, Dr. E. O. A. Aina, to make way for his designated successor, Alhaji R. Shaba, formerly of NARESCON. The six-month internal strife created in the public the impression that the Agency had become "the fiefdom of an individual". FEPA has resumed its functions again, even though its impact cannot be said to be strong in all the areas in which it is expected to operate. Programmes being highlighted by the agency today focus more on pollution control rather than environmental conservation.

It can be concluded that the large number of central environmental institutions that has erupted could be explained by two factors: firstly, the preference of the Nigerian political elite for centralised technical solutions to political problems, which continues as a legacy of colonialism in the Nigerian post-colonial state; secondly, the inclination of the political elite to create institutions for the purposes of satisfying the personal material aspirations of egoistic individuals. In spite of the myriad of environmental institutions, the Nigerian state has not hitherto achieved the monopoly on power to enable it to control the strong industrial enterprises, nor has it been able to enlist the masses in the environmental decision-making process. Nigeria, as the following part shows, has yet to use exhaustively the opportunity given by its federal state structure to build a viable legal and administrative capacity to face existing environmental challenges.

The Nigerian post-colonial state is, like its colonial predecessor, still overcentralised and not based on a legal-rational authority in the sense of Max Weber. Throughout the 1970s to the end of the crisis-ridden Second Republic (1983), local governments were dissolved by state governments at will because of alleged misallocation of the local government resources. Their functions were transferred to the so called "Caretaker Committees" or to totally new institutions established by the state governments. President General Sani Abacha, after the military coup d'état of November 1993, has used the Decree 15 of 1989, "Local Government Basic Constitutional and Transitional Provisions", to dissolve the local governments again. Political instability in the last two decades, which culminated in a change of federal government three times within only six months in 1993, has meant frequent structural and personnel changes in the environmental administrations. This situation does not augur well for long-term environmental policy and capacity-building either. In fact, for reasons of political instability, some planned environmental management institutions with approved World Bank loans have not taken off, several years after their approval. One grave consequence of the political instability at local level is the non-implementation of the clause in the national framework law on environment (Decree 58 of 1988) which directs that local government councils be encouraged to set up their own environmental protection bodies as soon as possible.

Courts have played virtually no role in Nigerian environmental management. Environmental authorities often avoid the prosecution of polluters because lack of personnel capacity has often resulted in long queues at the federal courts which have jurisdiction over environmental crimes. The government tactic of granting wide discretionary powers to control institutions generally inhibits effective judicial control of administrative actions. For instance, the Petroleum (Drilling and Production) Regulations of 1969 are defined in general terms, e.g. "good oil-field practice" without specific pollution standards. Environmental victims do not use the court for legal protection or redress, due to the time and money that a process entails and because of the judiciary's tendency to favour the polluters and deny awards of remedies to the victims. Wide executive discretionary powers have bred corruption: polluting companies, for fear of being portrayed negatively in the public, have often preferred to influence the officials by offering bribes or lucrative jobs in their industries (Akanle, 1991: 32; Ogbuigwe, 1989).

3.2 Green Organisations

The majority of victims of environmental degradation still have no effective formal organisations to give more weight to their agitation for a better environment. Many Nigerian communities have resorted to the strategy of non-compliance with the modern environmental laws, whereas traditional practices and laws on natural resources management which have a conservationary effect on natural resources are still heeded. Omobolaji Adewale found in a recent study that many communities in south-western Nigeria still abide by the regulations that allow hunting activities only in the dry season, during which time bush burning is also a punishable offence (Ajomo and Adewale, 1994: 157 ff.). Various attempts to establish an effective pressure group are currently being made by intellectuals all over the country. Donor organisations from industrially developed countries have aided this process, most especially in the years after the publication of the Brundtland Commission's report.

The Movement for the Survival of the Ogoni People (MOSOP). MOSOP formed by the people of the oil-producing area of Ogoni/Southeastern Nigeria in 1990 is a good example of an emerging environmental pressure group. MOSOP has calculated that Ogoniland has, since 1958, contributed an estimated $30 billion to the Nigerian national wealth, but has received "nothing" in return except an environment devastated to the extent of ecological disaster. Nigeria was, by 1993, in the opinion of MOSOP-leader Saro-Wiwa, "a federation only in name. In practice, it was a slave society, a rogue state with a very strong central government in which the ethnic majority oppressed the minorities, depriving them of political power and of their economic resources" (Saro-Wiwa, 1994: 18). MOSOP demands $4 billion compensation and the direct payment of $6 billion royalties and rent from the oil company Shell. In 1990 it submitted a "Bill of Rights" to the government, in which it demanded inter alia: political control of Ogoni affairs by Ogoni people, the right to control and use a fair proportion of Ogoni economic resources for Ogoni development, adequate and direct representation in all Nigerian national institutions and the right to protect the Ogoni environment and ecology from further degradation. MOSOP's strategy has changed radically since Ken Saro-Wiwa assumed the leadership of the organisation. The strategy was no longer limited to merely negotiating compensation, but included violent attacks and the mobilisation of international allies (Oyediran, 1994). The MOSOP's "Bill of Rights" illustrates the interrelationship of environmental and political issues in Nigeria.

Reactions of the government to Ogoni agitation have included military violence, as a result of which many Ogonis have been killed. The all-encompassing Civil Disturbances Decree (Special Tribunal) of 1987 was used to justify the trial of Saro-Wiwa and eight other MOSOP members and to sentence them to death in November 1995. According to the human rights activist, Gani Fawehinmi, cited in the Nigerian weekly magazine TELL, "the whole trial was a political inquisition" (TELL, 13 November 1995).

Urban-based non-governmental organisations. 27 urban-based NGOs formed by "concerned" Nigerian intellectuals have been identified in 1995. The activities of the urban-based environmental groups are intended to complement measures taken by the government to enhance its informational and personnel capacity. But fears that many urban-based NGOs are powerful members of the ruling elite, who negotiate agreements that could run counter to the actual needs of the grassroots are too important to be ignored.

Of the Nigerian urban-based NGOs, the "Nigeria Conservation Foundation" (NCF) registered in 1982 is very influential, because it has support from the big industries in Nigeria. About 75% of its funds is said to come from this source. The National Conservation Education Strategy for primary and secondary schools, developed by the federal government in cooperation with the NCF, was adopted in 1990 into the National Policy on Education. Another NGO with a growing impact is the "Foundation for Environmental Development and Education in Nigeria" (FEDEN) formed in 1991 and solely sponsored by the Rockefeller Foundation. FEDEN has through its "Leadership for Environment and Development Programme" (LEAD) trained 40 Nigerian environmental managers, including 7 from FEPA in the three years between 1991 and 1994. Other NGOs are: "Ecological Society of Nigeria" (ECOSON), "Nigerian Environmental Society" launched in 1985, and "Nigeria Study and Action Team" (NEST) founded in 1987. NEST published in 1991 a report on the ecological situation in the country entitled "Nigeria's Threatened Environment. A National Profile". Generally, the urban-based NGOs are seen to have "lost control of the research agenda as they mechanically respond to whatever the multilateral/bilateral donor agencies require by way of reports" (Bangura, 1994: 274). There are still no private research institutes for environmental studies.

Better Life Movement. Two women's environmental groups, the "Women's Environment and Development" (WEDO) and the "Country Women's Association of Nigeria" (COWAN), were registered in 1995. Starting from the general idea that sustainable agricultural development can be achieved through more participation of women in the government's development programme, the wife of the Nigerian President (1985-1993), Mrs Maryam Babangida founded in 1987 the "Better Life Movement". Its objective is to develop a "Better Life Programme" (BLP) to tackle the impediments to women's development in the rural areas. Major instruments employed by the "Better Life Movement" are multi-purpose cooperatives, created throughout the country under the patronage of the wives of the state military governors. Up to 200 women's cooperatives were created annually in some states between 1987 and 1990. The main function of this large pressure group is to further sustainable agricultural productivity.

The military government did not see any necessity for a green party when in 1989 it created two political parties per decree: the National Republican Convention (NRC) and the Social Democratic Party (SDP). However, it included in the manifestos drafted for the new parties a clause which makes it mandatory for the parties to continue the military government's strategy for environment protection.

3.3 The Media

The involvement of the media in environmental issues has a long tradition in Nigeria. Many ambitious health laws during the colonial days were enacted after the unrelenting pressure of newspapers on the government to remedy the bad sanitation conditions in some parts of the country (Echeruo, 1977: 23).

The influence of the Nigerian press on government policy cannot be denied, despite the strict press laws, especially since 1984. A case in point is the discovery of the illegal import of toxic chemical waste in 1988, which has been credited to the Nigerian press. The media participate actively in the environmental education of the country by providing detailed and in-depth information on Nigerian environmental issues. For instance, the fate of the "environmental refugees" – in the oil-destroyed areas of the Delta and in Bakolori, where a mechanised irrigation system has destroyed the environment – were given very wide press coverage in the 1980s. The weekly magazine "West Africa" publishes informative experts' comments on many important international developments in the area of the environment, in particular as they relate to Africa and the West African Subregion. Since the beginning of the 1990s the Nigerian daily paper "The Guardian" focuses at regular intervals on the effects of Nigerian environmental laws on property, while the "Nigerian Daily Times", among others, makes valuable contributions to the issue of the detrimental impact on health of environmental problems. Environmental institutions, especially FEPA since its inception, have made use of the press to mobilise the populace for their environmental programmes. To foster environmental education, FEPA sponsored in the early 1990s two television programmes: "Clean Talk" every Saturday and "Housing and Environment" every Tuesday. The NARESCON also started towards the end of 1992 a weekly programme on National Television where topics of conservation are discussed.

3.4 Economic Actors

FEPA found in a survey conducted in 1992 that as many as 463 firms have labelled themselves "environmental companies". The breakdown was as follows: 316 environmental consultants/contractors, 58 hazardous waste disposal companies, 89 municipal waste management companies. Of these, 118 deserved to be named "environmental companies" while only 30 could attain the standards set by FEPA for permits and accreditation to operate as such. More serious private entrepreneurs would have been enticed into the environmental industry if the demand for environmental technologies were high: only a total of about 5% of the Nigerian manufacturing companies have, according to the Lagos State government, installed some sort of treatment facility in the last ten years. The environmental industries have not been able to derive much economic gain from the government's environmental policy, because of their low degree of leverage coupled with the aforementioned low demand for environmental technologies and their small number. A prominent association of the environmental industries now emerging as a pressure group is the "Nigerian Industrial Cleaners Association". Willy Ikolodo, whose Lagos-based company Patovilki Industrial Planner Ltd. has been engaged in industrial cleaning since the 1970s, argued that the environmental problems in the country could be solved effectively if the environmental engineers from the private sector were consulted in the law planning phase. The "Nigerian Urban Forum" is new on the domestic environmental scene. Urban Forum was created in 1994 by the Urban Development Bank of Nigeria which itself was established through a decree in January 1992. The Urban Development Bank is mandated to provide resources for development in urban areas. It might be said optimistically that the Urban Forum epitomises a rethinking of development strategy in Nigeria, because it is created as a "multidisciplinary and cross cultural forum" to support and guide orderly urban growth and better management of the urban centres. Eleven oil-producing companies operating in Nigeria formed in 1981 a consortium, the "Clean Nigeria Associates" (CNA), with the aim of implementing the National Oil Contingency Plan.

3.5 Target Groups of Environmental Policy

Parastatals are the most prominent target groups, since the State still plays the major role in the economy. In 1988, the federal government announced 96 state-owned enterprises, including many in the manufacturing sector, that are to be commercialised or privatised. The NNPC, the largest industrial polluter, is still in the federal government's possession. State governments' parastatals such as the "Ogun State Property and Investment Company" (OPIC), the "Farmers Supply Company" (FASCOM), a parastatal of the Kaduna State in fertiliser operations (procurement and distribution), are, like federal government's parastatals, uncontrolled polluters.

The ongoing economic restructuring programme and the privatisation of some industries will bring some changes in the structure of target-enterprises. It is however doubtful whether less pollution will result from privatisation, as the proponents have argued. The environmental institutions are still not able to enforce environmental laws on the corporate large-scale polluters in the private sector, organised in strong associations such as the Manufacturer Association of Nigeria, the Nigerian Chamber of Commerce and Industry, the Nigerian Association of Chambers of Industries, Mines and Agriculture, and the Lagos Chambers of Commerce and Industry. A representative of Nigeria's biggest Cement industry once argued that "it would not be wise to start making stringent or unattainable laws at this point of our development. Any attempt to emulate the environmental protectionists laws of the USA or Europe will adversely affect the growth of our Gross Domestic Product" (Sowande, 1991).

Since the enactment of the Land Use Decree in 1978, big commercial enterprises and oil industries have gone into the agricultural business, and are instrumental in speeding up the growing tendency towards concentration in the agricultural sector as well as intensive agricultural methods. They constitute, together with the so-called gentleman-farmers – retired high military officers, retired superior permanent secretaries and staff members of academic institutions – important target groups, too powerful to be controlled.

The present forestry law, for example, dates back to colonial days and needs review. According to the Natural Resources Conservation Council in 1992, the Forestry Act of 1938 is unsuited to modern day requirements and appears to unduly protect the interests of forest exploiters to the detriment of forest resources conservation.

Transport sector. The Nigerian transport sector is dominated by a large number of private operators. 80% of all passenger and freight traffic is road-based. The uncontrolled importation of outmoded foreign vehicles for transport business is having adverse effects on the environment and human health because of maintenance problems, e.g. shortage of spare parts. The government announced in its 1994 budget the banning of the import of used cars which are more than ten years old. A positive effect from this measure is still to be seen.

4. Capacity-Building

4.1 Number, Strength, Professionalism and Independence of Proponents

Environmental Policy Institutions. The entire senior staff of FEPA comprises (in 1995) less than 200 graduates and technicians. They are responsible for a nation of 30 states, one Federal Capital Territory and 589 Local Government Areas. The number of staff members at the Head Office and four regional Conservation Monitoring Centres of NARESCON was 110 at the end of 1991, comprising 60 senior and 50 junior staff. Numbers of NARESCON staff increased later by a net total of 40 a year.

Knowledge. Comprehensive statistical data on the state of the Nigerian environment are not available because of a lack of coordination between the agencies responsible for information collection. Seminars and workshops have proven to be important sources of information. Various professional bodies and academic institutions collaborate with the government's environmental protection institutions to provide some information on the state and development of the Nigerian environment. The number of academic institutions has grown since the 1970s, providing well trained graduates in e.g. environmental law (e.g. the Faculty of Law of the University of Ibadan since 1984) and environmental science (e.g. the Institute of Ecology of the University of Ife since 1982). The general number of students increased from 14,360 in 1970 to 123,047 in 1987 (Bangura, 1994).

Participation. The government's proposal in the 1981 FEPA-Bill to establish institutional representation of interest groups (three persons from the public) through FEPA Governing Board (now Ministerial Governing Council) was omitted in the FEPA decree of 1988. Citizens are expected to influence the government's actions through their lobby organisations. The majority of the victims of environmental degradation, as discussed above, have hitherto no effective formal organisations to give more weight to their activities in support of a better environment. The degree to which members of the Nigerian private sector have been able to make their interests felt in environmental management varies considerably. Some economic interest groups, for example, have even participated as members of a Nigerian expert team in the preparation of the Natural Resources Conservation Action Plan. The Nigerian government and international donor organisations (e.g. UNDP) have given a positive response to the demands of hotel entrepreneurs that the tourism industry be promoted as an indirect instrument for environmental policy.

4.2 Increase of General Resources

According to World Bank estimates, environmental deterioration is causing health risks to 50 million Nigerians and would generate a defensive expenditure of 3-20% of the Gross Domestic Product (GDP) annually if left unchecked. Finance for environment measures still comes mainly from budgetary allocations and from the Ecological Fund established in 1981 to which 1% of the Federal Account (about 1.1 billion Naira) is paid for the purpose of ameliorating environmental degradation caused by natural disasters. The average annual growth rate of GDP dropped from 6% in 1965-1980 to 1.4% in 1980-1990, according to the World Development Report. What this amounts to is that the government's financial capability and commitment to address environment issues have decreased considerably.

On the whole, international donor organisations, both governments and non-governmental organisations, have increasingly become the major source of finance for federal and state environmental activities. 79% of the total cost of 37.9 million US$ for the Nigerian Environmental Management Programme initiated in 1990 is to be financed through external support. External financial dependence, according to a recent UNDP report, has led to at least three adverse effects on the environmental policy: a top-down communications approach, with little involvement of the affected citizens in designing environmental policy priorities, an increased debt burden, and a concentration on short-term environmental issues (UNDP, 1994: 8).

4.3 Degree of Policy Integration within Government

Environmental policy has also brought to light new areas of inter-governmental conflict. The cause of conflict is often the question of supremacy. One prominent example is the conflict between FEPA and NARESCON in 1993. Another case in point is the conflict between FEPA and Lagos State Waste Disposal Board (LSWDB) in 1990. FEPA recommended as a solution to a "garbage crisis" the cancellation of the compulsory monthly environmental sanitation exercise and the transfer of the responsibilities of waste management, including the building of solid waste depots across the federation to the FEPA. Such FEPA activities have been rejected by the LSWDB as usurpation of power, which is contrary to the government's intention as stated in the National Policy on the Environment and the Decree No. 58 of 1988. Conflicts have also arisen within the federal environmental institutions as a result of inconsistency of standards. For instance, under its enabling law, FEPA is empowered to monitor all imported chemicals for use in the country. This duty also falls under the responsibility of the National Agency for Food and Drugs Administration and Control (NAFDAC) of the Federal Ministry of Health and Human Services, thereby resulting in the conflicts that have many times inspired protests from manufacturers who claim they are being double-taxed and would only end up in passing the extra costs to consumers.

4.4 Development of Policy Paradigm

New "anticipate-and-prevent strategies", for achieving sustainable development were prepared after Rio in 1992 (see *Table 3*). The promulgation of the Environmental Impact Assessment Law (EIA) Decree No. 86 of 1992 is to be seen as an attempt to place the new preventive strategies on a legal footing, as the decree provides that a development project or activity that is likely to significantly affect the environment should be subjected to EIA studies prior to its execution. The EIA Decree enjoins the proponents of development projects to produce mitigation plans as well as future monitoring programmes. It vests the responsibility for the approval of EIA in the restructured FEPA, but makes provision for public review and hearing to interested parties and members of the public whose interests would be injuriously affected by the proposed projects (section 7). States or neighbouring local governments in the project areas are to be consulted (section 11). To improve the EIA process, some states have agreed with the private manufacturing enterprises on a comprehensive auditing and monitoring system for pollution control. At the federal level, the required legislation and guides for the EIA could be said to have been put in place for its enforcement, including, for instance: the establishment of a new department of Environmental Impact Assessment (1993), the creation of an EIA Secretariat and a Technical Committee in FEPA; publication of technical guides for the oil and gas, mining, manufacturing, agricultural and infrastructural sectors as well as an EIA practitioners and managers guide.

4.5 Public Awareness

The activities of the Nigerian environmental institutions and the lively debate on environmental issues since 1970 have impinged upon the standpoints of Nigerians towards environmental problems. A testimony to the growing environmental

Table 3. Institutional Arrangements and Policies in Nigeria.

Institutional Arrangement and Policies	*Pre-UNCED*	*Post-UNCED*
Institutional Arrangement	Old FEPA established (1988); Natural Resources Conservation established (1989); National Council of Environment (for vertical linkage – Federal/State/ Local Governments).	Restructured FEPA (Environmentally related agencies merged); Status of Governing Board raised to Ministerial Governing Council and assisted by a Technical Committee. Ministries of Budget & Planning and Finance, National Planning Commission, State Environmental Agencies (training and formulation of State Environmental Action Plans) to be strengthened.
Instruments	Command and Control; End-of-pipe Strategy; Evacuation.	Emphasising preventive strategy, e.g. Promulgation of Environmental Impact Assessment (1992); Incentives and other fiscal measures for industrial pollution control.
Policy	National Policy on Environment (1989); National Environmental Action Plan (1990); Natural Resources Conservation Action Plan (1992).	Proposal to review National Policy on Environment along with other sectoral policies; Conservation Policy (preparation in progress).
Resources Mobilisation	Budgetary allocation; Ecological Disaster Fund established in 1980 (1% of national revenue).	Budgetary allocation; Ecological Disaster Fund (increased in 1993, 25% of which will be statutorily allocated to the enlarged FEPA); Debt for Nature Swap (under negotiation); financial contributions of enterprises to the environmental measures on a voluntary basis (to be intensified), donations. Assistance from bilateral, multilateral agencies, e.g. UNDP, World Bank, FAO, UNEP, UNIDO.

consciousness is the high public response to the WAI Programme, which is shown in the large amount of refuse gathered on each Environmental Sanitation Day. Nevertheless, WAI reveals the inadequacy of the governments' infrastructure to cope with urban waste. Days after the monthly cleaning exercise, heaps of garbage collected by households are left uncleared by the roadside. In Lagos state, the military administration recently (1993) resorted to using soldiers to clear the mountains of refuse.

The first worldwide UNEP survey conducted towards the end of the 1980s, measuring public opinion and leadership attitudes in 14 nations, showed that the present economic recession has negatively influenced the attitudes of Nigerians towards environmental issues. In all countries, including Nigeria, there was more environmental concern among younger people and women. Worth mentioning is that a majority of both the public and leaders interviewed agreed on the need for determined government and international community actions and would be willing to pay higher taxes to government to protect the environment (Development Forum, 1989). In the Gallup Health of the Planet Survey conducted in 1992 in 24 countries of the world, Nigeria together with Hungary showed the lowest preference for the environment and a very high preference for the economy. The Philippines and Brasil, but also Great Britain ranked similarly low regarding preference for the environment (Dunlap et al., 1993).

4.6 Main Systemic Restrictions

Three factors limit the capacity for environmental policy: economic instability and dependency, the over-centralisation and the over-politicisation of the State's bureaucracy.

The dependence of the Nigerian economy on oil has made it vulnerable to international market demand and prices. At the same time, about 50% of raw materials used in Nigerian industries have to be imported. The economic crisis has reduced the capacity to act. The overreliance of Nigeria on imported technological expertise for its industrial development has exacerbated the environmental problems and still stands in the way of effective environmental solutions.

Many exemptions from established regulations are possible. For instance, section 15 of the Environmental Impact Assessment Decree provides that projects could be exempted from EIA on the order of the President or the FEPA council when the environmental effects of the proposed projects are deemed to be minimal or the projects are to be executed during national emergency situations or when they are in the interests of public health and safety. This gives wide political discretionary powers that can be misused by executives to protect certain industrial polluters.

The Structural Adjustment Programme prescribed for Nigeria by the World Bank and the International Monetary Fund has done a great deal to erode the administrative basis of environmental policy. Nigeria (along with Guinea) has one of the highest percentages of retrenchment in West Africa. According to a report of the Economic Community of West African States (ECOWAS), 156,550 or a retrenchment of 20% of the civil service work force has been registered for Nigeria between 1980 and 1990. The government is using the instrument of in-service training which, as indicated, is being complemented by the NGOs to enhance the capacity of the civil service. The above-mentioned restrictions could thwart any advantage for the building up of an efficient administrative capacity by way of training.

The high level of politicisation and centralisation in the political system are other restrictions. The legacy of the past (i.e. a centralist, overblown administrative structure) is being perpetuated.

Local government reforms since 1976 have led to a change in the self-perception of government officials at local levels. Many local officials no more see their main functions as restricted to the maintenance of law and order but increasingly in the provision of local amenities. Lack of autonomy for governments at the grassroots level has destroyed the image of local institutions and strengthened the proneness of Nigeria's educated and economic elites to direct their demands to central government. It should be stressed that the local governments have not been able to satisfy the increasing political aspirations of the people. In fact, the local government actors are also not involved in the international discussion on the implementation of Agenda 21 agreed upon at the Rio Conference, as is the case in the industrialised countries.

5 Policy Evaluation

A large number of central environmental institutions and regulations has sprung up in the last two decades. But Nigeria has yet to use exhaustively the opportunity given by its federal state structure to build an effective legal and administrative capacity for environmental protection. The absence of decentralisation and participation is a consequence of the political uncertainties and instability that have accompanied a long period of rule by the military which only sometimes played a modernising role. This has created a built-in limit on the capacity of the environmental institutions and has given rise to a situation where the involvement of affected citizens is no longer considered a prerequisite for viable environmental policy planning. The dependence of state and local governments on federal government funding for the implementation of environmental measures has mounted, while the federal government's financial commitment to environmental issues has diminished as a result of the country's economic crisis. All efforts to build a viable environmental capacity in Nigeria will be fruitless unless emphasis is placed on the democratisation and deconcentration of the political and administrative institutions (Salau, 1993).

References

Adamolekun, L., 1991: Federalism in Nigeria: Toward Federal Democracy, Publius, 21, 4.

Adewale, O., 1996: The role of NGOs in Implementing Environmental Treaties in the ECOWAS, Paper Presented at the Africa-Colloquium on "Regional Integration for Global Challenges: The Case of West Afrika", organised by the Society for International Development (SID), Munich.

Ajomo, M. and O. Adewale, 1994: Environmental Law and Sustainable Development in Nigeria. Lagos.

Akanle, O., 1991: The Constitution and the Nigerian Environment, in: E.O.A. Aina and N.O. Adedipe (eds.), The Making of the Nigerian Environmental Policy. Lagos.

Babangida, I., 1989: Our National Environmental Goals. Lagos.

Bangura, Y., 1994: Intellectuals, Economic Reform and Social Change: Constraints and Opportunities in the Formation of a Nigerian Technocracy, in: Development and Change, Vol. 25: 261-305.

Development Forum, 1989: Vol XVII, No. 4, July-August 1989.

Dunlap, R. E., G. H. Gallup Jr. and A. M. Gallup, 1993: Health of the Planet. Results of a 1992 International Environmental Opinion Survey of Citizens in 24 Nations. Princeton, N. J.

Echeruo, M., 1977: Victorian Lagos. Aspects of Nineteenth Century Lagos Life. Great Britain.

Etete, D. L., 1995: Nigeria. The Untold Ogoni Story, West Afrika, 27 November, pp. 1835-1836.

FEPA, 1991: Achieving Sustainable Development in Nigeria. National Report for the United Nations Conference on Environment and Development, Rio de Janeiro. Lagos.

Mabogunje, A. 1995: The Challenge of an Urban Forum in Nigeria, in: The Urban, Vol. 1, No. 1: 5-13.

Nigerian Environmental Study Team, 1991: Nigeria's Threatened Environment. A National Profile. Ibadan.

Ogbuigwe, A., 1989: Compensating Victims of Environmental and Industrial Pollution. Paper presented at the National Conference on Criminal Justice. Abuja, 28th-30th June 1989.

Okunola, O., 1995: FEPA, Capacity Building on Environmental Control and the EIA Law in Nigeria (unpublished paper).

Oyediran, O., 1994: Autonomy of Cummunes in Africa and German Speaking Countries in Europe. Ibadan

Salau, F. K., 1993: Kulturelle Identität und Umweltprobleme in Nigeria. München.

Salau, F. K., 1995: Die Ökologische Kultivierung der Ökonomie. Herausforderung für die Eine Welt. Fallbeispiel: Nigeria. Paper Presented at the Annual Conference of the Working Group of the German Development Aid Agencies (Arbeitskreis Kultur und Entwicklung), Oktober 1995.

Saro-Wiwa, K., 1994: Multi-Ethnic Demands in Nigeria: Minority versus Majority Interest, in: O. Oyediran, Autonomy of Communes in Africa and German Speaking Countries in Europe. Ibadan, 9-19.

Sowande, D., 1991: Control of Air Pollution WAPCCO Experience. Paper presented at the Lagos State Seminar on "Our Environment: The Dangers Ahead", 15th July 1991.

UNDP, 1994: Programme of the Federal Government of Nigeria. Programme Support Document. Lagos.

RUSSIA

Ivan Potravnyi and Ulrich Weißenburger

1 Environmental Problems and the Development of the Burden on the Environment

Russia's current environmental problems are first and foremost the result of the economic policy conducted under the system of a centralised planned economy. Independent businesses were non-existent under this system. The companies were each subordinate to their respective branch ministries and possessed only limited room for manoeuvre. Production volumes were laid down in the economic plan, prices and buyers by the State. There was no free trade in the means of production, environmental protection technology and environmentally friendlier technologies were almost unobtainable. Meeting or exceeding production plans, i.e. maximisation of output, had priority. Although environmental investment was financed predominantly out of the state budget, the running of environmental protection plants automatically had a negative impact on every significant indicator in the economic plan and an unwanted effect on company finances. Thus, for economic reasons, the companies were not interested in taking environmental protection measures or assigning them the appropriate finance. The pricing system also rewarded environmentally harmful behaviour. The low cost of energy promoted its wasteful use, natural resources such as land and water were used free of, or at a minimal charge. The "loose" budgetary restrictions, i.e. the guarantee of subsidies or credit to unprofitable companies, led to there being no economic pressure to use available resources sparingly.

A systemic opposition to innovation within the centralised planned economy had negative consequences for environmental protection. The effective life of capital stock in the USSR was longer by far than was usual in the West: in heavily polluting sectors such as the iron and steel industry, the average effective life of fixed assets was some 50 years. Capital spending on replacement was chronically inadequate. A considerable part of the environmental burden was thus caused by obsolete and seriously worn-out production plants with high consumption of materials, raw materials and energy, high repair costs and which brought with them a high risk of accidents. In a market economy, most of these plants would have had to be closed down for business reasons alone.

A decisive influence on pollution levels in Russia is industry's structurally intensive use of basic raw materials. Since the start of the process of transformation to a market economy, the importance of the fuel and energy complex and the basic materials industries has become even greater. There has been a particularly sharp increase in the electricity industry's share of industrial production (1991: 3%, 1994: 13.5%), while the fuel industry's share (1991: 7%, 1994: 16%) and that of the iron and steel industry (1991: 5%, 1994: 9%) have also increased. The engineering (1991: 24%, 1994: 19%), light industry (1991: 17%, 1994: 3%) and food (1991: 18%, 1994: 12%) sectors have all lost in importance.

Air pollution is caused primarily by industrial activity, and its emission levels have fallen continuously in recent years. In 1994, total industrial emissions were some 22 Mt, including 6.5 Mt SO_2 and 2.1 Mt NO_x (1993: total: 24.8 Mt, including 4.8 Mt dust, 7.2 Mt SO_2, 2.5 Mt NO_x, 5.9 Mt CO, 4.1 Mt hydrocarbons) (Goskomstat Rossii, 1994: 223, 1995: 123). However, this reduction in emission levels is not due to an improvement in environmental protection, but is entirely the result of a fall in industrial production, and production has fallen more sharply than emission levels during the transformation process:

Table 1. Industrial Production and Industrial Emissions in Russia.

	1991	*1992*	*1993*	*1994*
Industrial production	-8%	-18%	-14%	-21%
Industrial emissions	-7%	-11%	-12%	-12%

Among the reasons for this development are an increasing relative energy consumption due to lower use of capacity, the deteriorating condition of plant and the use of poorer quality raw materials and fuel.

A quarter of industrial emissions in 1993 came from the electricity industry, specifically 35% of SO_2 emissions, 56% of NO_x emissions and 39% of particulate emissions. The fuel industry, which is characterised by high levels of hydrocarbon emissions, gave rise to 18% of industrial emissions. Considerable air pollution is caused by burning off natural gas during crude oil extraction (1993: 7 billion m^3). The non-ferrous metal industries contributed 15% of industrial emissions in 1993 (SO_2: 39%), the iron and steel industry 13% (CO: 37%) (Minpriroda Rossii, 1994a: 14, 110 ff.). Regional blackspots for industrial air pollution are the industrial centres in the Urals and West and East Siberia.

The level of water conservation is extraordinarily low. Only 10% of waste water was clarified to the legal levels in 1994, some 25 billion m^3 (1993: 27 billion m^3) were emitted with inadmissible levels of pollution (cf. *Table 3*). In 1994, 51% of waste water classified as being polluted was caused by communal

Table 2. Industrial Pollutant Emissions in Russia 1981-1994 (in Mt).

	1981	*1985*	*1986*	*1987*	*1988*	*1989*	*1990*	*1991*	*1992*	*1993*	*1994*
Total[1)]	41.0	42.4	39.3	39.4	38.1	36.3	34.3	31.7	28.2	24.7	21.9
SO_2	12.1	11.7	11.2	11.1	10.5	9.9	9.4	9.2	8.2	7.2	6.5
NO_x	2.4	2.5	2.7	2.9	2.7	2.7	3.0	3.0	2.7	2.5	2.1
CO	9.1	8.5	8.7	9.2	8.9	8.4	.	7.6	6.7	5.9	
dust[2)]	9.1	9.1	8.7	8.7	8.4	7.9	7.1	6.4	5.6	4.7	
C_xH_x[2)]	6.7	6.7	7.0	7.6	6.7	6.7	.	5.2	4.2	4.1	
-VOC[2)]	.	.	.	.	.	.	.	1.7	1.6	1.6	
-other C_xH_x[2)]	.	.	.	.	.	.	.	3.5	2.6	2.5	

1) 1990 data from Goskomstat Rossii. According to Goskomstat SSSR 34.1 Mt.

2) 1991 from Ministry for the Environment figures. According to Goskomstat particulates 6.5 Mt; hydrocarbons 4.9 or 1.7 and 3.1 Mt.

Sources: Weißenburger (1995: 293), Goskomstat Rossii (1995: 123).

Table 3. Sewage Discharges in Russia 1985-1993 (in billion m^3).

	1985	*1986*	*1987*	*1988*	*1989*	*1990*	*1991*	*1992*	*1993*	*1994*
Total sewage volume[1)]	74.7	.	75.3	72.7	76.4	75.2	73.2	70.6	68.2	60.2
of which: slightly polluted[1)]	48.7	.	49.0	44.9	45.7	44.2	42.3	40.6	38.4	33.0
correctly purified	14.0	14.3	9.6	4.6	3.5	3.2	2.9	2.7	2.6	2.6
impermissi-bly polluted	12.0	11.3	16.7	23.3	27.1	27.7	28.0	27.1	27.2	24.6
of which: unclarified sewage	5.3	.	5.2	5.3	8.5	8.5	8.4	8.2	8.5	.

1) 1991 from environment ministry figures. According to Goskomstat 76.0 or. 45.1 billion m^3.

Sources: Weißenburger (1995: 294), Minpriroda Rossii (1995: 25).

and domestic activity (1992; 44%, 1993 45%), 35% came from industry (1992: 42%, 1993: 37%) and 13% from agriculture (1992: 13%, 1993: 17%). Pollution of numerous inland waters with crude oil products, phenols, organic pollutants, heavy metals and nitrogen frequently exceeds legal limits by ten times or more (Minpriroda Rossii, 1995: 25, 183 ff.), and the contamination of ground water resources gives great cause for concern. Ground water pollution has been shown at 1,200 locations, of which 80% are in the European part of Russia. Some of the contaminated ground water resources extend over an area of more than 100 km^2. Among the pollution sources are particularly crude oil extraction, inadequately or not at all secure sludge disposal, toxic industrial and domestic waste, livestock and unsafe storage of fertilisers and poisonous chemicals. In about a quarter of cases the ground water is contaminated with highly poisonous substances, including mercury, beryllium, lead, cadmium and arsenic. Ground water contamination is particularly extreme near chemicals industry facilities (Zelenyj mir no. 31, 1995: 7 f.; Soviet bezopasnosti Rossijskoj Federacii, 1994: 143 ff.).

Only little has been done up to now to reduce the risks arising from toxic waste deposits. In 1993 alone, 120 Mt toxic waste was produced by Russian industry (Germany; 15 Mt), of which only a quarter was recycled or rendered harmless by the companies concerned. The total mass of hazardous waste taken to landfill sites in 1993 was 250 Mt, and was thus double the amount of industrial toxic waste. There are no data on the origin of this difference, but it is probably a question of sludge, waste produced and temporarily stored in previous years, hazardous waste from other sectors or imported from abroad. The rate of recycling or destruction is particularly low, above all for the most hazardous waste (e.g. waste from electroplating, waste containing chrome and mercury), due to a lack of technology. And thus over 70 Mt industrial hazardous waste was deposited in domestic landfills in 1993. There are no legal regulations to guarantee adequate control of the extent and danger of the waste stored on business premises or of the safety of business disposal sites (Minpriroda Rossii, 1994a: 134 ff.; Soviet bezopasnosti Rossijskoj Federacii, 1994: 164).

A considerable environmental hazard comes from the numerous military and civilian nuclear installations. In its investigations in 1993, the state atomic control authority found some 20,000 infringements of safety regulations (Süddeutsche Zeitung, 17 February 1994). Even under normal working conditions, the radioactive burden on the environmental from nuclear installations is considerable. In Russia at this time there are 9 nuclear power stations with 29 reactors. Inadequate safety standards lead to two to three times as many malfunctions annually as in German nuclear power stations. The older type WWER-440 stations are virtually impossible to upgrade: they have now containment facility, the emergency cooling equipment is inadequate, fire and flood protection underdeveloped. The safety deficiencies in the type RBMK-1000 stations are also serious, especially the lack of a containment facility or leakage monitor, and the unsatisfactory fire protection (Atomwirtschaft, Oct. 1994: 709, Dec. 1994: 839 f.). The plutonium production and nuclear fuel reprocessing plants are at least an equally large source of danger. In the plants at the "Mayak" production group in

Oblast Chelyabinsk there have been a number of large accidents. Since 1948 this company emitted materials with a radioactivity of 150 million Curie, three times as much as was released in the Chernobyl disaster. A serious accident occurred in 1993 at the nuclear complex of the Siberian chemicals combine in Seversk (Tomsk-7) (Penjagin 1991: 24; Minpriroda Rossii, 1994a: 64 ff., 139).

The disposal of radioactive waste is associated with a high risk potential. The radioactivity of the total nuclear waste stored in state facilities (including burnt-out fuel rods) is over 5 billion Curie. The greater part of this waste is stored on the premises of fuel recycling companies (Chelyabinsk-65, Tomsk-7, Krasnoyarsk-26). Atomic waste is predominantly deposited in temporary storage facilities which fail to meet many of the current safety regulations. Some nuclear waste is stored in the open air or in waters, so that the release of radioactive particles into the environment cannot be ruled out. Some of the temporary storage facilities at the nuclear power stations are filled to capacity, and here too, accidents often occur. In the past, nuclear waste from the navy was usually sunk in the Arctic Ocean or in seas in the Far East. The reactors in sunken submarines and ships represent a great hazard, and the question of disposal of the reactors on 79 decommissioned nuclear submarines has still not been cleared up. Safety standards at navy disposal sites for burnt-out fuel rods and solid nuclear waste are thoroughly inadequate (Soviet bezopasnosti Rossijskoj Federacii, 1994: 58 ff.; Minpriroda Rossii, 1994a: 68 ff., 1995: 89 ff.).

2 Development of Environmental Policy

Environmental protection played no role within Soviet economic policy until the start of the 1970s. A nature conservation law was passed in 1960, but it was observed only loosely and was ineffective. Only in the face of increasing environmental problems, and after the 1972 Stockholm conference was environmental protection defined as a national goal. In the 1975 yearly plan and in the five year plan for 1976-1980 environmental protection became for the first time an element of national economic planning. Environmental planning was admittedly restricted to clean-up measures, achieving a specific environmental quality had no part in the planning process.

Institutional areas of responsibility in environmental protection were laid down in 1972, with the state environmental protection system characterised by much fragmentation and overlapping of areas of competence, and with no orientation around specific principles. At the level of the Union alone, 16 ministries and State committees were entrusted with environmental protection functions. Some institutions were responsible for protecting certain areas of the environment, such as the Ministry for Land Improvement and Water Management (water conservation), the State Committee for Hydrometeorology and Environmental Monitoring (air quality) and the Ministry of Agriculture (land and nature conservation). Other organs of state may have had full competence with respect to protection of the environmental area in question, but they were restricted in their scope for action. To this group belonged the State Planning Committee (environmental planning), the State Science and Technology Committee (environmental research) and the State Construction Committee (environmental investment). An additional hindrance to effective environmental protection was that decisive competencies were assigned to those organs of state for whom the environment was of only secondary importance, and who looked after primarily economic interests which sometimes conflicted with their functions as protectors of the environment. This was particularly the case with the Ministry for Land Improvement and Water Management, which was responsible for the many grandiose irrigation projects, and with the Ministry of Agriculture. Not until 1988 was an organ of state created with responsibility for the entire environmental policy, in the form of the State Committee for Environmental Protection, to which the environmental departments of the other ministries and state committees were assigned. The competencies of the new authority embraced directing research, monitoring the compliance with laws, preparing plans, determining quality standards, producing sustainability tests, issuing licences in respect of air and water emissions, waste disposal and the use of certain natural resources, and the coordination of environmental activities in other ministries and authorities[1] .

From the start of the 1970s, environmental policy concentrated on clean-up measures. In the period 1986-1990 environmental investment in Russia was some 40% higher than between 1976 and 1980, however the share of environmental investment in total investment remained virtually constant at 1.5%. Little changed in the structure of this investment either. Between 1986 and 1990, water conservation received 70% of environmental investment (1976-1980: 79%), air quality 13% (1976-1980: 9%) and land conservation 10% (1976-1980: 6%) (Goskomstat SSSR, 1991a: 310). Environmental investment fell by 40% between 1992 and 1994, investment in water conservation by 50% and air quality by over 20%. Although at a low level, investment in air quality did permit some more or less effective measures against dust emissions. Desulphurisation and denitrogenation could not be financed with so few resources. Strategies for emissions and waste avoidance had no real significance in environmental policy, although their primacy was often stressed. Because of low energy and raw materials prices and the lack of economic instruments for environmental policy, there was no stimulus to make appropriate investments.

In 1993, the Russian Ministry for the Environment presented a national plan of action on the environment, which is intended to guarantee the implementation of decisions made at the UN Conference on the Environment and Development (1992). This programme lays out the most important planned legislative measures (cf. section 4.2). In the course of converting the defence industry for civilian production, the manufacture of modern environmental technologies is given priority. In water conservation, the emphasis is laid on reducing heavy metal immisions, in air quality priority is given to cutting emissions of highly toxic

1 Sobranie postanovlenij pravitel'stva SSSR Nr.6, 1988, p. 83 ff.

substances such as chlorides, fluorides, hydrogen sulphide, carbon disulphide, mercury and lead (Minpriroda Rossii, 1993b).

3 Actors in Environmental Policy

3.1 State Institutions

The organisational structure of state environmental protection institutions has not changed decisively since the creation of the State Committee for Environmental Protection in 1988, despite the reforms carried out during the disintegration of the USSR and the market reform process. The Ministry for Environmental Protection and Natural Resources has assumed to a great extent the functions of the old Soviet State Committee. Its tasks are state direction of environmental protection as a whole, coordination between ministries and other organs of state, policy implementation in research and development, determining legal limits and other regulations, conducting sustainability tests, monitoring compliance with environmental laws, distributing information, and international cooperation. In general the Ministry for the Environment has limited competence in so far as in matters such as accident prevention and nuclear safety, responsibility lies with other organs of state (Federalnyi ekologicheskij fond Rossii, 1995: 4 ff.).

3.2 The Environmental Movement

Since the start of perestroika in 1985, Russia has seen the emergence en masse of non-governmental environmental organisations with a variety of organisational structures and fields of activities. The emergence of these organisations reflects the positive changes in the social and political life of the country, and a growing public consciousness and activism. At first, small-scale local organisations at local level predominated, formed as a rule in order to solve specific problems (e.g. the Baikal Protection Fund). In the following years, organisations at regional and national level began to appear, for example the Social and Environmental Union, which merged a number of different independent organisations from many regions in December 1988. The first practical steps they took were public demonstrations against the environmentally hazardous construction of the Volga-Don and Volga-Chograi canals. The Public Volga Rescue Committee was founded in 1989 with the aim of assisting the conservation of the Volga River and its tributaries as natural, cultural and economic resources, and of neutralising the negative impacts of economic activity. For this, the Committee formed the Volga Rescue Fund.

The changes begun in 1985 have had a strong influence on the public's attitudes to the environment. In many regions, there were public meetings and demonstrations to protest against environmentally hazardous production methods and the construction of new plants. Later on, these bodies evolved more organised structures. Non-governmental organisations (NGOs) have tended to turn their attention away from local issues to larger, more complex environmental and economic problems. In 1991, the Russian Green Party was registered as a political party. The green movement is a new actor in Russia's social and political arenas and has attracted the attention of experts, the mass media and the general public. Nonetheless, opinion on the development and activities of green organisations is divided. According to a survey of 362 experts – the heads of conservation and research bodies within the Ministry for Environmental Protection and Natural Resources – conducted in October 1992, 19.4% of them denied the existence of a green movement in Russia. In addition, 72.8% considered it to be without tangible social or political weight in their regions, while 17.5% were of the opposite opinion and 9.7% refused to answer. 34% of the experts questioned expressed their full support of the green movement, 35.9% qualified their support in some areas. A certain underestimation of the green movement is, in our opinion, connected with the current climate of confrontation between state environmental bodies and NGOs.

In 1993, over 800 environmental NGOs (non-profit and business organisations) were operating within Russia at local, regional, national and international levels. Green organisations are concentrating their efforts, among other things, on protecting specific natural resources, monitoring compliance with legislation, environmental assessment, training and promoting education.

An analysis of the number, activities and regional distribution of NGOs leads us to conclude that organised environmental forces do indeed exist in Russia. A common factor in their development was the politicisation of these forces during the run-up to the 1993 parliamentary elections. The most active were the Russian environmental movement "Kedr" ("Cedar") and the Russian Green Party. However, neither managed to gain the number of votes required to enter parliament (5%), which was evidence of their lack of integration and importance for the general public. Coordination between NGOs and cooperation and exchange of information with foreign green organisations are still at a very low level of development, although such cooperation does seem to have been given a stimulus recently, due to Greenpeace and the World Wide Fund for Nature (WWF) establishing their representatives in Russia. Lack of finance, up-to-date equipment and organisation are serious impediments faced by the green organisations.

3.3 Environmental Reporting and the Media

Environmental reporting in the Brezhnev era was extraordinarily restrictive. Data on pollution were kept under lock and key, statistical reporting restricted itself until 1986 to publishing information on environmental protection measures. In the statistical Yearbook for 1987, data on water and air pollution were published for the first time, and since 1988, an annual report on the environmental situation and the state of environmental policy has been published by the environmental ministry, in addition to the appearance of several statistical collections on pollution and protection measures carried out. The scope of environmental reporting has been extended in recent years to cover radioactive pollution and the environmental impact of the military-industrial complex. That being said, public faith in official reporting is not great, which goes back not least to the public relations exercise after the Chernobyl disaster.

At present 50 weekly and monthly magazines are published in over 30 regions of Russia, but due to their limited target audience – mainly specialists – their actual importance is relatively low. Many newspapers, such as *Argumenty i Fakty* or *Izvestya* run stories on environmental topics, and regional newspapers have special columns for environmental information, often with supplementary comment from the relevant officials.

According to surveys, about 80% of the population gets its news from TV and radio. The TV channel *Ostankino* has developed a programme on the environment produced with cooperation between journalists and conservation bodies. In many regions there are regular TV and radio programmes on the environment which are made possible by financial support from regional environmental funds, as a major difficulty faced by the mass media is the lack of financial support and rising publishing or transmission costs.

3.4 Other Actors in Environmental Protection

Environmental business exists mainly in fields such as eco-tourism, secondary resources and waste recycling. The first businesses emerged in the late 1980s, and by 1991 they numbered 17,000. Forty commercial banks and two exchanges (Commodity and Raw Materials Exchange, Secondary Resources Exchange) support the development of the environmental goods and services market. The Ecoprombank (Perm), the Volga Regional Environmental Protection Bank (Ecobank-Volga), and the Federal Econatsbank (formed in 1994) operate in environmentally threatened regions.

Specialist departments at universities and technical colleges are involved in research and education in environmental protection.

The interdepartmental GATT commission deals with environmental certification and standards, trade issues, licences and quotas. The Russian Chamber of Commerce and Industry provides, among other things, an information service and help with marketing environmental products.

3.5 Important Restrictions

The most significant restriction on environmental policy under the planned economic system was the strong position of the branch ministries. Individual companies were bound by their directives, which were themselves based on the national economic plan. Thus, in their attempts to uphold the regulations, the environmental authorities were immediately confronted by the branch ministries, which could always cite the economic plan in case of conflict. Particularly against the military-industrial complex and its suppliers the environmental authorities were virtually powerless, all the more since emissions monitoring in defence-related areas could only take place with the permission of the relevant ministries. This constellation of powers was evident in the conflict over production by the cellulose combine at Lake Baikal, which was also of military importance, and whose construction was pushed through against the resistance of environmental protectionists.

Although the military sector remained outside their control, the position of environmental authorities in relation to that of the branch ministries was strengthened during Gorbachev's term of office. This was shown in the decommissioning of a series of heavily polluting production facilities, particularly in the chemicals, cellulose and iron and steel industries, which began in 1986. The decisive factor in this change of course by the new Party leadership was most probably the fact that the decommissioning of obsolete production facilities was also necessary for the modernisation of the national economy and the intensification of economic development that were being striven for at the time. Correspondingly, as supply problems increased towards the end of perestroika, the government's readiness to shut down businesses was significantly reduced.

The start of the economic transformation process marked a watershed in the framework conditions for environmental protection. The sharp fall in production has reduced the burden on the environment, but companies' financial difficulties and the high budget deficit have had a negative impact. The collapse in investment in production (a fall of 75% between 1991 and 1994) has led to an increase in the rate of obsolescence of capital stock and in the risk of accidents. At the same time, the State has also reduced its levels of environmental investment, and environmental protection has lost much of its importance in the scale of political priorities as a result of the crisis in the economy. The State's capacity for effective action on the behaviour of business is small above all in those sectors whose exports guarantee a positive balance of trade: the fuel and energy, metals and basic chemicals industries.

4 Capacity-Building

4.1 Proponents

The current state of capacity-building in environmental protection is marked by a clear discrepancy between the relatively highly developed institutional structure of the environmental authorities and the existence of formal legal and administrative instruments on the one side, and the extraordinarily weak implementation of environmental policy goals on the other. These deficiencies are extremely large in comparison to Western countries, due to the lack of a democratic tradition based on the rule of law, so that the primary problem lies in the implementation of environmental regulations.

Russia has at its disposal environmental policy institutions at all administrative levels. There are even environmental departments in most other ministries.

Monitoring is however inadequate, due to insufficient (e.g. measuring) equipment. Personnel deficits have been reduced somewhat, through the introduction of environmental qualifications at universities and technical colleges and the creation of new specialist educational institutions. In comparison with the perestroika period, the influence of environmental groups has fallen, which can be attributed first and foremost to the change in political priorities caused by the economic crisis. Although many new companies may have sprung up to exploit the possibilities for action in environmental protection, their total economic importance is, however, relatively low. The development of these businesses has been affected very strongly by the real fall in expenditure on environmental protection.

The disintegrative tendencies which, after the fall of the Soviet Union, have also engulfed the Russian Federation, have impacted negatively on the efficiency of environmental protection in some places. On many occasions, environmental and conservation regulations issued by the authorities are ignored for economic reasons, and the environmental boards are not in a position to uphold them (Kotov and Nikitina, 1993: 13 f.).

4.2 Legal Opportunities

Since 1991, charges have been levied for pollution and the use of natural resources in the Russian Federation. These charges are intended to provide an economic stimulus for improving pollution levels and for sparing use of natural resources. Pollution charges, designed to uphold the Polluter Pays Principle, are levied for emissions of over 200 atmospheric pollutants, 125 water pollutants and for the depositing of waste. The rates of these charges are supposed to be oriented around the hazards posed by the pollutants or the damage they cause. They are regionally differentiated, i.e. higher charges are to be paid in heavily polluted environmentally sensitive areas. They are levied on permitted emissions and immissions, and for exceeding legal limits or in cases of environmental damage caused by negligence higher charges are levied. Payment of these charges does not free businesses from their obligations to pay compensation or fines for environmental damage through negligence or breaking environmental regulations. The charges for use of natural resources, i.e. land, water, mineral deposits, forestry, hunting and fishing, are composed of a fee for rights of use (rent) and a contribution towards covering the costs of conservation and restocking (e.g. reforestation). In case of excessive use, fines are imposed.

In 1992 the law on "Protection of the Natural Environment" came into force. This law comprehensively puts in order the entire range of environmental protection measures for the first time. A major aim of the new law is the creation of economic instruments for achieving environmental policy goals. The use of natural resources will be regulated through contracts between regional authorities and the user, or through the issuing of licences by the Federation. The pollution and resource use charges, introduced by presidential decree, were embodied in law. Special funds, managed independently from the general budget, are being set up at all administrative levels for the solution of urgent environmental problems. Among the many sources of finance for these ringfenced funds are 90% of the pollution and resource use charges, all compensation for environmental damage and fines for environmental offences; 60% of the funds' resources go to the local level, 30% to the regional administrative level and 10% to the federation. The law provides for the introduction of an environmental insurance scheme. Environmental protection measures will be promoted through tax concessions, and all investment projects must undergo sustainability tests (environmental assessments). Only after a positive report will the implementation and financing of the project be permitted. Areas with serious environmental problems can obtain special status, by which economic activity can be restricted or prohibited (Zakon no. 3, 1993: 4 ff).

However, as a result of the high rates of inflation, the environmental charges lost any real significance soon after their introduction: although industrial prices rose by 240% in 1991 and 3,300% in 1992, rates for environmental charges were not raised until the beginning of 1993, and then by only 400% (Soviet Ministrov RSFSR, 1991: 1 ff.; Minpriroda Rossii, 1994c: 287 ff.). Revenue from charges thus remained insignificant, and was sufficient to finance only a fraction of expenditure on environmental protection:

Table 4. Environmental Protection Expenditure and Revenue (Financial Sums in Billion Rouble).

	1991	*1992*	*1993*
Total expenditure on environmental protection	18.2	299.6	2803
Revenue from pollution charges	1.1	4.4	79.8
Share of charges revenue in financing environmental protection expenditure	6%	1.5%	2.8%

Sources: Minpriroda Rossii (1994 a: 162, 165), Goskomstat Rossii (1992: 5), Goskomstat Rossii (1993a: 5).

Since the start of 1994, regional authorities have been permitted to index-link pollution charges (Minpriroda Rossii, 1994c: 344).

The effectiveness of environmental charges has for a long time been negatively influenced not only by high rates of inflation, but also by the continuingly "loose" budget restrictions. It is also doubtful whether an efficient monitoring of emissions is at all possible in the face of the large number of pollutants for which there is a charge. It would be sensible to restrict the levying of a charge to only the most important pollutants, such as SO_2, NO_x, dust, CO and hydrocarbons. Equally unproductive is the levying of a charge for highly toxic substances such as mercury, whose large-scale emission or immision is impermissible in any case, and cannot be sanctioned by the payment of higher charges.

Current planning at the Russian Ministry for the Environment foresees complementing economic instruments with permitting trade in emission rights, and there is a proposal to introduce special taxes for the use of environmentally threatening production processes.

In September 1994, the draft of a new environmental protection law was published. The 1992 environment law still oriented itself in many places around centralised and planned economic state structures. The new law is intended to suit environmental legislation to the federal structure embodied in the constitution, and make possible a greater level of public participation in decisions on environmentally important questions. Environmental protection is viewed as a common task of the Federation and other authorities, so that the present draft legislation has the function of a framework law, whose regulations may be complemented with legislative activity at other administrative levels. In decisions on environmentally relevant economic matters, such as the licensing of investment projects, for example, the already binding environmental assessments, whose subject is compliance by the project with environmental regulations, will be complemented with a comprehensive assessment of the environmental impact of the project. Also new is the follow-up assessment of projects with sustainability tests. The importance of licensing procedures, for example in respect of permissible emissions or sewage discharge is to be enhanced in future. The draft law's arrangements for economic instruments, in particular the pollution and natural resources use charges, are more or less identical to those in the current law. Regulations on the environmental funds were however made more precise, so that they must now restrict their activities to financing environmental protection measures (Zelenyj mir no. 22, 1994: 5 ff.).

As before, there are considerable loopholes in the law, particularly as regards waste disposal, the safety of industrial technology and nuclear safety. Still within the legislative process are laws on, among other things, production and domestic waste, industrial safety, hazardous substances, the transport of hazardous goods and dealing with radioactive waste (Soviet bezopasnosti Rossijskoj Federacii, 1994: 191; Minpriroda, 1995: 223). Monitoring of waste volumes and disposal is to be improved through the creation of a state-wide information system. The introduction of licensing procedures as to the extent and type of toxic waste stored on business premises, the length of their storage and the obligatory safety measures is also foreseen (Minpriroda, 1994a: 136 f.; Zelenyj mir no. 28, 1994: 10 ff.). Of particular importance among the legal instruments intended to bring about an improvement in safety standards is the introduction of permission procedures (licensing) for potentially hazardous areas of production, and the laying down of compulsory safety parameters (environmental certification) for technological processes and production facilities (Soviet bezopasnosti Rossijskoj Federacii, 1994: 29; Minpriroda, 1994c: 183 ff.).

4.3 Financial Resources

Among the financial resources available for environmental protection measures, the extra-budgetary special funds play only a subordinate role. Their share of investment financing was only 2.4% in 1994. Half of all environmental investment in 1994 came from the resources of partially privatised companies, 25% came from state companies and institutions, 19% from the communal sector and 4% from the private sector (Minpriroda, 1995: 255).

The decisive impasse in the implementation of the present environmental law lies in the small scope for sanctions against offenders. The punishments provided for are extraordinarily slight, and the criminal prosecution of environmental offences is only possible in exceptional cases. This can be seen in the discrepancy between the damage caused by offences and the fines and compensation demanded in each case. In 1994 for example, the fines and compensation demanded for water offences came to only 4% of the total damage caused in those cases (Minpriroda, 1995: 243 f.). In the face of the public budget deficit, and as a result of the fall in production, insufficient sources of finance are a considerable obstacle to improvement in environmental protection.

4.4 International Cooperation

International cooperation is, all in all, a positive factor in capacity-building, since carrying out joint projects assists in the solution of international problems, and since Russia's capacity to guarantee a more efficient protection of the environment is improved through international aid programmes. These joint projects and aid programmes, which include the 1993 Lucerne Environmental Action Programme, the 1992 Helsinki Baltic Sea Action Programme and other international programmes, comprise joint investment projects, making technical equipment available, training environmental specialists and advice on setting legal standards.

Seventeen joint Russian-German projects are currently underway, such as: a project on environmentally sound oil and gas extraction and the creation of an environmental monitoring system in Tver. German firms are participating in the construction of clean-up facilities in St.Petersburg, based on jointly developed designs, and with the financial support of Germany and the EC, the International Environmental Bureau was established. In the development of eco-tourism in the Baikal watershed area German businesses, technology and equipment are widely involved. The automated radiation control system installed at the nuclear power plant in Smolensk was developed as part of the IRIS project.

The United States is providing technical assistance and support for some Russian environmental projects, in particular on air quality management (Volgograd), water quality control (Moscow) and personnel training. Twenty projects are being executed jointly by Russia and Denmark, such as the construction of waste water clean-up facilities in Novgorod and Gvardeisk (Kaliningrad region). The Russian-Finnish cooperation programme, developed to fulfil obligations under the Declaration on Baltic Sea Environmental Protection, is aimed at pollution reduction in bordering territories (Republic of Carelia, Leningrad, Murmansk, Novgorod, Pskow regions and St.Petersburg) and includes the construction of water clean-up facilities, a composite waste reprocessing plant, installation of equipment for desulphurisation at a mining complex and the introduction of clean technologies.

In 1995 the International Bank for Reconstruction and Development granted Russia a credit of US$ 110 million over 17 years for the project on technical assistance in environmental process management at federal and local levels and stabilisation of the natural environment.

4.5 Policy Integration

In February 1994 the Fundamentals of the Russian State Strategy on Environmental Protection and Sustainable Development was approved by the Russian President. The document aims to provide a mechanism for government decision-making and a basis for drawing up federal, regional and industrial development programmes.

As for integration between the government and the other main actors, it must be described as weak and without noticeable impact on government decisions. Confrontation takes place when sharing out financial resources between regions or economic sectors.

4.6 Development of the Policy Paradigm

End-of-pipe and high-chimney strategies prevailed through the 1970s and 1980s in dealing with pollution. The 1990s showed a shift towards precaution and prevention, and away from compensation and dilution, at least in terms of policy aims. State environmental assessment is practised as a pollution prevention instrument and comprises the following areas: assessment of feasibility studies and design documentation for industrial projects; assessment of inspection data on certain areas with the aim of granting special environmental status; and assessment of the environmental usefulness of licenses and certificates. The success of ecological modernisation strategies has remained modest, however. Only 15% of the total reduction in industrial emissions can be attributed to modernisation of technological processes.

4.7 Main Restrictions

The major, most far-reaching factor restricting policy realisation is the deep and prolonged crisis in the national economy. Technological modernisation and the transition to environmentally clean production processes, as well as the construction of effective clean-up facilities at major polluting plants, management of material inputs in production all require considerable investment, and can therefore hardly be implemented at the moment, considering the general dearth of finance. For instance, Russia is experiencing difficulty in meeting the requirements of the Montreal Convention on the ozone layer.

Tough financial conditions, inflation, chronic non-payment between businesses have all led to a fall in demand for new equipment and technologies, and obstruct the use of economic instruments such as pollution charges. Lack of efficient legislation and effective mechanisms for reliable and comprehensive data collection on the state of the environmental and natural resources can be summed up as one restrictive factor. Besides this, the mechanisms of environmental policy regulation are at a formative stage and require recasting to fit in with market economic reforms. One consequence in particular is the unchecked pollution of the environment by small businesses, against which environmental protection institutions have at present no means of regulation.

Another restriction is weak motivation and poor public environmental awareness, bound up with deficits in environmental education as well as social-economic living standards. Thus in 1993, environmental protection was ranked fourth among urgent problems in public opinion, after "decline in standards of living", "living conditions", and "price rises".

5 Conclusion

In spite of a certain reduction in environmental pressure recently, the transformation process has yet to lead to a real improvement in environmental protection. The reduction in air pollution was linked almost exclusively to the fall in production, thus a rise in industrial pollution would presumably lead to a fresh rise in industrial emission of pollutants. Discharge of polluted waste water has remained at a virtually constant level, the level of water purification remains extraordinarily low. Permanent improvements in air quality and water conservation will only be achieved with a fundamental modernisation of capital stock, with integrated conservation measures, and with the decommissioning of the obsolete production facilities with high energy consumption and levels of pollution. However, because of the steep fall in investment, the problem of obsolescent fixed assets has become still more serious. Not least as a result of low capital spending on replacement and the inadequate plant maintenance, the number of serious accidents rose by 20% in 1993. Still more difficult to manage are the problems arising from the disposal of toxic and radioactive waste, especially since stocktaking of the inherited environmental burden has not yet been completed. Clean-up of such inherited problems is, as the East German experience shows, associated with extraordinarily high capital expenditure. In order to reduce excessive consumption of energy and raw materials, appropriate pricing is required. Among the positive developments in recent years can be mentioned the development of a market for environmental goods and services.

In general, the case of Russia shows clearly that there are two indispensable preconditions for a successful environmental policy:

- a stable, functioning political system which is able to influence at least the most significant developments in the country, and
- a stable, functioning economic system which can provide the material foundation and a minimum of technical possibilities for environmental protection.

Both preconditions at present are still absent in Russia. Therefore the introduction of a new environmental policy in the early 1990s, the existence of the appropriate institutions and of an active ecology movement, and critical reports in the media all lack the elementary political and economic foundation for them to be in any way effective. Economic instruments such as charges are a useful tool for environmental policy, but they cannot be effective if the large majority of enterprises is in a state of economic crisis. Good laws cannot be implemented if the government is weak and has no authority in the country.

The situation is aggravated by the heavy burden of environmental damage and continuing risks carried over from earlier times – an additional demand on finances that cannot be met.

There are some hopeful signs and there has been some help from abroad. But even after overcoming the current political and economic crisis, there will still be a long road to travel towards sustainability in Russia.

References

Gneveckow,J. and I. Potrawnij, 1992: Umweltpolitik am Baikalsee – Bestandsaufnahme, Taxonomie der Umweltprobleme und Entwicklungslinien für ein Handlungsprogramm, Osteuropa, no. 9.

Jaroshenko, V. (ed.), 1994: Ekologicheskaja ekonomika. Perspektivy primenenija ekonomicheskich instrumentov v oblasti ochrany okruzhajushchej sredy v Germanii, Rossii i Ukraine. Moscow.

Kotov, V. and E. Nikitina, 1993: Russia in Transition. Obstacles to Environment, Environment 35, 10.

Lebedinskij, J., I. Potrawnij and B. Krasnjanskij, n.y.: Umweltschutz in der Sowjetunion. Was bedeuten "Perestroika" und "Glasnost" für die Ökologie? Hamburg: VSA-Verlag.

Penjagin, A.N. (ed.), 1991: Rezonans: Juzhno-Uralskaja atomnaja: Byt' ili ne byt'. Tscheljabinsk.

Potrawnij, I., 1991: Umweltorientierte Unternehmertätigkeit: Regelungs- und Leitungsmechanismen, in: Umwelt in Europa: Umweltprobleme kennen keine Grenzen. Konstanz: Univ.-Verlag.

Potrawnij, I., 1993: Ökonomische Instrumente der Umweltpolitik im Transformationsprozeß Rußlands. Köln: Bundesinstitut für ostwissenschaftliche und internationale Studien..

Potravnyi, I.,1993: Economic Strategies for Avoiding Waste and Abatement of Prior Damages in Russia, in: Fourth International KfK/TNO Conference on Contaminated Soil, May 3-7 1993, Berlin, vol.1. Dordrecht/Boston/London: Kluwer Academic Publishers, 139-140.

Weissenburger, U., 1994a: Umweltschutz in der Russischen Föderation, Wochenbericht des DIW no.5/1994.

Weissenburger, U., 1994b: Ökonomische Instrumente im Umweltschutz in den Nachfolgestaaten der Sowjetunion, Vierteljahrshefte zur Wirtschaftsforschung no. 4/1994.

Weissenburger, U., 1995: Zur Entwicklung des Umweltschutzes in der Russischen Föderation, Wochenbericht des DIW no. 14/1995.

Official documents:

Federalnyj ekologicheskij fond Rossii, 1995: Ekologija Rossii. Spravochnik – ezhegodnik. Moscow.

Federal'nyj zakon, 1994: Ob ochrane okruzajushchej sredy v Rossijskoj Federacii. Proekt, Zelenyj mir no. 22/1994.

Goskompriroda SSSR (Gosudarstvennyj komitet SSSR po ochrane prirody), 1989: Sostojanie prirodnoj sredy v SSSR v 1988 godu. Moscow.

Goskompriroda SSSR (Gosudarstvennyj komitet SSSR po ochrane prirody), 1990: Sostojanie prirodnoj sredy i prirodoochrannaja dejatel'nost' v SSSR v 1989 godu: Gosudarstvennyj doklad. Moscow.

Goskomstat Rossii (Gosudarstvennyj komitet Rossijskoj Federacii po statistike), 1991: Narodnoe chozjajstvo RSFSR v 1990 g. Moscow.

Goskomstat Rossii (Gosudarstvennyj komitet Rossijskoj Federacii po statistike), 1992: Ochrana okruzhajushchej sredy v Rossijskoj Federacii. Moscow.

Goskomstat Rossii (Gosudarstvennyj komitet Rossijskoj Federacii po statistike), 1993 a: Ochrana okruzhajushchej sredy v Rossijskoj Federacii v 1992 godu. Statisticheskij sbornik. Moscow.

Goskomstat Rossii (Gosudarstvennyj komitet Rossijskoj Federacii po statistike), 1993 b: Narodnoe chozjajstvo Rossijskoj Federacii 1992. Statisticheskij ezhegodnik. Moscow.

Goskomstat Rossii (Gosudarstvennyj komitet Rossijskoj Federacii po statistike), 1994: Rossijskij statisticheskij ezhegodnik 1994. Moscow.

Goskomstat Rossii (Gosudarstvennyj komitet Rossijskoj Federacii po statistike), 1995: Rossija v cifrach 1995. Kratkij statisticheskij sbornik. Moscow.

Goskomstat SSSR (Gosudarstvennyj komitet SSSR po statistike), 1989: Ochrana okruzhajushchej sredy i racional'noe ispol'zovanie prirodnych resursov v SSSR. Moscow.

Goskomstat SSSR (Gosudarstvennyj komitet SSSR po statistike), 1991a: Narodnoe chozjajstvo SSSR v 1990 g.. Moscow.

Goskomstat SSSR (Gosudarstvennyj komitet SSSR po statistike), 1991b: Ochrana okruzhajushchej sredy i racional'noe ispol'zovanie prirodnych resursov. Moscow.

Minpriroda Rossii (Ministerstvo ochrany okruzhajushchej sredy i prirodnych resursov Rossijskoj Federacii), 1992: Gosudarstvennyj doklad o sostojanii okruzhajushchej prirodnoj sredy Rossijskoj Federacii v 1991 godu. Moscow.

Minpriroda Rossii (Ministerstvo ochrany okruzhajushchej sredy i prirodnych resursov Rossijskoj Federacii), 1993a: Gosudarstvennyj doklad o sostojanii okruzhajushchej prirodnoj sredy Rossijskoj Federacii v 1992 godu. Moscow.

Minpriroda Rossii (Ministerstvo ochrany okruzhajushchej sredy i prirodnych resursov Rossijskoj Federacii), 1993b: Nacional'nyj plan po realizacii reshenii konferencii OON po okruzhajushchej srede i razvitiju. In: Zelenyj mir no. 19/1993, no. 20/1993, no. 21/1993 und no. 22/1993.

Minpriroda Rossii (Ministerstvo ochrany okruzhajushchej sredy i prirodnych resursov Rossijskoj Federacii), 1994a: Gosudarstvennyj doklad o sostojanii okruzhajushchej prirodnoj sredy Rossijskoj Federacii v 1993 godu. Moscow.

Minpriroda Rossii (Ministerstvo ochrany okruzhajushchej sredy i prirodnych resursov Rossijskoj Federacii), 1994b: Sostojanie okruzhajushchej sredy i prirodoochrannaja dejatel'nost' na territorii vyvshego SSSR – ot Stokgol'ma k Rio. Spravochnoe posobie. Tom I, II. Moscow.

Minpriroda Rossii (Ministerstvo ochrany okruzhajushchej sredy i prirodnych resursov Rossijskoj Federacii), 1994c: Ekonomika prirodopol'zovanija. Analiticheskie i normativno-metodicheskie materialy. Moscow.

Minpriroda Rossii (Ministerstvo ochrany okruzhajushchej sredy i prirodnych resursov Rossijskoj Federacii), 1994d: Koncepcija sozdanija gosudarstvennoj sistemy ochrany okruzhajushchej sredy ot otchodov proizvodstva i potreblenija. In: Zelenyj mir no. 28/1994.

Minpriroda Rossii (Ministerstvo ochrany okruzhajushchej sredy i prirodnych resursov Rossijskoj Federacii), 1995: Gosudarstvennyj doklad o sostojanii okruzhajushchej prirodnoj sredy Rossijskoj Federacii v 1994 godu. Moscow.

Plan dejstvij Pravitel'stva Rossijskoj Federacii po ochrane okruzhajushchej sredy na 1994-1995 gody, in: Sobranie zakonodatel'stva Rossijskoj Federacii. no. 4/1994.

Postanovlenie Central'nogo Komiteta KPSS i Soveta Ministrov SSSR: O korennoj perestrojke dela ochrany prirody v strane, in: Sobranie postanovlenij pravitel'stva SSSR no.6/1988, S. 83 ff.

Pravitel'stva Rossijskoj Federacii: Programma uglublenija ekonomicheskich reform, in: Voprosy ekonomiki no. 8/1992.

Sovet bezopasnosti Rossijskoj Federacii, 1994: Ekologicheskaja bezopasnost' Rossii. Vypusk I. Moscow.

Sovet Ministrov RSFSR 1991: Postanovlenie no. 13 ot 9 janvarja 1991.

Ukaz presidenta Rossijskoj Federacii 1994: O gosudarstvennoj strategii Rossijskoj Federacii po ochrane okruzhajushchej sredy i obespecheniju ustojchivogo razvitija. In: Sobranie aktov Prezidenta i Pravitel'stva Rossijskoj Federacii. no. 6, 7. 2. 1994.

Zakon Rossijskoj Sovetskoj Federativnoj Socialisticheskoj Respubliki: Ob ochrane okruzhajushchej prirodnoj sredy. In: Rossijskaja gazeta vom 3rd March 1992.

SUMMARY: GLOBAL ENVIRONMENTAL POLICY LEARNING

Martin Jänicke and Helmut Weidner

Introduction

In this book the national environmental policy of a number of different countries has been characterised. In this, the method of "structured" case studies was used, to take into account both the complexity of the individual historical development and any possible general or cross-national characteristics. Thirteen case studies is a relatively large number, but they still represent a somewhat narrow basis for any systematic generalisations. The cases, however, possess similarities which cannot be overlooked: even though the analyses were strictly limited to domestic development, the authors reveal important international influences and similarities. In this summary we will focus especially on this global dimension of the book.

1 The Profile of Global Environmental Change

To start with the profile of environmental change, one important feature can be drawn from the environmental statistics in the appendix as well as from the case studies: the close relationship between per capita GNP and environmental indicators. The wealthier countries both need and can afford more environmental protection. They enjoy a higher level of electrification and motorisation, more intensive production of chemical products and utilisation of concrete. As well as increasing the burden on the environment, this affluence also offers better economic, institutional and informational options for countering pollution (Jänicke, 1992).

However, it can be seen from the data in the appendix that the relationship between affluence and environmentalism is significant, but very much contradictory: economic development leads to both improvements and deterioration in the state of the environment. SO_2 (or dust) emissions have been reduced, sewerage systems have been extended at a rate more or less parallel to per capita GDP growth. A similar (but weaker) link exists with respect to nature conservation. CO_2 emissions as well as fertiliser consumption appear to increase up to a certain per capita GDP, after which a stabilisation or a slow decrease begins. Road traffic emissions such as NO_x (or VOC) are nonetheless still higher in most rich countries, even if measures are being taken to reduce their levels. Additionally, the volume of waste produced (and the area per capita of built-up land) is increasing at a high level in the more advanced industrial countries. There are more than 100,000 severely polluted sites in the Netherlands, 170,000 in Germany, and approximately 10,000 in Switzerland, while many other countries (e.g. Japan) may well have similar problem areas, but do not present similarly systematic information on them. Since the more advanced countries are also the older industrial societies, accumulated pollutants are generally a crucial problem for them. And the accumulation is increasing, even in cases where their annual flow is tending to decrease. With respect to bio-diversity, the rich countries are also often worse off in fact (UNEP, 1993). The list of environmental indicators deteriorating in parallel with economic development (e.g. groundwater contamination) could go on for some time.

The less developed countries exhibit a general deterioration in the environmental situation and high pollution levels, especially in cities. Many of them have, in addition, particular environmental problems, such as desertification, deforestation, water shortage etc. (which could not be taken into account in the appendix, but are mentioned in the case studies). But the developing countries are generally starting from lower per capita emission levels and still may not have accumulated as much pollution in their environments as the richer countries.

In general, there is no tendency to "get rich and get clean" (for a partly different view, see *Mao* in this volume). Even in the economically more advanced countries success is mainly restricted to certain specific forms of pollution, while many relevant environmental problems remain unsolved (see also Jänicke, Mönch and Binder, 1996). Thus the need for capacity-building in environmental protection is by no means restricted to the developing world, as the debate up to now has seemed to suggest.

2 Institutionalisation: Policy Innovation and Diffusion

Capacity-building in environmental policy started in most industrialised countries in the late 1960s or early 1970s with the creation of specialised government institutions. This can be seen as the necessary – though not sufficient – condition for formulating and implementing environmental policy. Comparison between our case studies provides some important insights into the general features of the process of institutionalisation in environmental policy. A systematic overview of the dates when relevant institutions in the new policy area were created in selected countries is given in *Table 1* in the appendix to this volume.

Environmental policy innovation and institution building at national level was initiated by frontrunners such as the United States or Sweden. From there, the general idea of creating specialised regulatory and advisory bodies, as well as comprehensive framework laws or constitutional principles in the new field of environmental policy rapidly spread to most industrial, and even some developing countries (see also Jörgens, 1996). At that time, even in Germany relevant environmental ideas and expertise were imported from the United States, although they were subsequently adapted during their actual institutionalisation (*Jänicke* and *Weidner* in this volume). The German government started introducing national institutions well before environmental groups became a significant political factor in the country. Thus it can be said that the impulse and general concept of environmental policy institutionalisation in Germany came from *abroad* and the initiative to react to it was then taken up by the politico-administrative system ("*top-down initiative*"). This "*abroad/top-down*" type of institutionalisation dominated the starting phase of environmental policy in most industrialised countries. The national governments played a crucial role in the beginning. This pattern can be contrasted with a few cases characterised as "*domestic*" (policy innovation developed in the country itself) and "*bottom-up*" (initiatives coming mainly from societal forces). The United States and to some extent Japan (in the late 1960s) belong to this group. Sweden may be characterised as "*domestic/top-down*", as the main ideas were developed within the country, while "environmental interest groups had little influence on the formation of the new environmental policy" (*Lundqvist* in this volume). In Chile, the start of an institutionalised environmental policy was strongly influenced by societal forces ("*bottom-up*"), the relevant information and ideas coming from *abroad*. Here, as in Russia (and other parts of Eastern Europe), the environment figured in the policy agenda of a victorious democratic opposition to the old regime.

The emulation of institutional arrangements, as can be seen from the above, was the first mechanism for international policy diffusion in respect of the environment. At that time, the UN Conference on the Human Environment held in Stockholm in 1972 provided the most important institutional mechanism for global policy learning for the advanced industrial countries analysed in this volume. In the four developing countries, on the other hand, the core institutions for environmental policy were established relatively late on, mainly towards the end of the 1980s. Here, as in most developing countries, the UNCED Conference in Rio de Janeiro (1992) – including its preparatory phase, which begins with the publication of the Brundtland Report in 1987 – gave an important impetus. This first wave of environmental policy innovation in developing countries was paralleled by a remarkable second wave within the industrialised world, as many countries created new environmental framework laws (Great Britain, Japan, Denmark) or reorganised their environmental administration (Great Britain, Switzerland, but also France, New Zealand and others; see also *Table 1* from the appendix).

The close connection between national innovation and global diffusion could again be observed with respect to the new institution of a national environmental policy plan as described in Agenda 21 (1992). The policy innovation here lies in the emphasis placed on long-term goal-setting with a broad political and societal basis, the integration of environmental policy objectives into other policy areas (intersectoral integration), a cooperative target group policy and the mobilisation of additional decentralised capacities. The frontrunners of national environmental policy planning were such countries as the Netherlands or Denmark. But South Korea also followed suit rapidly, institutionalising national environmental planning in its Basic Environmental Policy Act of 1990. Again, the UN Conference and preparations for it acted as the main catalyst for policy diffusion. This can be illustrated by the fact that UNCED is referred to explicitly in the Japanese Environmental Plan of 1995. But the rate of diffusion now seems to be higher, and its scope broader than in 1972. Russia and even Nigeria have – in contrast to some highly developed countries such as Germany or the United States – drawn up a national environmental action plan.

In Europe additional mechanisms for policy diffusion, such as the EU (or ECE) have been influential. In Great Britain, "without doubt the principal international influence on the development of pollution control policy has been the European Union", as new legislation has been passed under EC/EU directives and standards have been tightened (*Weale* in this book).

Rapid international diffusion of policy innovations can also be observed with respect to instruments such as environmental impact assessments, liability rules, insurance regulations, emission management through "bubble" concepts, eco-labelling, "cradle-to-grave" waste management concepts etc. All our case studies report a preference for administrative command-and-control measures. Nevertheless, in nearly all countries the policy style is changing and moving towards softer approaches and negotiated regulation. Networking, informational instruments and voluntary agreements are gaining more and more in importance, at least in the highly developed countries. Environmental charges and taxes still play a minor role in most countries. It is, however, interesting to note that they are applied more frequently in countries with a relatively interventionist regulatory style such as Denmark, Sweden and also South Korea, while they are less important in the more liberal and market-oriented states (e.g. UK, US). Ironically, there seems to be no country with more environmental charges than Russia.

It is above all some of the smaller countries in the group of frontrunners, especially the Netherlands, that have developed the institutional and scientific preconditions to play an active and cooperative role in the international arena. This openness and orientation towards foreign developments may also be used to overcome internal resistance to change, as can be illustrated by the way in which the Dutch minister of the environment "cleverly used the report 'Our Common Future' to lure other ministers into accepting certain basic principles which eventually and inevitably resulted in the National Environmental Policy Plan" (*Bressers* and *Plettenburg* in this volume). In other countries also, we sometimes find a tendency for national environmental policy institutions to use the international stage in order to influence domestic policy.

Not all countries possess the ability to learn from international developments to the same extent. The United States – itself often the starting point for policy learning in other countries – clearly finds it difficult to learn from developments elsewhere, or to give a pro-active response to the initiatives of international organisations, a criticism made by *Andrews* in this publication.

3 The Change in Pioneer Countries

The global process of capacity-building in environmental protection is heavily dependent on national innovators. And here the role of leading nation can be seen to pass periodically from one country to another. Great Britain, the world's first industrialised country, suffered from the world's first heavy smog crisis (1952). Its reaction, the clean air policy in the late 1950s and 1960s (as well as its still earlier developed industrial air pollution control policy) may have played a certain pioneering role at the time. The British policy of water protection also reached a high level of development, when compared internationally. But in the 1980s, after a severe economic crisis and radical political change under the Thatcher government, nobody could find much that was innovative in British environmental policy any more. At EU level however, Great Britain has had some innovative influence with its initiatives on the introduction of eco-auditing and other instruments. The United States, the next pioneer nation in environmental policy, underwent a similar change when Reagan became President, although the activities of US civil society remained remarkable (as did the influence of scientists), even in the international arena. Japan, after a severe ecological crisis, was a pioneer in the 1970s, influencing environmental policy abroad (especially in Germany), developing new instruments (e.g. a compensation system linked to an SO_2 charge) and stimulating discussion of and challenges to established legal doctrines which favoured polluters (such as the burden of proof). For a long time it was also the leading exporter of environmental technology. But in the late 1980s the alliance of environmental modernisers was weakened by extremely low energy and raw material prices, and by other economic and (in the 1990s) political problems. Germany, which in the 1980s unexpectedly became a leader in environmental policy and the main exporter of environmental technology, lost much of its dynamism after the 1994 federal elections (with a considerably weaker minister for the environment) and as a result of the most severe recession in its post-war history. However Sweden, the pioneer of the early 1970s, seems unwilling to relinquish its global role: in 1996 the new government at least announced a new "ecological" model for its economy. The Netherlands too have been environmental policy innovators for more than two decades, while Denmark, partly due to its green parliamentary majority in the 1980s, has developed into one (see *Andersen* in this volume).

Backward steps in domestic environmental policy can be described as a loss of national capacity if, for instance, participation rights are limited (as in Germany recently), if conditions for raising and promoting public awareness worsen (e.g. as a result of changes in the media), or if the economy comes into structural crisis. It may be that domestic retrograde steps are limited by the established international environmental capacity, especially the present influence of the global public.

In some cases the retrogression may also be explained as under-utilisation of existing institutional, economic or informational capacities by political and societal actors who are lacking the adequate "will and skill". The remarkable capacity of environmental organisations, for example,. is sometimes in stark contrast to their day-to-day routines, and to the loss of strategic orientation, which may also encompass the "exit option" if cooperation does not lead to significant change. In general it is wise not to under-estimate this "subjective factor" in environmental policy, as has been demonstrated in a positive sense by the pioneering actions and sophisticated political strategies of organisations such as Greenpeace, or of some politicians, for example the first director-general of the Japanese Environmental Agency, the Dutch Environment Ministers Winsemius and Alders, the German Environment Minister Töpfer, or President Carter.

4 Non-governmental Proponents of Environmental Protection

Our case studies show that it is not only government institutions which constitute the capacities for fostering progress in environmental policy. In most countries, organised non-governmental proponents of environmental interests were influential in the early years of environmental policy, and their influence has grown almost everywhere over time. The organisational strength and professional competence of the environmentalist movement today in countries such as the Netherlands, Switzerland, the United States or Germany is remarkable. Their memberships today are usually larger than those of the political parties. In Switzerland it ranks close and in the Netherlands it is even larger than that of the trade unions (cf. *Knoepfel*; *Bressers* and *Plettenburg* in this volume).

The other important non-governmental actor is the mass media. They are mentioned as a relevant factor in all case studies. Critical reports in the media have been a significant factor in increasing public concern everywhere. Especially in Japan, but also in Germany, the successes of the environmental movement could not have been achieved without committed reporting on the part of the media. They have been important even in countries like China (which has the largest environmental newspaper in the world, with circulation figures of 400,000), or Chile under the dictatorship. In Nigeria also, a press campaign in 1988 against the illegal import of toxic waste was important for the institutionalisation of environmental policy.

In Japan the influence of environmental organisations and the mass media shrank noticeably after the confrontational period of the late 1960s and early 1970s, where they were highly successful in pushing government and industry towards an – in international terms – extremely strict new environmental policy (see *Imura* in this volume; Tsuru and Weidner, 1989). This period of decline in activity correlates closely with a period of stagnation in government environmental policy. However, in the last few years the mass media seem once more to be playing a significant role in increasing public awareness of global environmental challenges, and a process of reform can be discerned within the highly decentralised environmentalist movement (there are thousands of citizens' action groups active at local level), intended to gain greater influence at national level.

The impetus now comes to a large extent from the activities of international organisations concerned with global environmental challenges.

In countries with unfavourable systemic conditions for environmental policy-making, non-governmental organisations are usually the main driving force. This is the case in China and Russia, and the authors in these two cases see growth in autonomous NGOs as a basic precondition for better environmental performance. In Nigeria, the connection of local environmental groups to the global network of Greenpeace and other organisations was a significant factor in the heavy conflict with the military government. An impressive example for the importance of ecological interest groups in environmental capacity-building is reported in the Chilean case study. Here, the NGOs not only helped the cognitive, political and instrumental capacities already gained to survive the long period of dictatorship, but they were also able to outmanoeuvre strong government censorship – for example by organising scientific congresses in the full glare of the media – thereby increasing public awareness and setting up a network which served as the basis for rebuilding governmental environmental capacities after the end of the dictatorship (*Silva* in this volume).

Particularly in the early stages of environmental policy-making, the influence of environmentalist organisations consists especially in their use of unconventional methods from outside the established channels of influence. Later on, they appear to gain more influence in the structuring of environmental policy through professional activities and cooperation with other actors. In some countries, especially the Netherlands, Denmark, Germany, Switzerland, and to some extent also the United States, most of the larger organisations have abandoned their heavy reliance on extra-parliamentary action and "developed very much into an institutional and professional actor", as *Bressers* and *Plettenburg* characterise the Dutch situation (in this volume).

In these countries it can be seen that there is an increasing tendency not only towards bilateral cooperation between the environmental organisations and government or business, but also towards trilateral cooperation between all three actors. At least in the Netherlands, Germany, and Switzerland, this form of cooperation now takes place to such an extent that an "eco-corporatist" interest mediation system appears to be taking shape. At any rate, some of the established policy networks, which were dominated in the past by government and economic actors and their scientific consultants, have opened up increasingly to environmental organisations or green research institutes. In the case of Switzerland, *Knoepfel* (in this volume) calls the actor configuration a new "iron triangle". In both the Netherlands and Denmark, the government has also created new capacities for exerting influence on the international environmental policy arena by means of financial support and strategic cooperation with environmental organisations.

In countries such as the Netherlands, Denmark, Sweden, Switzerland, Germany or the United States the influence of "green businesses" – i.e. enterprises committed to environmentally sound management practices and sustainable development – is important and on the increase, now being an important factor of ecological modernisation. In Germany, for example, green businesses have set up their own associations (*B.A.U.M.*, *future*, *UnternehmensGrün*). Only recently (in October 1995) these organisations joined together as the "*Unternehmerinitiative Umwelt*" ("Enterprise for the Environment"), in order to achieve greater influence in the political arena and on traditional trade associations. Environmentally-minded enterprises and their interest associations are not only important as coalition partners for other proponents in concerted lobbying, but also as "catalysts of innovation", promoting environmentally sound products and processes. As a rule, the environmental measures taken by these actors are observed by other proponents all over the world and used in the national debate.

The most important preconditions for developing and strengthening the influence of non-governmental proponents are the presence of democratic rules providing for the formation of autonomous societal organisations (certainly not the case in China), and some measure of financial support, whether in the form of direct subsidies (as in the Netherlands, for example) or preferential tax status (as in Germany, the US and Denmark, among others).

Finally, legal provision for participation, information rights and civil court action (most developed in the US) are important. On the other hand, the role of the courts themselves varies so strongly from country to country in this study that it does not appear meaningful to view them as a general resource for non-governmental proponents. In some countries (the US, Japan and to some extent Germany), their current or former significance is or was very great, in others – among them equally progressive countries such as Sweden and Denmark – their importance has been only slight. In Japan, the lower courts, through pioneering innovative judgements, were an important trigger for the environmental policy change of heart in national government and business in the 1960s and 1970s. Later on, this commitment of the courts to environmental politics fell back sharply, the higher courts in particular applying the brakes. Recently however, in the context of compensation claims, court judgements in favour of those affected are on the increase. In the United States heavy involvement in court battles, more or less as an end in themselves, has weakened to some extent the ability of different parties to cooperate for common environmental protection measures.

With very few, short-lived, exceptions (Germany and Japan), trade unions have played no part as proponents of environmental protection. The unions in the Netherlands, who made a case for strong environmental protection even during a period of recession (1994), appear to have developed furthest as supporters of environmental protection measures. In the majority of cases, though, they are brakemen to new policy initiatives, or even influential members of "anti-environmental coalitions".

According to our case studies (especially those of the United States, Sweden, Denmark, Japan and South Korea) green parties seem to be no necessary condition for environmental policy innovations. Despite relatively high voting potential for green parties, the electoral conditions in Great Britain, Japan and the US are unfavourable to them. In Germany, on the other hand – with its federal structure and its broadly proportional voting system, combined with public financial support for political parties – they have been an important factor since 1983. There is no country in the world where so many green party members are ministers at state level or hold key jobs in government institutions.

The globalisation of environmental policy has generally been underpinned by the development of a "global environmental public", encompassing a broader spectrum of actor groups than merely those belonging to the so-called epistemic community (Haas, 1990). It is constituted out of international NGOs (such as Greenpeace, Friends of the Earth and the Worldwide Fund for Nature), innovative business groups, the global communications network of the mass media, and the scientific organisations, the latter often with strong links to international institutions and authorities.

5 The Changing Environmental Policy Paradigm

The developmental approach to environmental policy used in this book can also be applied to the changing structure of the public discourse on ecological questions. The debate usually starts with questions of environmental health, pollution of ambient air and water (alongside an older tradition of nature conservation). This leads first to a strategy of dilution (high chimney policy), then to a demand for add-on clean-up technologies. Environmental policy at this stage is preoccupied with mobilising financial resources and formulating standards relating to particular environmental media such as water or air. In highly developed countries, such as the Netherlands, Germany, or Japan, the debate has started to switch to the overall resource input to industrial production. In the developing countries and Russia (and Eastern Europe), however, the paradigm of end-of-pipe treatment still dominates the environmental discourse, and dilution is often the only solution. The extremely grave environmental and economic situation in these countries may be the best explanation for this. But even in the more advanced countries the policy paradigm has undergone only gradual change. A broad and stable consensus for a strategy of resource input reduction has not yet been reached.

The main reason for this may be the strong position of opposing economic interests. This structural restriction seems nonetheless to have stimulated a new wave of global environmental policy learning, which is still continuing. The new policy debate is about strategies aimed at internalising environmental responsibility within the polluting industries (and "their" policies), at enlarging the scope for action of relevant extra-governmental actors and at mobilising societal activities towards a new type of cooperative long-term planning. Such strategies were developed in the smaller West European countries, but their global diffusion took place remarkably quickly.

6 Systemic Resources and Restrictions

A country's capacity to identify and solve environmental problems depends not only on the abilities of government and private proponents. They act within a national context, which provides systemic opportunities and restrictions. From the Russian and Nigerian case studies we learn that the elementary precondition for environmental policy-making and management is a minimum of political and economic stability. Both cases also demonstrate the importance of democracy as a precondition for effective strategies. This can also be deduced from the Chilean and South Korean cases. In Russia and South Korea, the process of democratisation was accompanied by a rapid growth in environmental organisations but, due to its better economic position, only in South Korea was the process of capacity-building a stable one.

In highly developed democracies the importance of the opportunity structure of the political system is not at issue. As *Andrews* puts it in his case study: "One of the most powerful resources available to U.S. green organisations (...) is their extensive formal rights of access to government information and decision processes".

Another systemic precondition of policy success is the internal integration of relevant government environmental activities. This is a difficult and time-consuming task (e.g. in South Korea, Russia, but also still in Japan, even with its long tradition of environmental policy-making). Intersectoral integration of different policies is even more complicated. At present the integration of environmental and transport policy does not seem to be functioning anywhere. The same can be said of energy policy, although the Danish example of merging the Ministry of the Environment with the (smaller) Energy Ministry is most promising in this respect (*Andersen* in this book). Other political areas where the integration of environmental policy is problematic are usually agriculture and construction. An ambitious attempt to integrate environmental concerns into relevant polluting sectors has been made with the Dutch National Environmental Policy Plan, which had to be signed by four ministers (environment, transport, agriculture, and economic affairs) before coming into force (*Bressers* and *Plettenburg* in this volume). However, it still seems to be too early for a final evaluation of this approach.

The "external" integration of new societal actors is another task for capacity-building. Successful environmental policy and management depends more and more upon the participation of a broad spectrum of actors and the voluntary cooperation of polluters (target groups). Our case studies show a general learning process in this direction. Countries with strong traditions of government-industry cooperation (Japan) or neo-corporatist interest mediation (Sweden, the Netherlands, Switzerland, Denmark, and Germany) seem to enjoy better conditions for the establishment of creative networks. The Swedish case is explicitly described in terms of "ecological corporatism", but this refers to public-private partnership in the field of waste management, without any participation of green organisations (see *Lundqvist* in this book).

In the US, the pronounced fragmentation of political responsibilities within the system (the so-called "bidding-up strategies" of central actors) has led to environmental organisations being confronted with what are admittedly open input structures (policy formulation), but having little influence on the output (implementation, enforcement) functions, or having only conflicting means – primarily litigation – at their disposal in exercising these functions. The "swinging pendulum" of US environmental policy may also be caused by the contradiction between "openness" and "fragmentation" (see *Andrews* in this book).

The available knowledge about problems or options, and the capacity to learn and to apply that learning to environmental policy processes is the third important systemic resource of environmental policy and management. For a long time global policy-learning in this field was heavily dependent on the scientific capacity and experience of the United States. Only later on did countries such as Germany develop a broad scientific infrastructure to provide independent environmental expertise of a standing which could not be challenged by industrial opponents. The pro-active role of Sweden and the Netherlands cannot be explained without reference to the excellent material and institutional infrastructure of the environmental sciences. Despite its present political role in global environmental policy, the scientific influence of the US is still very important for other countries. The development of the global environmental "epistemic community" was also strongly influenced by the Netherlands, Sweden, Germany and Japan.

The public media also needed time to become a significant factor in promoting public awareness. Political restrictions on them have been a significant obstacle to a "normal" development of the policy area. This demonstrates the importance of press freedoms (and of the media being independent of economic or political interests) in environmental capacity-building. The media, being an important factor in policy diffusion, were also largely responsible for generating a public environmental consciousness on a global scale.

Surveys carried out in the Netherlands show that strong media attention to the environment – especially from the press – results in greater public awareness, and that information about imminent catastrophe is the most influential form (*Bressers* and *Plettenburg* in this book). Similar results have been gathered from other countries. This means that the less spectacular forms of ecological degradation – which are becoming all the more important in advanced countries – create a need for additional capacity-building in this field, i.e. for an environmental journalism which is able to anticipate dangerous but "silent" developments.

The overall economic situation has had considerable effects on the opportunities of environmental proponents. Environmental policy comes under particular pressure in times of recession, even though it has been demonstrated that an ambitious environmental policy also has benefits on economic development and can become a competitive advantage in the global market economy. There are nevertheless some remarkable exceptions to this rule. In Japan in the 1970s, despite severe economic upheavals following the oil crisis, strict environmental protection measures were pushed through, without any resulting economic disadvantage. Indeed, the measures contributed to a modernisation of the economic structure and a dominant role in the world market for environmental technology. Germany started similarly in 1983 with an active environmental strategy just after a heavy recession, and ended up as the leading exporter of environmental technology. In the early 1990s, Denmark and Sweden introduced "green" taxes/charges within a recession.

Once again it is economic performance which decides whether such economic instruments are effective or not. Neither in China nor in Russia are existing environmental charges helped by the prevailing economic conditions, for which reason they have hardly any effect. In Russia especially, the system of pollution and resource use charges (which was introduced in the early 1990s) has been neutralised by rapid inflation, chronic non-payment by businesses, low investment rates, inefficient monitoring of emissions etc. (see *Potravnyi* and *Weißenburger* in this volume). Non-availability of clean-up technology, mostly due to a lack of financial resources or an inadequate scientific and technological infrastructure, is another restriction, and the international transfer of technology is seen as helpful (see also *Salau* in this book).

7 Capacities for still Unsolved Problems

According to the relevant case studies, capacity-building for the environment in developing countries depends mainly on material and human resources, scientific, technological and administrative knowledge, open political access and policy integration, public awareness and the strength or competence of environmental organisations. But as mentioned above, even in the advanced industrial countries success in environmental protection is restricted to only some specific problems.

The conceptual framework presented in the first chapter stressed the character of the environmental problem to be solved as one of the main factors influencing policy outcomes. Systematic models of environmental policy-making have often ignored this category. Our case studies, however, give valuable indications of the importance of this factor. It is no coincidence that certain problems, such as traffic emissions, waste production, soil contamination or extensive land use, have not undergone significant improvement in most "average" countries, while other pollution problems, such as SO_2 emissions, municipal sewerage or industrial waste water (e.g. the emissions from Swedish pulp and paper industry), show impressive improvements almost everywhere. The missing link here could well be the structure of the problem, in terms of its visibility and urgency, the availability of a standard technological solution and the societal importance and composition of the relevant target groups (*Jänicke* in this volume).

For *Lundqvist*, traditional environmental policy was adequate for point sources, while the ecological problems caused by non-point sources have a more complicated structure. *Bressers* and *Plettenburg* (in this volume) argue similarly that "target groups consisting of large numbers of small units (such as households, farmers, car drivers, retailers and other small firms) are much more difficult to reach by means of a target group policy". In these cases economic approaches, especially the use of eco-taxes, seem to be a more promising strategy. The reduction of energy consumption in Danish households, as reported by *Andersen* (in this volume), is however an example where a heterogeneous polluter group was successfully steered towards environmentally friendlier behaviour. In Japan, where the early and widespread use of environmental protection technologies has led to remarkable success in fighting industrial pollution, the necessity of changing consumer behaviour is now seen to be becoming increasingly important (*Imura* in this book).

A more important social dimension of the structure of ecological problems may be the power to ignore environmental imperatives from the public. The broad alliances and the closed policy networks of transport, energy, construction and agro-business nearly everywhere have been strong enough to obstruct preventive measures that go beyond end-of-pipe treatment.

Here changes are taking place only very slowly: the opening-up of the powerful policy network of power supply in the US (especially in California) was an important first step in this direction. The weakening of the energy and coal industry in the UK was an interesting side-effect of privatisation, which may also improve the chances for integrated solutions (see *Weale* in this volume). The Netherlands and Denmark have succeeded in weakening the influence of the agricultural lobby through sophisticated political management, thereby opening this policy arena up to environmentalists. A similar approach was the creation of the Transport Council in Denmark to counteract the influence of the traditional policy network (see *Andersen* in this volume). There is still a vast untapped reservoir of possible improvements in this respect. The development of strategies against the privilege of powerful polluters of not having to learn about environmental matters may well be the crucial task of the future.

8 Conclusion

From the above summary of the main findings in this book, we can draw the following conclusions:

- There has been a remarkable increase in capacities for environmental policy and management in the last 25 years which could not have been anticipated at the outset. The rapid development of specialised policy institutions, including even a (cautious) penetration of other policy fields, seems to be without precedence in the history of the modern state. The most important characteristic of this development is nevertheless "societal" capacity-building beyond government institutions.

- With innovations such as new forms of participation, information rights, self-regulation, networking or cooperative planning, the ecological question has contributed to the general process of political modernisation. In more advanced countries, such as Denmark, Sweden, the Netherlands, Switzerland, Japan or Germany, it has also become a motor of economic modernisation and innovation.
- This development can be described as a process of global environmental policy learning. We can thus see from our case studies that the process of globalisation not only leads to greater ecological stress world-wide. It also significantly enlarges domestic capacities for environmental protection in most countries.
- The communicative mechanisms of this global learning process seem to be bilateral imitation at first, then diffusion through international institutions – especially the UN system, later also the OECD – and finally learning within a broad international network of environmental government institutions and non-governmental organisations, the media, research institutions and "green business" organisations.
- Throughout the countries included in this volume, a global environmental policy convergence can be observed, irrespective of national differences in e.g. policy style. The patterns of environmental policy have become similar even under extremely varying domestic conditions. Even the change of the dominant environmental policy paradigm – from end-of-pipe treatment and government command-and-control to softer instruments and more sustainable approaches – takes place on a global scale.
- Thus it is not so much the general policy pattern but the problems, capacities and outcomes of environmental policy and management which differ significantly in global terms. In some parts of the world, the new "standard" environmental policy institutions and laws were only introduced on paper, and have no effective basis within government or society. China, as well as Russia, possess a fairly comprehensive and differentiated system of environmental regulation. The main problem lies in the field of implementation. To a lesser extent this contradiction can also be observed in South Korea (see *Mao*, *Potravny* and *Weißenburger*, and *Nam* in this volume).
- It is questionable whether a relatively uniform pattern of environmental policy can provide adequate solutions for so many different countries. In particular, it must be questioned whether strategies so strongly influenced by the rich industrialised countries can be implemented effectively in the poorer developing countries. In any case, the search for new, more appropriate and cheaper solutions, especially ones to prevent the occurrence of problems already experienced by highly industrialised countries must continue.
- Up to now, existing environmental capacity, even in the most advanced countries, has been sufficient only for – predominantly technological – standard solutions, such as end-of-pipe treatment of air and water pollution, waste collection, nature conservation, the control (substitution) of toxic substances, or more efficient use of energy or water. For the development of

effective strategies for land use or material flow management, for soil or climate protection, additional capacities are needed, and here there are still structural restrictions to be overcome.

References

Binder, M., 1996: Die Operationalisierung umweltpolitischen Erfolges: Probleme und Lösungsansätze, in: M. Jänicke (ed.), Umweltpolitik der Industrieländer. Entwicklung – Bilanz – Erfolgsbedingungen. Berlin: edition sigma, 133-152.

Haas, P. M., 1990, Saving the Mediterranean: The Politics of International Environmental Cooperation, New York: Columbia University Press.

Jänicke, M., 1992: Conditions for Environmental Policy Success: An International Comparison, *The Environmentalist* 12, 1, 47-58.

Jänicke, M. (ed.), 1996: Umweltpolitik der Industrieländer. Entwicklung – Bilanz – Erfolgsbedingungen. Berlin: edition sigma.

Jänicke, M., H. Mönch and M. Binder, 1996: Getting Rich - Getting Clean? Umweltindikatorenprofile im Industrieländervergleich, Zeitschrift für angewandte Umweltforschung (ZAU) 9, 1, 41-55.

Jörgens, H., 1996: Die Institutionalisierung von Umweltpolitik im internationalen Vergleich, in: M. Jänicke (ed.), Umweltpolitik der Industrieländer. Entwicklung – Bilanz – Erfolgsbedingungen. Berlin: edition sigma, 59-111.

UNEP (United Nations Environmental Programme) 1993: Environmental Data Report 1993-1994.Oxford: Blackwell Publishers.

Weidner, H., 1996, Basiselemente einer erfolgreichen Umweltpolitik. Eine Analyse und Evaluation der Instrumente der japanischen Umweltpolitik, Berlin: edition sigma.

APPENDIX

International Profiles of Change since 1970

Note:

Sources for the environmental data given in this appendix are mainly OECD and UNEP (in particular the latest OECD Data Compendium and UNEP Environmental Data Report), supplemented with data from other international statistical sources (such as the FAO Yearbook) or national environmental reports. Source for the data on institution-building in environmental policy: Jörgens, H., 1996: Die Institutionalisierung von Umweltpolitik im internationalen Vergleich, in: M. Jänicke (ed.), Umweltpolitik der Industrieländer. Entwicklung - Bilanz - Erfolgsbedingungen. Berlin: edition sigma, 59-111.

National signs for cars are used to indicate the industrialised countries. CHI = Chile, CHN = China, NIG = Nigeria, SKO = South Korea, THA = Thailand.

Table 1. Dates of Institutionalisation in Environmental Policy.

Countries[1]	Ministry of the Environment	National Environmental Agency	National Environmental Report	Environmental Framework Law	Article in the Constitution	Council of Environmental Experts	National Environmental Plan[3]
Japan	•	1971	1969	1967/93	•	1967	(1995)
Sweden	1986	1967	1977	1969	1974	1968	•
USA	•	1970	1970	1969	•	1971	•
GDR (until 1990)	1971	1988	1990	1970	1968	•	•
Great Britain	1970	1972/95	1978	1974/90	•	1970	(1990)
Denmark	1971	1971	1983	1973/91	•	•	•
Netherlands	1971/82	1984	1973	1979/93	1983	1974	1989
France	1971/84	1991	1973	•	•	1975	1990
Poland	1972	1980/91	1972	1980	1976-89	1993	(1992)
Austria	1972	1985	1978	•	1984	1971	1995
FRG	1986	1974	1976	1974[2]	1994	1971	•
Hungary	1990	1974	1975	1976	1972/89	•	(1992)
Switzerland	•	1971	1990	1983	1971	•	•
Australia	1971	1988	1980	1974	•	•	•
Portugal	1979/90	1971/87	1987	1987	1976	•	1994
Yugoslavia	1984	1974	1978	•	1974	•	(1991)
Canada	1971	•	1986	1988	•	1972	1990
Norway	1972	•	1976	1981	1992	1990	•
Greece	1980	1976	•	1986	1975	•	•
Bulgaria	1990	1976	1989	1991	1968/91	•	(1992)
Romania	1989	1973	•	1973	1991	•	(1992)
Belgium	1975	1988	1979	•	•	•	•
New Zealand	1972/86	•	1993	1986/91	•	1970-88	(1994)
Finland	1983	1986	1984	•	•	1977	(1990)
Turkey	1991	•	1984	1983	1982	1982	•
Italy	1973/83	•	1989	1986	•	•	•
CSSR (until 1989)	•	•	•	•	1971	•	•
• Czech. Rep.	1989	1992	1990	1992	1992	•	(1992)
• Slovakia	1990	•	1992	•	1992	•	(1992)
Spain	1996	•	1977	•	1978	•	•
China	•	1984	1989	1979/89	•	1991	(1994)
USSR/Russia	1988	•	1988	1991	1977/93	•	(1993)
Ireland	1978	1993	1985	•	•	1992	•
Korea	1990/94	•	1991	1990	980/87	•	1991
Chile	•	1990/94	1992	1994	1980	•	•
Nigeria	•	1988	1992	1988	•	•	(1990)

1) The ranking of the countries is based on the average of the first three stable and unrestricted steps of institutionalisation.
2) Cannot be fully counted as an environmental policy institution.
3) Years in brackets represent plans which contain primarily action planning without quantitative goals.

Figure 1. SO_2-Emissions.

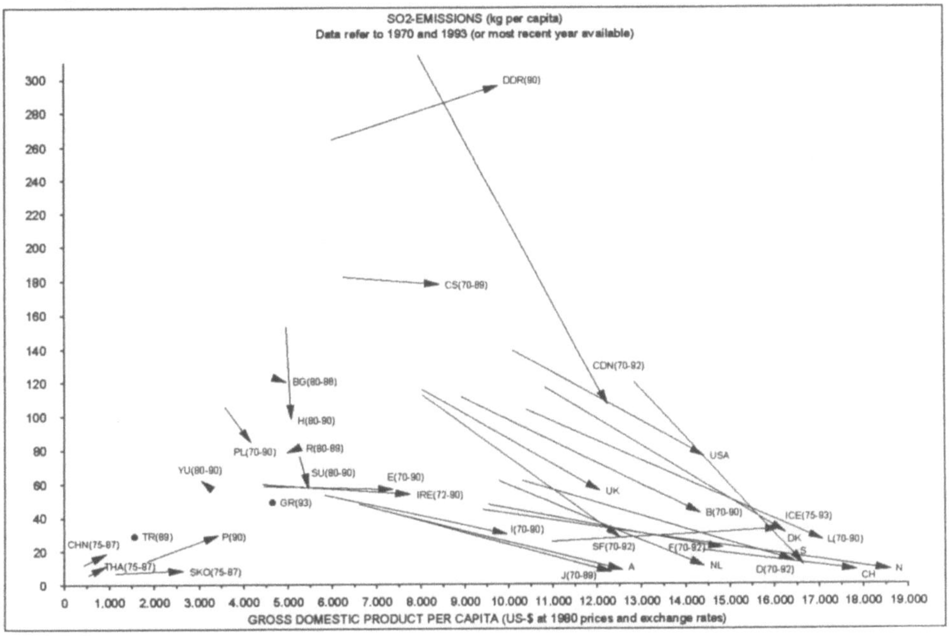

Figure 2. NO_x-Emissions.

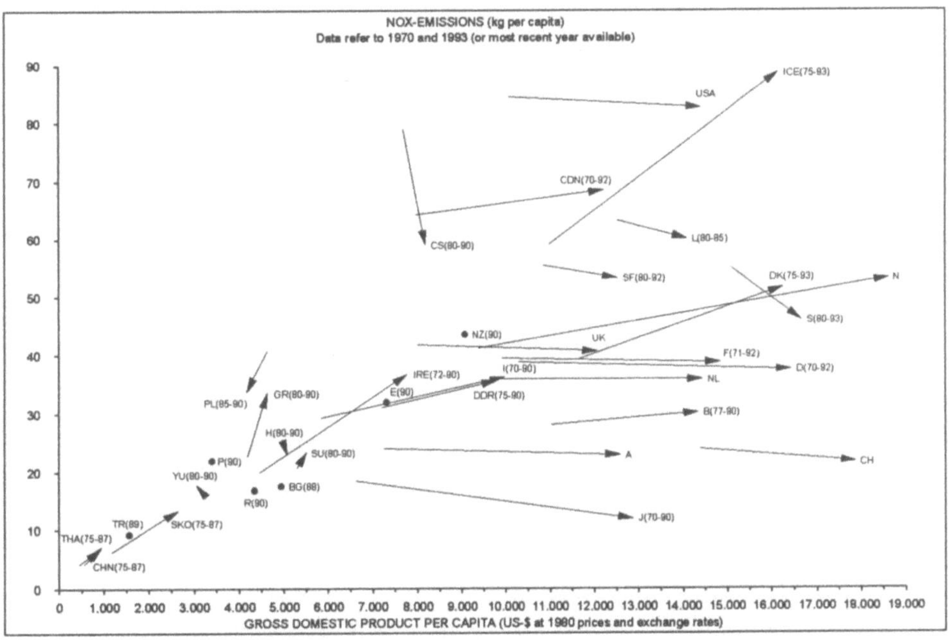

Figure 3. CO_2-Emissions.

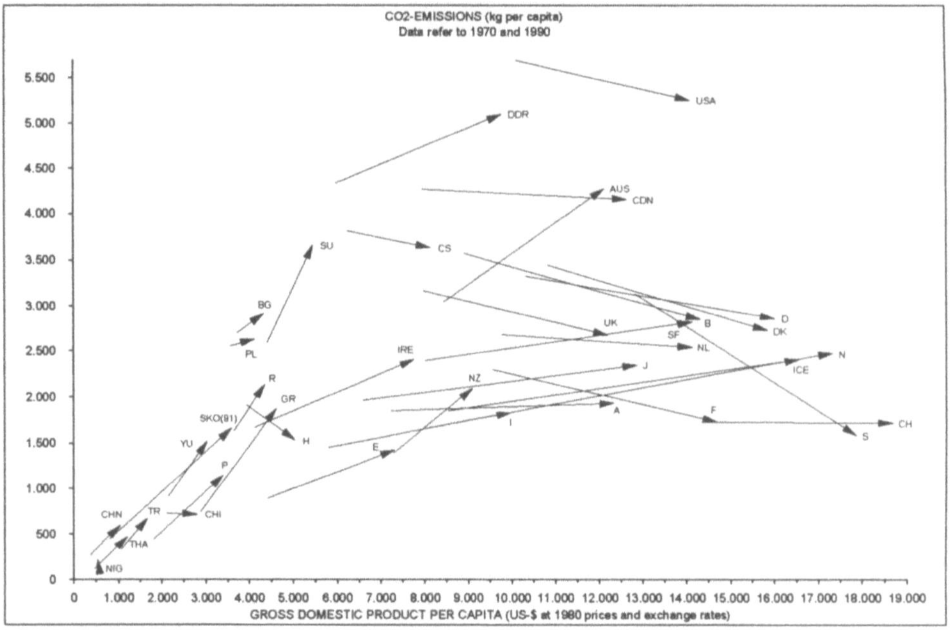

Figure 4. Population Served by Waste Water Treatment Plants.

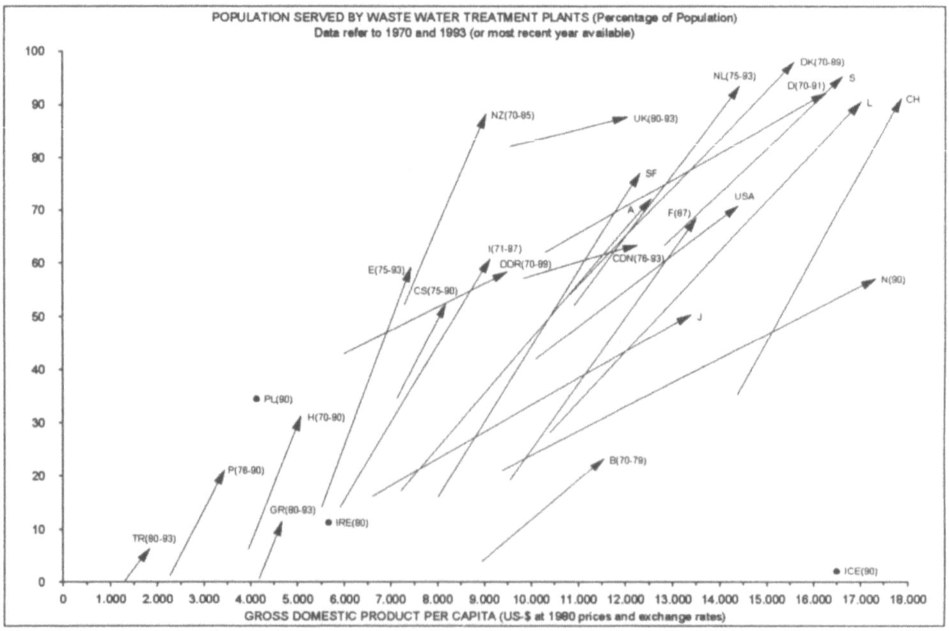

Figure 5. Amounts of Municipal Waste.

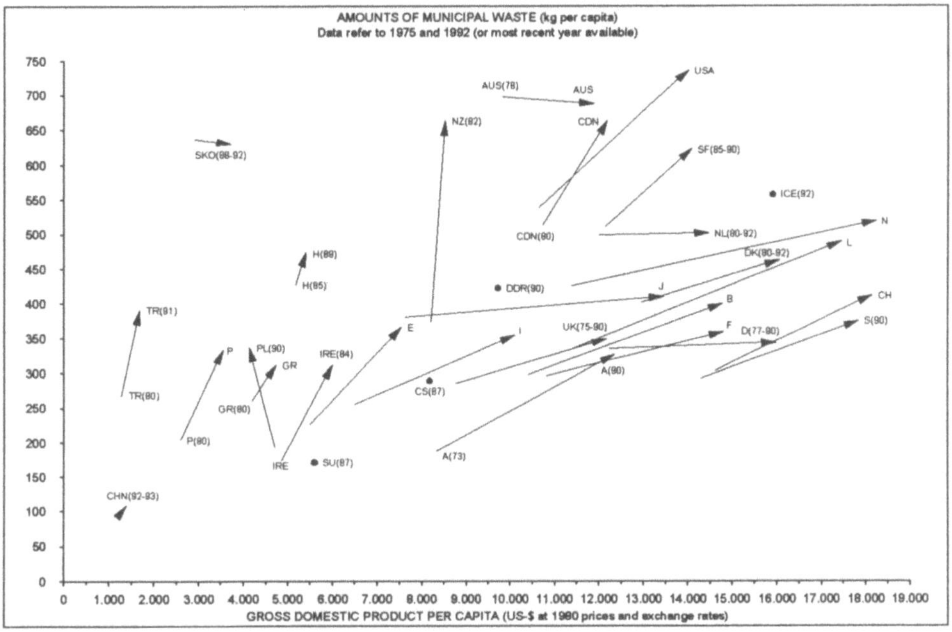

Figure 6. Fertilizer Consumption.

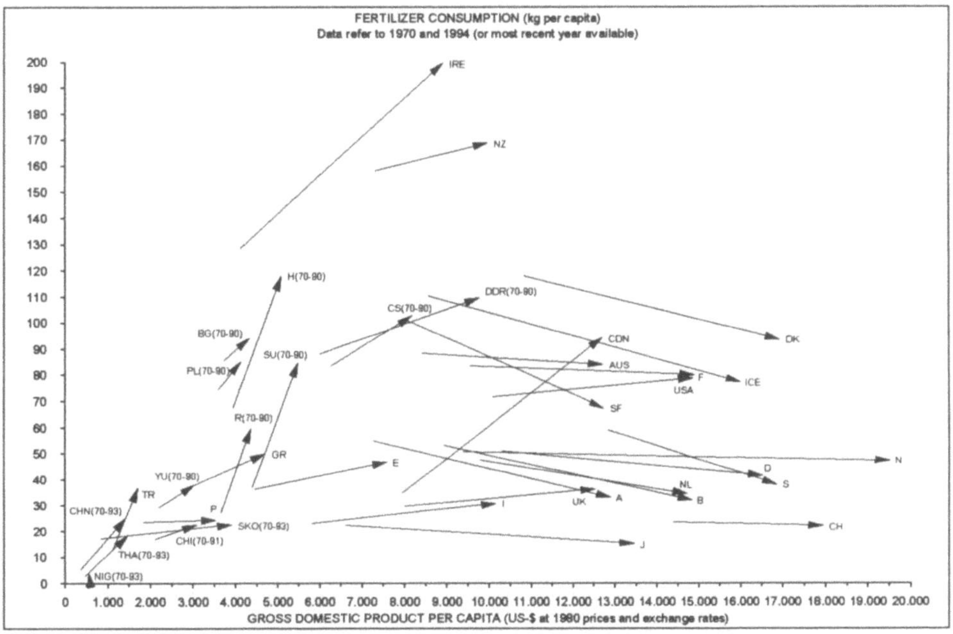

Figure 7. Major Protected Areas.

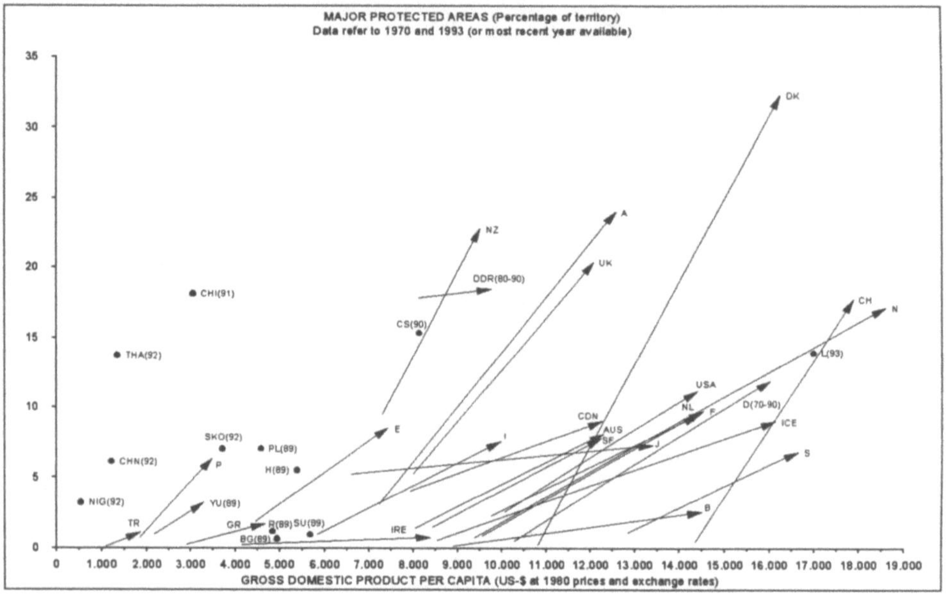

MIX
Papier aus verantwortungsvollen Quellen
Paper from responsible sources
FSC® C105338

If you have any concerns about our products,
you can contact us on
ProductSafety@springernature.com

In case Publisher is established outside the EU,
the EU authorized representative is:
Springer Nature Customer Service Center GmbH
Europaplatz 3, 69115 Heidelberg, Germany

Printed by Libri Plureos GmbH
in Hamburg, Germany